# EPILEPSY TOPICS

Edited by **Mark D. Holmes**

**Epilepsy Topics**
http://dx.doi.org/10.5772/57017
Edited by Mark D. Holmes

**Contributors**

Azra Alajbegovic, Jasminka Đelilović-Vranić, Salem Alajbegović, Boulenouar Mesraoua, Dirk Deleu, Heinz Gregor Wieser, Hassan Al Hail, Gayane Melikyan, Manjari Tripathi, Jane McCagh, Raidah Saleem Albaradie, Karen Gale, Patrick Forcelli, Lee Hively, Todd McDonald, William Ashbee, Lorena Vega-Zelaya, Jesus Pastor, Marta Alvarez, Elena Ezquiaga, Maria De Toledo, Jaime Nogueras, Rafael G Sola, Charles Akos Szabo, Russell Pella, Lakshmi Mukundan, Batool Kirmani, Diana Mungall, Hyang Woon Lee, Reese Terry

**Published by InTech**

Janeza Trdine 9, 51000 Rijeka, Croatia



**Publishing Process Manager** Iva Lipovic

**Technical Editor** InTech DTP team

**Cover** InTech Design team

First published July, 2014

Printed in Croatia

Additional hard copies can be obtained from orders@intechopen.com

Epilepsy Topics, Edited by Mark D. Holmes

p. cm.

ISBN 978-953-51-1630-1

For EU product safety concerns:
IN TECH d.o.o., Prolaz Marije Krucifikse Kozulić 3, 51000 Rijeka, Croatia,
info@intechopen.com or visit our website at intechopen.com.

# Meet the editor

Mark D. Holmes MD is currently Professor of Neurology and Associate Director of the Regional Epilepsy Center at the University of Washington in Seattle, Washington, USA. He received his medical degree at the Ohio State University, medical internship at the Cleveland Clinic, neurology training at Letterman Army Medical Center in San Francisco, California, and fellowships in epileptology and clinical neurophysiology at the National Institutes of Health in Bethesda, Maryland. He is the author of numerous papers and book chapters on various aspects of epilepsy. His current research focuses on understanding the epileptic network through high density EEG recordings in combination with neuroimaging modalities.

# Contents

**Preface IX**

Chapter 1 **Brain Circuits Responsible for Seizure Generation, Propagation, and Control: Insights from Preclinical Research 1**
Patrick A. Forcelli and Karen Gale

Chapter 2 **Role of EEG in Epilepsy 15**
Manjari Tripathi and Man Mohan Mehendiratta

Chapter 3 **Nonlinear Epilepsy Forewarning by Support Vector Machines 29**
W.S. Ashbee, L.M. Hively and J.T. McDonald

Chapter 4 **Juvenile Myoclonic Epilepsy: An Update 53**
Boulenouar Mesraoua, Dirk Deleu, Hassan Al Hail, Gayane Melikyan and Heinz Gregor Wieser

Chapter 5 **Juvenile Myoclonic Epilepsy — A Maturation Syndrome Coming of Age 77**
Russell D. Pella, Lakshmi Mukundan and C. Akos Szabo

Chapter 6 **Reflex Epilepsy 129**
Raidah Saleem AlBaradie

Chapter 7 **Vagus Nerve Stimulation Therapy for Epilepsy 139**
Reese S. Terry Jr

Chapter 8 **Neuromodulation Therapy: Nonmedical, Nonsurgical Treatment for Intractable Epilepsy 161**
Hyang Woon Lee

Chapter 9 **Psychogenic Non-Epileptic Spells 173**
Batool F. Kirmani, Diana Mungall Robinson, Jose Aceves, David Gavito, Richard Phenis and Daniel Cruz

Chapter 10 **Psychogenic Non-Epileptic Seizures in a Surgical Epilepsy Unit: Experience and a Comprehensive Review 193**
Lorena Vega-Zelaya, Marta Alvarez, Elena Ezquiaga, Jaime Nogeiras, María Toledo, Rafael G. Sola and Jesús Pastor

Chapter 11 **Quality of Life Issues in Epilepsy 225**
Jane McCagh

Chapter 12 **Marriage of Epileptic Patients 241**
Azra Alajbegovic, Jasminka Djelilović-Vranić, Salem Alajbegović and Lejla Alajbegović

# Preface

Epilepsy is recognized as of one of the most common primary disorders of the brain. It affects all ages, from infancy to old age, and afflicts at least one person in fifty during the course of his or her lifetime. The eminent 19st century British neurologist, J. Hughlings Jackson, conceived of an epileptic seizure as an abnormal, paroxysmal and excessive electrical discharge in cortical neurons. Today, nearly a century and a half later, modern neurologists agree that Jackson's formulation is essentially correct. Today, neurologists understand that epileptic seizures may be a final common pathway of virtually any one of the myriad processes that can potentially result in brain injury, ranging from genetic, vascular, traumatic, infectious and degenerative etiologies.

Just as the causes of epilepsy may be protean, so may be the clinical manifestations of epileptic seizures. The signs and symptoms of seizures are determined by the location of ictal origin, the distribution and spread of discharges within an epileptic network, and the duration of the seizure. The spectrum of adverse effects on brain function that result from a seizure is extensive. These may range from temporary aberration of motor or sensory function, to collapse, convulsive activity, and the complete loss of consciousness which transiently, but completely, robs the affected individual of his or her essential humanity. Particularly when mental status is impaired during seizures, the overall impact on quality of life may be profound, with deleterious effects on nearly all aspects of the patient's life. Despite the significant advances that have been made over the last several decades by basic scientists in understanding the fundamental nature of epileptic seizures, and by clinical researchers in improving diagnosis and treatment, at least one-third of affected subjects remain medically refractory. Epilepsy thus remains a major world-wide public health problem.

Although the management of patients with epilepsy usually falls under the purview of neurologists or epileptologists, other specialists are frequently involved in epilepsy patient evaluation and care as well, including neurosurgeons, neuropsychologists, neuroradiologists, psychiatrists, social workers, nurses, emergency physicians, and primary care providers, including obstetrician-gynecologists. In fact, nearly any health care provider who has direct patient contact will most likely, at some point, be involved with the assessment or treatment of patients with seizures. For this reason, widespread dissemination of knowledge regarding fundamental issues surrounding the patient with epilepsy is vitally important.

In this volume, an international group of recognized epilepsy researchers and clinicians has been assembled to discuss a variety of topics on epilepsy. The subject matter is diverse, and includes new concepts in brain circuitry involved in seizure generation, a discussion on reflex epilepsy, updates on juvenile myoclonic epilepsy, the role of EEG in epilepsy evaluation, the novel possibility of seizure prediction from scalp EEG, the roles of vagus nerve stimulation and other neuromodulatory therapies in epilepsy management, non-epileptic seizures, and, no less important to the individual, some of the psychosocial issues that confront the patient and his or her family. This volume is not intended be a comprehensive overview of the field of epilepsy, but each discussion is focused and will be valuable to both investigators and practitioners.

**Mark D. Holmes MD**
Regional Epilepsy Center
Department of Neurology
University of Washington
Seattle, Washington, USA

Chapter 1

# Brain Circuits Responsible for Seizure Generation, Propagation, and Control: Insights from Preclinical Research

Patrick A. Forcelli and Karen Gale

Additional information is available at the end of the chapter

http://dx.doi.org/10.5772/58584

## 1. Introduction

In the early 1870s, John Hughlings Jackson, the father of modern epileptology wrote, that a seizure is "a symptom, and implies only that there is an occasional, an excessive, and a disorderly discharge of nerve tissue" [1]. When one considers that he wrote this more than 50 years before the first human electroencephalographic recordings [2], his level of insight is quite remarkable. Indeed, his later definition of epilepsy as "the name for occasional, sudden, excessive, rapid, and local discharge of grey matter" [3] could be used without alteration today.

There is a key difference between Jackson's two definitions: his later definition no longer included the concept of seizures as "disorderly". While seizures are a symptom of a disorder, the temporal pattern of signs and symptoms of seizures are far from disorderly or disorganized; this was evident to Jackson in the march of seizure activity through somatosensory cortex [1,4]. Today, relying not only on seizure semiology, but also electroencephalographic, neuroimaging, and animal models, we can without hesitation state that seizure activity does not spread randomly through the brain, but moves through anatomically constrained pathways and networks.

These pathways are the focus of this chapter; we will discuss specific brain networks that are capable of seizure generation, seizure propagation, and seizure suppression. From the perspective of preclinical research, we will emphasize several points:

1. How do we identify seizure circuits?

2. What is the importance of animal models for understanding seizures (with an emphasis on circuit-level manipulations and species-specific features)?

3. How do emerging technologies enable translation of network-level manipulations to the clinic?

## 2. Identifying seizure circuits

Seizure semiology can provide insight into the brain networks impacted for a given seizure type: for example, the "fencing posture" seen in patients with frontal lobe seizures involving pre-motor cortex can be recapitulated by selective stimulation of pre-motor cortex [5,6]. Similarly, sensory-specific auras e.g., odors in temporal lobe epilepsy can be localized to piriform cortex, [7,8], complex visual hallucinations in anteromedial temporal lobe, occipito-temporal and occipital epilepsy [9]. These symptoms provide an index of regions impacted by seizures, and the temporal order of the occurrence of these symptoms can provide a measure of seizure spread. However, working backwards from these symptoms to identify the path and origin of seizure propagation is a near impossible challenge.

Take, for example, electrical wiring in a house as an anology: a surge of power may cause the lights to flicker in the living room, but that does not necessitate (or even indicate) that the surge started in the living room. Indeed, we know that both parallel and serial wires exist in the house, connecting power sources to fuse boxes to distribution nodes. Various signs and symptoms (burnt wiring, a tripped circuit breaker, etc.) may represent primary causes or secondary effects. Troubleshooting a circuit problem in the house, as complex as it may be, is feasible because there are wiring diagrams to guide you. Without these wiring diagrams tracing a problem would be much more complicated.

At the present, we are working, at best, with very incomplete wiring diagrams for the brain. Thus, we assert that understanding how seizure networks are wired in the "normal" brain is essential to determine how faults in this wiring leads to chronic seizures.

A variety of "mapping" approaches have been employed to identify brain regions engaged by seizures, including electrographic, metabolic (e.g., 2-deoxyglucose), immediate early gene (e.g., fos, zif), and functional magnetic resonance imaging [10–19]. While informative, these approaches, in isolation, only identify areas activated by seizures. Mapping approaches alone cannot determine the role of a region in initiation, propagation, or seizure suppression; these determinations can only be made on the basis of circuit manipulations. The need for circuit-level manipulations is one of several reasons that animal models are vital for deciphering seizure circuitry.

## 3. Importance of preclinical research using animal models

Studies in human patients have provided many valuable insights into the networks supporting seizures, but the conclusions that can be drawn from these studies are limited by the following:

http://dx.doi.org/10.5772/58584

1. Changes observed in association with repeated or recurrent seizures cannot be readily identified as cause, effect, or compensation.

2. The great deal of variability across patients and studies with respect to diagnosis, etiology, and treatment.

3. The inability to use matched controls for invasive procedures.

Animal models overcome these limitations. For example, it is only by directly manipulating a brain pathway or region that one can determine whether the structure is necessary for seizure initiation, amplification, distribution, or inhibitory (feedback) control. These direct manipulations include circumscribed lesions, electrical stimulation, pharmacological inactivation/silencing, and optogenetic approaches.

When these techniques are applied to intact, normal animals, their impact on the circuitry can be evaluated uncompromised by preexisting pathologies. Moreover, the effect of the manipulation can be studied in both animals with a seizure profile and in control animals that are seizure naïve, allowing one to determine how pathology changes circuit function.

Four major types of animal models have been used in epilepsy research: genetic (naturally-occurring and engineered), evoked epileptogenesis, and evoked seizures. Entire texts have been written on this subject (see for example, [20]), so our discussion below is by no means intended to be comprehensive.

Naturally-occurring and inbred models are seen in a variety of species, ranging from mouse (e.g., the El mouse [21–24]; and others [25]), rats (e.g., GEPR rats [26–28]; Wistar Audiogenic Rats [29,30]), gerbils [31], dogs [32], and non-human primates (e.g., baboons [33]). The truly spontaneous seizures that occur in these cases suggest that the circuitry that produces epilepsy has been highly conserved over phylogeny.

Transgenic models of epilepsy are of increasing importance as new mutations for inherited epilepsies are discovered. These models have been used to identify abnormalities at the microcircuit level (e.g., interneurons in the SCN1A knockout mice [34]), but abnormalities at the macrocircuit level still require investigation for most of these models.

Models that evoke epileptogenesis are vital when the goal is to identify what neuroplastic changes, if any, lead to epilepsy. However, if the goal is to delineate networks through which seizures preferentially propagate, then the use of an acute or subacute seizure model is most appropriate, especially a model that does not cause brain injury. It may be worthwhile to compare the pattern of seizure propagation in an injured vs uninjured brain, but for this purpose, the injury should be highly controlled and reproducible. Unfortunately, models such as status-epilepticus (SE) induced spontaneous seizures suffer from some of the one of the same drawbacks associated with studies in patient populations, e.g., heterogeneity of injury. Moreover, SE can cause severe and widespread damage that often exceeds the level of damage seen clinically [35]. The need for highly reproducible and focal epileptogenic insults may potentially be filled by controlled models of traumatic brain injury, which provide greater control over the location and extent of damage [36-38].

## 4. Types of models and manifestations: what is seizure-related and what is due to compensatory mechanisms?

Determining how seizure networks are changed by epileptogenesis is a necessary step in understanding epilepsy, however, this can only be understood in the context of a comparison between the "normal" and "disease" state. The need to examine seizure propagation in a "normal" network is one of several reasons that evoked seizure models are a powerful tool in modern preclinical epileptology. In addition to this utility, evoked seizure models may be preferable for examining network mechanisms because they offer experimental control of seizure timing, severity, etc. This contrasts with most models of epileptogenesis, in which seizures occur spontaneously and unpredictably.

## 5. Seizure models evoked by pharmacological agents

In rats and mice, systemic administration of GABA-A receptor antagonists (bicuculline, pentylenetetrazole, picrotoxin, beta-carbolines) trigger, in a dose-dependent manner, myoclonic, clonic (complex partial/limbic-motor), and tonic-clonic seizures (for a review see: [39]). These compounds have been used to screen virtually every anticonvulsant drug currently available for clinical use. At least one of these compounds (pentylenetetrazole, Metrazol) has been used to trigger tonic-clonic seizures in human patients. In the non-human primate, most of these compounds trigger generalized tonic-clonic response at the lowest effective dose; this may reflect higher sensitivity of hindbrain seizure networks as compared to limbic forebrain networks in the monkey (discussed below).

Other chemoconvulsants (e.g., pilocarpine, kainate) have been widely used for modeling epileptogenesis, and have also been used to examine seizure circuitry [40,41]. Non-convulsant seizure triggering agents (e.g., gammabutyrolactone) have been used to evaluate circuitry underlying thalamocortical spike-and-wave seizures [42].

Focal application of drugs or electrical stimulation of discrete brain nuclei allows for highly controlled and reproducibly evoked seizures of focal or partial onset. This approach also allows for multiple sites within a network to be manipulated. An example of an especially sensitive and circumscribed site in the forebrain effective for triggering complex partial seizures is "Area Tempestas". This functionally defined region is located in the anterior deep piriform cortex and has been identified in rodents and non-human primates [43–48]. Interestingly, fMRI and PET data suggest that an anatomically homologous area exists in human patients with epilepsy [49]. Moreover, unruptured aneurysms of the middle cerebral artery, located in close proximity to piriform cortex, have been associated with unilateral olfactory auras and complex partial seizures (e.g., [7,8]).

http://dx.doi.org/10.5772/58584

## 6. Multiple seizure networks

While a seizure can appear to "progress" on a continuum from complex partial to generalized tonic-clonic, the progression actually results from successive engagement of independent and dissociable seizure networks: one network supporting complex partial seizures and another supporting tonic-clonic seizures [50–52].

For example, complex partial seizures can be evoked by stimulation of the piriform cortex [43], while activation of the inferior colliculus [53] and/or reticular nuclei [54] triggers tonic-clonic seizures. While it is striking that such focal manipulations can trigger these seizures, the independence of these seizure networks is even more impressive. In both the cat and rat, disconnection of the forebrain from the hindbrain via precollicular transections does not impede the ability of the forebrain to show characteristic EEG seizure responses to focal or systemic chemoconvulsant treatment [50,52,55]. Thus, communication with the hindbrain is not necessary for forebrain seizures. Moreover, these animals are still capable of demonstrating normal tonic-clonic and running/bouncing clonus. Thus, communication with the forebrain is not necessary for hindbrain seizures. These data provide a compelling argument for the independence of these seizure networks, an observation that has been supported by localization of focal trigger zones and circuits for these various seizure types. This leads us to the question, are these seizure "trigger zones" necessarily the same as a "seizure focus"?

## 7. Insights into seizure foci from animal models

It is often assumed that the first site to show ictal activity is the site of seizure initiation. By focally evoking seizures from piriform cortex in the rat, we have found that this is not necessarily the case. Shortly after bicuculline microinjection, piriform cortex displays an interictal like pattern, while ictal activity can be seen in other limbic brain regions. Thus the first ictal activity can appear in a site distal to the site that triggers a seizure.

Clinically, sites of histopathology are often examined as presumptive seizure foci. While in some cases the site(s) of pathology may indeed be the site(s) of seizure onset, animal models have demonstrated that this is not true in all cases. For example, in the tish rat (a model of cortical heterotopia), the *normotopic* neurons, not the heterotopic neurons, are more likely to display epileptiform activity [56]. Moreover, suppression of activity within the heterotopias reduces epileptiform activity *only within the heterotopia* and not within normotopic cortex; conversely, suppression of activity within normotopic cortex suppresses epileptiform activity in both normotopic and heterotopic cortex.

Indeed, even in a highly controlled animal model (e.g., electrically-induced self-sustained status epilepticus), the site within the limbic network showing earliest ictal electrographic activity can vary both between and within subjects [57]. Together, these findings suggest that pathology is not by necessity a clear indicator of the site of seizure initiation. While this does not preclude the possibility that a seizure *can* begin at the site of pathology, it underscores that this is not necessarily the case.

## 8. Translating semiology and terminology across species

Much of the terminology that is used to describe seizures in animal models has been borrowed from the clinic. However, because seizure semiology differs across species, accurate mapping of terms presents a challenge.

For example, there are behavioral differences in seizures evoked from area tempestas in the monkey as compared to rodents. In the monkey, these seizures are characterized by facial automatisms and arm posturing – behaviors that are strikingly similar to those seen during complex partial seizures in humans. These seizures have high face validity. AT-evoked seizures in the rat are typical limbic-motor seizures, similar to those seen after low doses of systemically-administered bicuculline, pentylenetetrazole, kainate, or after electrical kindling [43,58]. These seizures are characterized by facial clonus (perhaps akin to lip smacking seen in patients and monkeys), forelimb clonus (perhaps akin to arm posturing), and rearing with loss of balance. The rearing and loss of balance seen in rats is strikingly different than the behaviors observed in primate species.

Thus, by examining AT-evoked seizures across species, it has become clear that complex partial seizures have species-specific behavioral manifestations but share the qualities of focal automatisms and engage the same brain network.

## 9. Species specific nature of seizure spread: What is a generalized seizure?

In human patients, complex partial seizures that secondarily generalize have two characteristic features: 1) involvement of the whole brain when the seizure generalizes and 2) tonic-clonic manifestations when the seizure generalizes (as compared to automatisms prior to generalization).

In the monkey, AT-evoked seizures can secondarily generalize showing bilaterally asynchronous tonic-clonic and electrographic features. This pattern fits both the electrographic and behavioral definitions used for secondary generalization in humans. In contrast, in the rat, AT-evoked seizures do not show tonic-clonic (brainstem) manifestations, but rapidly show bilateral synchronization of the limbic/cortical EEG and associated motor automatisms with rearing and loss of balance. Thus, in the rat, it appears that the "path of least resistance" for seizure propagation is transcallosal or commissural (hence bilaterally synchronized limbic motor and electrographic manifestations) whereas in the primate it appears to be down the neuraxis (hence the involvement of brainstem seizure networks).

Can, then, the limbic motor seizure with rearing and loss of balance in the rat (i.e., a Racine Stage 5 seizure) be considered a secondarily generalized tonic-clonic seizure? Some have suggested that because these seizures engage basal ganglia, they should be considered generalized [59,60]. However, seizure activity in limbic-evoked motor seizures (i.e., Stage 5 amygdala kindled) engages basal ganglia substrates (substantia nigra pars reticulata) even before other limbic structures (such as hippocampus) [61–63]. Moreover, because these seizures lack the prominent tonic-extensor phase (which requires brainstem engagement) seen in secondarily generalized seizures in monkeys and humans, we suggest that the repeated clonus and rearing/loss of balance that is characteristic of these seizures should not be considered tonic-clonic. On this basis, we suggest that focal limbic seizures (e.g., seizures early in kindling that do not engage the basal ganglia) are akin to simple partial seizures, seizures that spread to the basal ganglia are akin to complex partial seizures (as they are still confined to the forebrain), and only when hindbrain circuits are engaged should these seizures be considered truly generalized.

## 10. Manipulating circuits with focal stimulation as a therapeutic intervention

With the success of deep brain stimulation trials in epilepsy (e.g., stimulation of the anterior nucleus of the thalamus), focal manipulation of circuitry for the control of epilepsy has become a reality. However, identifying the best locations for targeting is a work-in-progress. Continued circuit analysis in animals is essential, not only for identifying targets, but also for examining newer approaches (e.g., optogenetics, chemical-genetics) that offer exciting levels of specificity in cell-type and pathway-specific targeting [64-70].

One approach that remains underexplored clinically is enhancing the function of seizure-suppressive network nodes that have been identified in animal models. One such node is the substantia nigra pars reticulata [71–76]. Suppression of activity within this region is potently anticonvulsant in a variety of seizure models, and across several species. This structure is particularly compelling for further investigation because it is positioned at the interface of two different seizure networks (i.e., the forebrain network, with heavy interconnections to limbic structures, and the hindbrain network, projections to colliculus and brainstem targets). This anatomical position may underlie the anticonvulsant effects that this region exerts across seizure types: it decreases the duration of tonic hindlimb extension triggered by maximal electroshock and it decreases seizures focally evoked from piriform cortex. As we continue to refine our circuit maps, we open the door to therapeutic approaches such as suppressing activity in seizure "distribution" nodes or activation of endogenous "surge suppressors". These possibilities that can only be realized through the use of appropriate animal models.

## Author details

Patrick A. Forcelli* and Karen Gale

*Address all correspondence to: paf22@georgetown.edu

Department of Pharmacology & Physiology, Georgetown University, Washington DC, USA

## References


[1] Jackson JH. A study of convulsions. Trans St Andrews Med Grad Assoc. 1870;1870(3):162–204.

[2] Berger H. Über das Elektrenkephalogramm des Menschen. Arch Für Psychiatr Nervenkrankh. 1929 Dec;87(1):527–70.

[3] Jackson JH. On the anatomical, physiological, and pathological investigations of epilepsies. West Rid Lunatic Asylum Med Rep. 1873;3:315–49.

[4] Jackson JH. Notes on the physiology and pathology of the nervous system. Med Times Gaz. 1868.

[5] McGonigal A, Chauvel P. frontal lobe epilepsy: seizure semiology and presurgical evaluation. Pract Neurol. 2004 Oct;4(5):260–73.

[6] Penfield W, Welch K. Instability of response to stimulation of the sensorimotor cortex of man. J Physiol. 1949 Sep;109(3-4):358–365, illust.

[7] Mizobuchi M, Ito N, Tanaka C, Sako K, Sumi Y, Sasaki T. Unidirectional olfactory hallucination associated with ipsilateral unruptured intracranial aneurysm. Epilepsia. 1999 Apr;40(4):516–9.

[8] Miele VJ, Bendok BR, Batjer HH. Unruptured aneurysm of the middle cerebral artery presenting with psychomotor seizures: case study and review of the literature. Epilepsy Behav EB. 2004 Jun;5(3):420–8.

[9] Bien CG, Benninger FO, Urbach H, Schramm J, Kurthen M, Elger CE. Localizing value of epileptic visual auras. Brain. 2000 Feb 1;123(2):244–53.

[10] Cassidy RM, Gale K. Mediodorsal thalamus plays a critical role in the development of limbic motor seizures. J Neurosci Off J Soc Neurosci. 1998 Nov 1;18(21):9002–9.

[11] Maggio R, Lanaud P, Grayson DR, Gale K. Expression of c-fos mRNA following seizures evoked from an epileptogenic site in the deep prepiriform cortex: regional distribution in brain as shown by in situ hybridization. Exp Neurol. 1993 Jan;119(1):11–9.

[12] Lanaud P, Maggio R, Gale K, Grayson DR. Temporal and spatial patterns of expression of c-fos, zif/268, c-jun and jun-B mRNAs in rat brain following seizures evoked focally from the deep prepiriform cortex. Exp Neurol. 1993 Jan;119(1):20–31.

[13] DeSalvo MN, Schridde U, Mishra AM, Motelow JE, Purcaro MJ, Danielson N, et al. Focal BOLD fMRI changes in bicuculline-induced tonic-clonic seizures in the rat. NeuroImage. 2010 Apr 15;50(3):902–9.

[14] Englot DJ, Modi B, Mishra AM, DeSalvo M, Hyder F, Blumenfeld H. Cortical deactivation induced by subcortical network dysfunction in limbic seizures. J Neurosci Off J Soc Neurosci. 2009 Oct 14;29(41):13006–18.

[15] Mishra AM, Ellens DJ, Schridde U, Motelow JE, Purcaro MJ, DeSalvo MN, et al. Where fMRI and electrophysiology agree to disagree: corticothalamic and striatal activity patterns in the WAG/Rij rat. J Neurosci Off J Soc Neurosci. 2011 Oct 19;31(42): 15053–64.

[16] Kato M, Malamut BL, Caveness WF, Hosokawa S, Wakisaka S, O'Neill RR. Local cerebral glucose utilization in newborn and pubescent monkeys during focal motor seizures. Ann Neurol. 1980 Mar;7(3):204–12, 230–2.

[17] Engel J Jr, Kuhl DE, Phelps ME. Patterns of human local cerebral glucose metabolism during epileptic seizures. Science. 1982 Oct 1;218(4567):64–6.

[18] Foster JA, Puchowicz MJ, McIntyre DC, Herkenham M. Activin mRNA induced during amygdala kindling shows a spatiotemporal progression that tracks the spread of seizures. J Comp Neurol. 2004 Aug 9;476(1):91–102.

[19] Dwyer BE, Fujikawa DG, Wasterlain CG. Metabolic anatomy of generalized bicuculline seizures in the newborn marmoset monkey. Exp Neurol. 1986 Oct;94(1):213–27.

[20] Pitkänen A, Schwartzkroin PA, Moshé SL. Models of seizures and epilepsy. Amsterdam; Boston: Elsevier Academic; 2006.

[21] Suzuki J, Nakamoto Y. Abnormal plastic phenomena of sensory-precipitated epilepsy in the mutant El mouse. Exp Neurol. 1982 Feb;75(2):440–52.

[22] Todorova MT, Burwell TJ, Seyfried TN. Environmental risk factors for multifactorial epilepsy in EL mice. Epilepsia. 1999 Dec;40(12):1697–707.

[23] Forcelli PA, Orefice LL, Heinrichs SC. Neural, endocrine and electroencephalographic hyperreactivity to human contact: a diathesis-stress model of seizure susceptibility in El mice. Brain Res. 2007 May 4;1144:248–56.

[24] Frankel WN, Valenzuela A, Lutz CM, Johnson EW, Dietrich WF, Coffin JM. New seizure frequency QTL and the complex genetics of epilepsy in EL mice. Mamm Genome Off J Int Mamm Genome Soc. 1995 Dec;6(12):830–8.

[25] Noebels JL. Spontaneous Epileptic Mutations in the Mouse. Models of Seizures and Epilepsy. Elsevier; 2006. p. 223–32.

[26] Faingold CL. The genetically epilepsy-prone rat. Gen Pharmacol. 1988;19(3):331–8.

[27] Dailey JW, Yan QS, Adams-Curtis LE, Ryu JR, Ko KH, Mishra PK, et al. Neurochemical correlates of antiepileptic drugs in the genetically epilepsy-prone rat (GEPR). Life Sci. 1996;58(4):259–66.

[28] Faingold CL. Neuronal networks in the genetically epilepsy-prone rat. Adv Neurol. 1999;79:311–21.

[29] Doretto MC, Fonseca CG, Lôbo RB, Terra VC, Oliveira JAC, Garcia-Cairasco N. Quantitative study of the response to genetic selection of the Wistar audiogenic rat strain (WAR). Behav Genet. 2003 Jan;33(1):33–42.

[30] Galvis-Alonso OY, Cortes De Oliveira JA, Garcia-Cairasco N. Limbic epileptogenicity, cell loss and axonal reorganization induced by audiogenic and amygdala kindling in wistar audiogenic rats (WAR strain). Neuroscience. 2004;125(3):787–802.

[31] Buckmaster PS. Inherited Epilepsy in Mongolian Gerbils. Models of Seizures and Epilepsy. Elsevier; 2006. p. 273–94.

[32] Ekenstedt KJ, Patterson EE, Mickelson JR. Canine epilepsy genetics. Mamm Genome Off J Int Mamm Genome Soc. 2012 Feb;23(1-2):28–39.

[33] Menini C, Silva-Barrat C. The photosensitive epilepsy of the baboon. A model of generalized reflex epilepsy. Adv Neurol. 1998;75:29–47.

[34] Yu FH, Mantegazza M, Westenbroek RE, Robbins CA, Kalume F, Burton KA, et al. Reduced sodium current in GABAergic interneurons in a mouse model of severe myoclonic epilepsy in infancy. Nat Neurosci. 2006 Sep;9(9):1142–9.

[35] Covolan L, Mello LE. Temporal profile of neuronal injury following pilocarpine or kainic acid-induced status epilepticus. Epilepsy Res. 2000 Apr;39(2):133–52.

[36] Pitkänen A, McIntosh TK. Animal models of post-traumatic epilepsy. J Neurotrauma. 2006 Feb;23(2):241–61.

[37] Pitkänen A, Immonen RJ, Gröhn OHJ, Kharatishvili I. From traumatic brain injury to posttraumatic epilepsy: what animal models tell us about the process and treatment options. Epilepsia. 2009 Feb;50 Suppl 2:21–9.

[38] Curia G, Levitt M, Fender JS, Miller JW, Ojemann J, D'Ambrosio R. Impact of injury location and severity on posttraumatic epilepsy in the rat: role of frontal neocortex. Cereb Cortex N Y N 1991. 2011 Jul;21(7):1574–92.

[39] Velíšek L. Models of Chemically-Induced Acute Seizures. Models of Seizures and Epilepsy. Elsevier; 2006. p. 127–52.

[40] Cavalheiro EA, Naffah-Mazzacoratti MG, Mello LE, Leite JP. The Pilocarpine Model of Seizures. Models of Seizures and Epilepsy. Elsevier; 2006. p. 433–48.

[41] Dudek FE, Clark S, Williams PA, Grabenstatter HL. Kainate-Induced Status Epilepticus. Models of Seizures and Epilepsy. Elsevier; 2006. p. 415–32.

[42] Snead OC 3rd, Depaulis A, Vergnes M, Marescaux C. Absence epilepsy: advances in experimental animal models. Adv Neurol. 1999;79:253–78.

[43] Piredda S, Gale K. A crucial epileptogenic site in the deep prepiriform cortex. Nature. 1985;317(6038):623–5.

[44] Maggio R, Liminga U, Gale K. Selective stimulation of kainate but not quisqualate or NMDA receptors in substantia nigra evokes limbic motor seizures. Brain Res. 1990;528(2):223–30.

[45] Halonen T, Tortorella A, Zrebeet H, Gale K. Posterior piriform and perirhinal cortex relay seizures evoked from the area tempestas: role of excitatory and inhibitory amino acid receptors. Brain Res. 1994 Jul 25;652(1):145–8.

[46] Fornai F, Busceti CL, Kondratyev A, Gale K. AMPA receptor desensitization as a determinant of vulnerability to focally evoked status epilepticus. Eur J Neurosci. 2005 Jan;21(2):455–63.

[47] Wardas J, Graham J, Gale K. Evidence for a role of glycine in area tempestas for triggering convulsive seizures. Eur J Pharmacol. 1990 Oct 2;187(1):59–66.

[48] Gale K, Dubach M. Localization of area tempestas in the piriform cortex of the monkey. Society for Neuroscience, Annual Meeting. 1993.

[49] Laufs H, Richardson MP, Salek-Haddadi A, Vollmar C, Duncan JS, Gale K, et al. Converging PET and fMRI evidence for a common area involved in human focal epilepsies. Neurology. 2011 Aug 30;77(9):904–10.

[50] Browning RA, Nelson DK. Modification of electroshock and pentylenetetrazol seizure patterns in rats after precollicular transections. Exp Neurol. 1986 Sep;93(3):546–56.

[51] Browning RA, Wang C, Nelson DK, Jobe PC. Effect of precollicular transection on audiogenic seizures in genetically epilepsy-prone rats. Exp Neurol. 1999 Feb;155(2):295–301.

[52] Browning R, Maggio R, Sahibzada N, Gale K. Role of brainstem structures in seizures initiated from the deep prepiriform cortex of rats. Epilepsia. 1993 Jun;34(3):393–407.

[53] Millan MH, Meldrum BS, Faingold CL. Induction of audiogenic seizure susceptibility by focal infusion of excitant amino acid or bicuculline into the inferior colliculus of normal rats. Exp Neurol. 1986 Mar;91(3):634–9.

[54] Peterson SL. The effect on maximal electroshock seizures induced by GABA agents and antiepileptic drugs microinfused into the nucleus reticularis pontis oralis. Epilepsy Res. 1996 Nov;25(3):161–7.

[55] Magistris MR, Mouradian MS, Gloor P. Generalized convulsions induced by pentylenetetrazol in the cat: participation of forebrain, brainstem, and spinal cord. Epilepsia. 1988 Aug;29(4):379–88.

[56] Chen ZF, Schottler F, Bertram E, Gall CM, Anzivino MJ, Lee KS. Distribution and initiation of seizure activity in a rat brain with subcortical band heterotopia. Epilepsia. 2000 May;41(5):493–501.

[57] Bertram EH. Functional anatomy of spontaneous seizures in a rat model of limbic epilepsy. Epilepsia. 1997 Jan;38(1):95–105.

[58] Racine RJ. Modification of seizure activity by electrical stimulation. II. Motor seizure. Electroen Clin Neuro. 1972;32(3):281–94.

[59] Velíšková J. Behavioral Characterization of Seizures in Rats. Models of Seizures and Epilepsy. Elsevier; 2006. p. 601–11.

[60] Englot DJ, Mishra AM, Mansuripur PK, Herman P, Hyder F, Blumenfeld H. Remote effects of focal hippocampal seizures on the rat neocortex. J Neurosci Off J Soc Neurosci. 2008 Sep 3;28(36):9066–81.

[61] Bonhaus DW, Walters JR, McNamara JO. Activation of substantia nigra neurons: role in the propagation of seizures in kindled rats. J Neurosci Off J Soc Neurosci. 1986 Oct;6(10):3024–30.

[62] Bonhaus DW, Russell RD, McNamara JO. Activation of substantia nigra pars reticulata neurons: role in the initiation and behavioral expression of kindled seizures. Brain Res. 1991 Apr 5;545(1-2):41–8.

[63] Shi L-H, Luo F, Woodward DJ, McIntyre DC, Chang J-Y. Temporal sequence of ictal discharges propagation in the corticolimbic basal ganglia system during amygdala kindled seizures in freely moving rats. Epilepsy Res. 2007 Jan;73(1):85–97.

[64] Paz JT, Davidson TJ, Frechette ES, Delord B, Parada I, Peng K, et al. Closed-loop optogenetic control of thalamus as a tool for interrupting seizures after cortical injury. Nat Neurosci. 2013 Jan;16(1):64–70.

[65] Wykes RC, Heeroma JH, Mantoan L, Zheng K, MacDonald DC, Deisseroth K, et al. Optogenetic and potassium channel gene therapy in a rodent model of focal neocortical epilepsy. Sci Transl Med. 2012 Nov 21;4(161):161ra152.

[66] Sukhotinsky I, Chan AM, Ahmed OJ, Rao VR, Gradinaru V, Ramakrishnan C, et al. Optogenetic delay of status epilepticus onset in an in vivo rodent epilepsy model. PloS One. 2013;8(4):e62013.

[67] Armstrong C, Krook-Magnuson E, Oijala M, Soltesz I. Closed-loop optogenetic intervention in mice. Nat Protoc. 2013 Aug;8(8):1475–93.

[68] Berglind F, Ledri M, Sørensen AT, Nikitidou L, Melis M, Bielefeld P, et al. Optogenetic inhibition of chemically induced hypersynchronized bursting in mice. Neurobiol Dis. 2014 May;65:133–41.

[69] Kätzel D, Nicholson E, Schorge S, Walker MC, Kullmann DM. Chemical-genetic attenuation of focal neocortical seizures. Nat Commun. 2014;5:3847.

[70] Forcelli P, Soper C, Gale K. Optogenetic stimulation of the superior colliculus attenuates seizures evoked by pentylenetetrazole. 2013. Available from: https://www.aesnet.org/meetings_events/annual_meeting_abstracts/view/1750955

[71] Garant DS, Gale K. Substantia nigra-mediated anticonvulsant actions: role of nigral output pathways. Exp Neurol. 1987 Jul;97(1):143–59.

[72] Maggio R, Gale K. Seizures evoked from area tempestas are subject to control by GABA and glutamate receptors in substantia nigra. Exp Neurol. 1989;105(2):184–8.

[73] Gale K, Iadarola MJ. Seizure protection and increased nerve-terminal GABA: delayed effects of GABA transaminase inhibition. Science. 1980 Apr 18;208(4441):288–91.

[74] Iadarola MJ, Gale K. Substantia nigra: site of anticonvulsant activity mediated by gamma-aminobutyric acid. Science. 1982;218(4578):1237–40.

[75] Depaulis A, Vergnes M, Marescaux C. Endogenous control of epilepsy: the nigral inhibitory system. Prog Neurobiol. 1994;42(1):33–52.

[76] Veliskova J, Moshe SL. Update on the role of substantia nigra pars reticulata in the regulation of seizures. Epilepsy Curr. 2006;6(3):83–7.

Chapter 2

# Role of EEG in Epilepsy

Manjari Tripathi and Man Mohan Mehendiratta

Additional information is available at the end of the chapter

http://dx.doi.org/10.5772/57430

## 1. Introduction

The human electroencephalogram (EEG) was discovered by the German psychiatrist, Hans Berger, in 1929. Its potential applications in epilepsy rapidly became clear, when Gibbs and colleagues in Boston demonstrated 3 per second spike wave discharge in what was then termed petit mal epilepsy. EEG continues to play a central role in diagnosis and management of patients with seizure disorders—in conjunction with the now remarkable variety of other diagnostic techniques developed over the last 30 or so years – because it is a convenient and relatively inexpensive way to demonstrate the physiological manifestations of abnormal cortical excitability that underlie epilepsy (Smith, 2005) [1].

The electroencephalograph records spontaneous electrical activity generated in cerebral cortex. This activity reflects the electrical currents that flow in the extracellular spaces of the brain, and these reflect the summated effects of innumerable excitatory and inhibitory synaptic potentials upon cortical neurons. This spontaneous activity of cortical neurons is much influenced and synchronized by subcortical structures, particularly the thalamus and high brainstem reticular formation. Afferent impulses from these deep structures are probably responsible for entraining cortical neurons to produce characteristic rhythmic brain wave patterns, such as alpha rhythm and sleep spindles.

The EEG provides confirmation of Hughlings Jackson's concept of epilepsy – that it represents a recurrent, sudden, excessive discharge of cortical neurons; but like other ancillary tests, it must be used in conjunction with clinical data.

## 2. How can EEG help in epilepsy?

- Diagnosis of epilepsy

  - Differential diagnosis of paroxysmal neurological events
  - Distinction between a focal and generalised seizure disorder
  - Identification of syndrome specific changes
  - Recognition of photosensitivity
- Management of epilepsy
  - Assessing risk of recurrence after an unprovoked seizure
  - Selection of antiepileptic treatment
  - Likelihood of seizure relapse if medication is withdrawn
  - Identification of epileptogenic region in epilepsy surgery candidates
  - Investigation of cognitive decline
  - Detection of non-convulsive status
  - Monitoring in convulsive status

Although the diagnosis of seizures and epileptic syndromes is primarily made from careful history and examination, the EEG remains an important investigative tool. The EEG often provides supportive evidence of seizure disorder and assists with classification of seizures and epilepsy. Moreover, EEG findings are important for determination of seizure focus and may also help with prognosis under certain circumstances. (Sundaram M et al, 1999) [2]

## 3. Technical considerations

### 3.1. Electrodes

The international ten-twenty system of electrode placement, originally proposed in 1958 (Jasper, 1958) [3], is now widely used and is the recommended standard method for recording scalp EEG. The American EEG Society has recently advocated slight modifications to the original alphanumeric nomenclature (American EEG Society) [4] The original T3, T4, T5 and T6 are now referred to as T7, T8, P7 and P8 respectively. This modification allows standardized extension of electrode placement in the sub-temporal region (e.g., F9, T9, P9, F10, T10, P10) and designates named electrode positions in the intermediate coronal lines between the standard coronal lines (e.g., AF7, AF3, FT9, FT7, FC5, FC3, FC1, TP9, TP7, CP5, CP3, CP1, PO7, PO3 and so on). Additional and more closely spaced scalp electrodes placed midway between the standard electrodes of the 10-20 system often provide further localization of epileptiform discharges in patients with partial seizures (Morris et al, 1987) [5]. Several electrodes are available for demonstrating temporal lobe activity. Sphenoidal electrodes are particularly useful for detecting mediobasal temporal discharges and are inserted under the mandibular notch (app. 2.5 to 3 cm anterior to the tragus) and directed posterosuperiorly towards the foramen ovale (Rovit et al, 1961). [6] Anterior "cheek" electrodes (placed on the maxilla

http://dx.doi.org/10.5772/57430

approximately 2 cm anterior to the site of entry of the sphenoidal electrode) and anterior temporal electrodes (placed 1 cm above one third the distance from the external auditory meatus to the external canthus) are also useful for demonstrating epileptiform discharges (ED) from the temporal lobe and the yield appears comparable to that from sphenoidal electrodes (Krauss et al, 1992) [7].

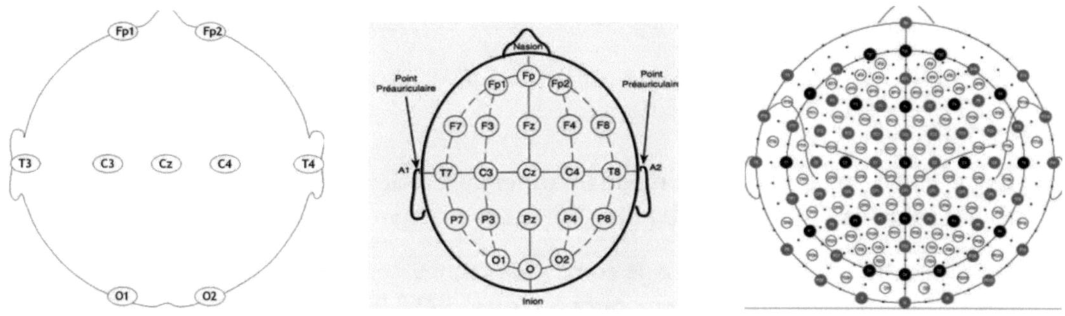


**Figure 1.** Figure of 10x20 system, 10x10, 10x5; neonatal-10-20, 10-5 placement

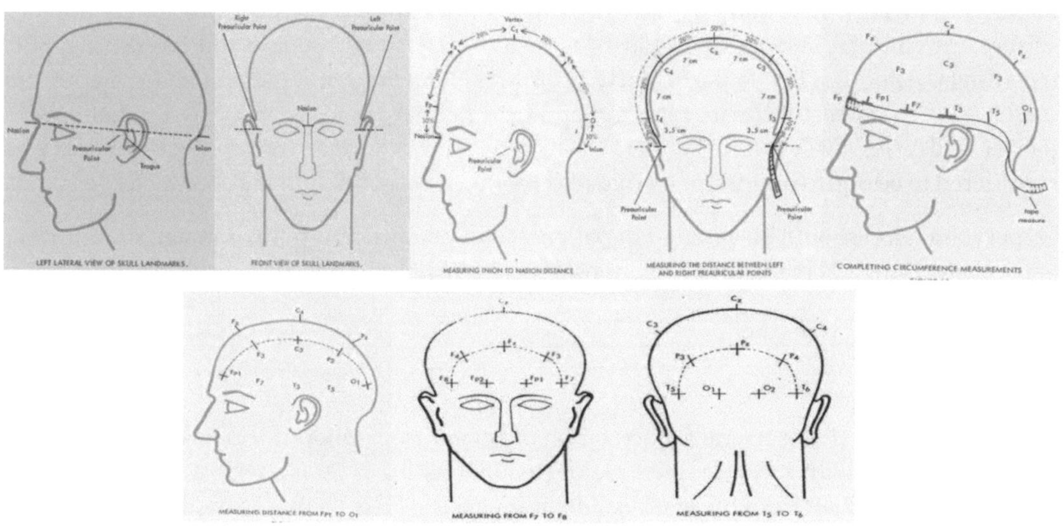

**Figure 2.** Measurement landmarks10-20 landmarks, measurements, deformity adjustment

## 3.2. Digital EEG

Digital recording machines are rapidly replacing the traditional "paper" systems.

Advantages:

- Digital EEG is particularly useful for detecting and analyzing ED as the waveforms in question can be reformatted in various montages after the recording is completed.
- Very little storage space requirement, elimination of paper costs, automatic event detection and the ability to network different recording stations.
- Filter and paper speed settings with digital recordings are accurate and automatic, thereby avoiding technician oversight.
- Problems due to pen alignment and curvilinear effect are not seen with digital systems.

Disadvantages:

- The incompatibility of systems made by different vendors, often forcing one to resort to paper printouts for transmission of EEG data between two centers.
- Comparing two separate epochs is somewhat cumbersome, as only limited data can be observed simultaneously on the monitor (Gorney, 1992) [8].

### 3.3. Activation procedures

#### *3.3.1. Hyperventilation*

Forster [9], in 1924, first demonstrated that hyperventilation (HV) may precipitate absence seizures in children and this method of activation has since become routine during EEG recordings. Although HV is particularly useful for demonstrating generalized epileptiform discharges, it may also activate focal epileptiform discharges in up to 10% of patients with partial epilepsies (Miley and Forster, 1977) [10]. The neuronal irritability during HV is considered to be due to brainstem mediated cerebral vasoconstriction induced by hypocapnia.

Hyperventilation should be avoided in patients with potential for brain damage from further vasoconstriction, e.g. malignant hypertension, subarachnoid hemorrhage, sickle cell disease or trait.

#### *3.3.2. Photic stimulation*

Photic stimulation (PS) is useful for activation of generalized epileptiform discharges. Testing is generally done with stepwise increase of frequencies up to 30 Hz with a strobe light at a distance of 20 to 30 cm from the eyes. At low frequencies, PS is recommended with eyes open and then closed. At medium and higher frequencies, stimulation should start with the eyes open, and the patient is asked to close the eyes during PS, thereby continuing with PS for a few more seconds with the eyes remaining closed. Eye closure during PS is particularly useful for augmenting ED and should routinely be used. ED outlasting PS strongly suggest generalized seizure disorder, whereas those confined to the train of PS may be an incidental finding in nonepileptic subjects, especially in the setting of drug withdrawal or toxic metabolic encephalopathy, or simply represent a genetic trait (Newmark and Penry, 1979) [11].

Photic stimulation is particularly useful in primary generalized epilepsy and ED may occur during PS in up to 40% of these patients (Gastaut et al, 1958) [12]. Recent evidence indicates that approximately a quarter to a third of EEGs with photic related ED also contain spontaneous focal or generalized ED elsewhere in the records (Gilliam and Chiappa, 1995) [13].

#### *3.3.3. Sleep Deprivation (SD)*

When the first EEG fails to show ED in patients with epilepsy, sleep deprived recording often helps. Several studies have convincingly documented that the chances of finding ED increase with sleep deprived recordings in both partial and generalized seizure patients of all ages (Degan, 1980) [14]. Epileptiform discharges following sleep deprivation occur both in the awake and sleep portions of the EEG. Moreover, Rowan and co-workers (1982) [15] have shown that EEGs following sleep deprivation are more likely to contain ED than the recordings of similar length done following sedation.

## 4. Clinical Significance of Interictal Epileptiform Discharges (ED)

### 4.1. ED in nonepileptic subjects

Although the presence of interictal ED generally supports the diagnosis of seizure disorder, caution is necessary in interpreting the clinical significance as ED may occur in subjects without seizures. Among healthy adults without seizure history, the frequency of ED is approximately 0.5% (Robin et al, 1978) [16]. Practically none of these healthy subjects subsequently develops seizures. "Incidental" ED occur slightly more often (app. 2%) in subjects with a history of previous neurological insults such as trauma, stroke, craniotomy, infections, cerebral palsy or during migraine (Zivin et al, 1968) [17]. Up to 14% of these patients subsequently develop seizures. In children without prior seizures, ED may occur in up to 5% and this may be as high as 8% if adequate sleep is recorded (Okubo et al, 1994) [18]; these tend to be benign rolandic or occipital spikes or generalized 3 Hz spike-wave discharges and likely represent incidental genetic trait. Risk of subsequent seizures in these children is around 6% (Cavazzuti et al, 1980) [19]. Certain EEG patterns, however, almost always indicate associated clinical seizures and these include hypsarrhythmia and 1 or 2 Hz generalized slow spike-wave complexes.

### 4.2. ED in the first and serial EEGs

First standard EEGs in patients with a reasonably certain diagnosis of seizure disorder contain ED in approximately 50% (Ajmone-Marsan et al, 1970) [20]. Yield from the first EEG in children with absence seizures, however, is higher, around 75% (Goodin and Aminoff, 1984) [21]. Apart from sleep, several other factors have been shown to increase the likelihood of ED and these include i) recording within 48 hours of a seizure and ii) ongoing seizure frequency of at least one attack per month (Sundaram et al, 1990) [22]. The yield, however, is not significantly altered by neurological status, etiology of seizures, age of the patient and anti-epileptic drug therapy (Sundaram et al, 1990) [22].

Serial EEGs are often necessary for demonstrating ED. Most patients who eventually show ED do so by the fourth EEG. Recordings are persistently negative in only 8% of epileptics although there is evidence that a higher proportion of patients with partial seizures may have persistently negative serial EEGs (Sundaram et al, 1990) [22].

The above observations suggest that –

**i.** the ideal time for obtaining an EEG is the first day or two after a seizure,

**ii.** one should consider long-term monitoring if four routine recordings have remained negative in patients with ongoing "seizures".

## 5. Ictal EEG

While interictal ED generally provides support for the diagnosis of seizure disorder, electrographic or clinical seizures during EEG confirm seizures. The scalp EEG may not reflect all of the ictal activity as this depends on –

**i.** the frequency-filtering properties of the skull and scalp,

**ii.** the distance and orientation of the focus from the recording electrode and,

**iii.** the surface area of the focus with respect to the recording electrode. In spite of these limitations, scalp recorded seizures provide valuable information regarding the seizure type and focus.

### 5.1. Partial seizures

Partial seizures, in scalp EEGs, are metamorphic, i.e., they show two or more distinct phases (Sharbrough, 1993) [23]. The most common patterns consist of a series of rhythmic waves, sequential spikes/sharp waves, a mixture of spikes and rhythmic waves or regional voltage attenuation. Most often the initial frequency of temporal lobe seizures is in the alpha or theta range with slower frequencies occurring in a lesser proportion (Geiger and Harner, 1978) [24]. Extra temporal seizures, however, often start in the beta frequencies rather than slower frequencies. With scalp EEG, the frequency may diminish or augment, but as the seizure ends, rhythmic waves or sequential spikes change to a slow spike-wave pattern that gradually decreases in frequency. Focal electrodecremental events are of excellent localizing value, reflecting intense neuronal depolarization or high frequency firing (Sharbrough, 1993) [23]. Following metamorphic seizures, there is often postictal delta slowing, suppression or activation of focal spikes. These postictal changes also have good localizing value for seizure origin and should be carefully sought (Kaibara and Blume, 1988) [25].

It is important to recognize that simple partial seizures, especially those with sensory rather than motor symptoms, may not be associated with discernable changes in routine scalp EEG in up to 80% of seizures (Devinsky et al, 1988) [26]. However, the yield in these patients may be augmented by using additional closely spaced electrodes (Bare et al, 1994) [27].

### 5.2. Generalized seizures

Typical absence seizures are characterized by isomorphic and stereotyped patterns that do not evolve as partial seizures. However, the spike-wave discharges may change from 3.5 or 4 Hz at the onset to 2 or 3 Hz as the seizure progresses. Also, the spike amplitude may decrease during the later part of the seizure. Atypical absence attacks frequently show gradual onset and offset with spike-wave discharges occurring at frequencies less than 3 Hz.

Generalized tonic-clonic seizures may be preceded by diffuse polyspike-wave complexes. Ictal recordings during the tonic phase typically shows generalized attenuation with or without high frequency rhythmic waves that gradually increase in voltage ("epileptic recruiting rhythm") and evolve into polyspikes. The clonic phase is characterized by paroxysmal spike activity mixed with slow waves and the post-ictal period shows generalized attenuation followed by gradual recovery of rhythms (Gastaut et al, 1972) [28].

Myoclonic seizures are associated with 10 to 15 Hz polyspikes with or without slow waves, whereas tonic seizures show generalized paroxysmal fast activity or diffuse voltage attenuation preceded or followed by sharp and slow wave complexes. Generalized atonic seizures may show 2-3 Hz spike-wave discharges or may not be associated with any scalp EEG change.

### 5.3. PLEDs (Periodic lateralized epileptiform discharges)

Although PLEDs have traditionally been considered "interictal" (Young et al, 1988) [29], there is some evidence that this pattern in some patients may be "ictal" in nature, especially when seen following traditional ictal patterns (Handforth et al, 1994) [30]. Moreover, Reiher and co-workers (1991) [31] reported that the "PLEDs plus pattern", consisting of periodic epileptiform activity closely followed by brief, low amplitude, stereotyped rhythmic discharges, is often associated with clinical seizures and may indeed be a foreteller of imminent seizures. However, whether the PLEDs or PLEDs plus pattern requires aggressive treatment similar to status epilepticus remains unclear (Treiman, 1995) [32].

## 6. Prognosis of epilepsy

Routine EEG is useful for prognostic purposes in at least three situations:

**i.** assisting in epilepsy syndrome classification,

**ii.** predicting recurrence after the first seizure and

**iii.** providing information on seizure relapse after anticonvulsant withdrawal.

### 6.1. Classification of epilepsy

The EEG provides important information for classification of various epileptic syndromes and thereby assists in predicting the natural history of the syndrome. For example, a child with normal neurological examination and rolandic spikes in EEG has a high probability of "outgrowing" seizures and may not even need treatment following isolated, infrequent seizures. Similarly, generalized 4-6 Hz spike-wave and polyspike discharges in an adolescent with seizures suggest juvenile myoclonic epilepsy of Janz: a condition with a high response rate to valproic acid.

### 6.2. First seizure

Prediction of recurrence after a single seizure is clinically important and many studies have addressed this question. However, differences in methodology make comparison of these studies difficult and the results still remain somewhat controversial. A meta analysis of sixteen published reports suggests that EEG abnormalities may increase the risk of recurrence after first seizure (Berg and Shinnar, 1991) [33].

A recent large prospective study of children with single unprovoked seizure (Shinnar et al, 1994) [34] showed that, in those without obvious etiology ("idiopathic"), the presence of epileptiform discharges in the EEG was associated with a recurrence rate of 54% whereas the rate was only 25% when the first EEG was normal. In the above study, the EEG was not of any predictive value in children with remote symptomatic seizures.

Several recent prospective studies suggest that the EEG is useful in adults with first seizure, especially among those with idiopathic seizures (Van Donselaar et al, 1992) [35]. The Dutch workers (Van Donselaar et al, 1992) [35] showed that when two EEGs (one baseline and one sleep deprived recording) are normal, the recurrence rate was 12% at two years, whereas in those with one or both EEGs containing ED, recurrence rate increased to 83%. The Italian first seizure trial group (1993) [36] also showed a 1.7 fold increase in seizure recurrence when the EEG contained ED. Some controversy still exists in this area as some authors maintain that the EEG findings are of no predictive value after first seizure (Hopkins et al, 1988) [37].

### 6.3. Anti-epileptic drug withdrawal

The role of EEG in predicting relapses after anti-epileptic drug withdrawal remains more controversial. A recent meta analysis discussing in depth various factors in predicting relapses after anti-epileptic drug withdrawal indicates that any EEG abnormality (epileptiform activity or slowing) is associated with a relative relapse risk of 1.45 (Berg and Shinnar, 1994) [38]. Other factors found to increase the relapse rate in the above meta analysis were adolescent or adult epilepsy onset (rather than childhood onset) and known remote etiology.

## 7. Video EEG monitoring

Although the EEG remains the gold standard for confirming seizures, an actual attack or event is rare during a standard 20 to 30 minute recording. Even serial EEGs may fail to reveal ED in up to 10% of epileptics (Ajmone-Marsan et al, 1970) [20]. When the nature of attacks or the exact seizure focus cannot be ascertained with several routine EEGs, telemetry monitoring often provides necessary additional information. With current telemetry systems, EEG data may be collected continuously for several days or even weeks. This may be done as an inpatient procedure using VEEG or at home/work environment with ambulatory EEG. The equipment also has video capability and provides an opportunity to analyze physical changes during the ictus. Most of the equipment available today is highly sophisticated and digitised and portable.

VEEG monitoring is useful:

**i.** for confirming the nature of epileptic attacks and nonepileptic events such as pseudoseizure, paroxysmal movement disorders, and sleep disorders

**ii.** for determination of seizure focus in patients with atypical features (e.g. frontal lobe seizures, gelastic seizures), and for presurgical evaluation

**iii.** for exact classification of seizures prior to appropriate therapy

**iv.** for assessing the response to anticonvulsant therapy, and

**v.** for research purposes, e.g. analysis of the relationship between the quantity of interictal spikes and clinical seizures, sleep etc.

Video telemetry is generally indicated when visual analysis of physical changes during the event is necessary as in pseudoseizure, frontal lobe seizures, and paroxysmal movement disorders. Ambulatory monitoring without video may be sufficient for confirming the nature of events such as syncope or absence attacks. VEEG is a labour intensive and costly method of investigating patients with difficult to control epilepsy. It involves continuous video and synchronised EEG recording done over many hours usually more than 24 hrs with documentation of at least 3 events or more (specially if discordant). The VEEG is also used in the differential diagnosis of the epilepsy specially when nonepileptic events are suspected. A short term VEEG (3-6 hrs) may be performed in patients where psychogenic non epileptiform events are suspected. It is also useful when the number of episodes are several in a day.

All procedures should be carried out ONLY by trained technicians and Neurologists trained in epilepsy and epilepsy monitoring. Since VEEG does carry a risk a standard operating procedure and manual should be available in all centres carrying this out.

**Pearl**

A normal EEG does not rule out the diagnosis of epilepsy.

Often the more the number of EEGs more the chance of picking up an abnormality.

The yield inceases with performing an EEG with both sleep and awake state.

An abnormal EEG should be interpreted according to the clinical situation.

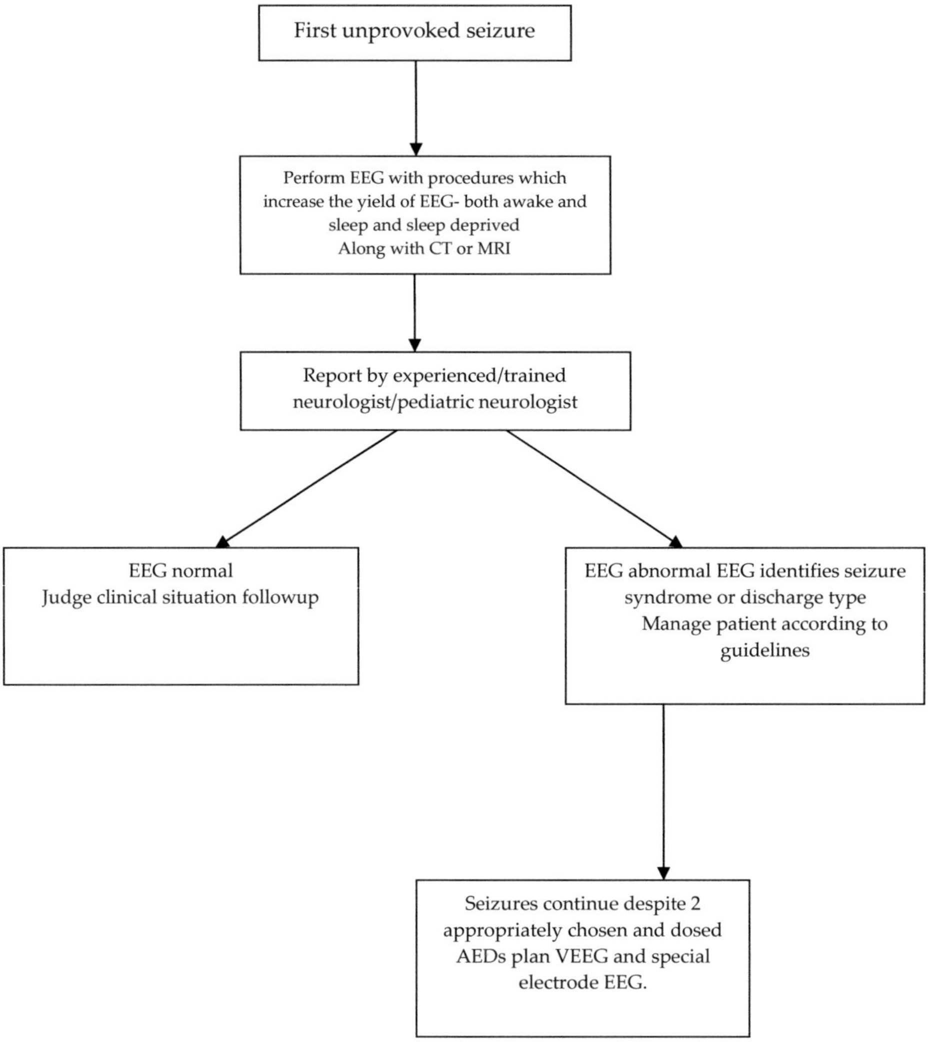


**Figure 3.** Documentation before recording: Patient demographic details such as name, age, clinical diagnosis or indication for EEG, state of the patient, medication details, test number and comments have to be entered.

## Author details

Manjari Tripathi[1*] and Man Mohan Mehendiratta[2]

*Address all correspondence to: manjari.tripathi@gmail.com

1 Neurology, Neurosciences Centre, AIIMS, New Delhi, India

2 Neurology, GB Pant Hospital, Delhi, India

## References

[1] Smith S J M. EEG in the diagnosis, classification, and management of patients with epilepsy. Journal of Neurol Neurosurg and Psychiatry 2005;76:ii2-ii7

[2] Sundaram M, Sadler RM, Young GB, Pillay N. EEG in Epilepsy: Current Perspectives. Can. J. Neurol. Sci. 1999; 26: 255-262

[3] Jasper H. Report of committee on methods of clinical exam in EEG. Electroencephalogr Clin Neurophysiol 1958; 10:370-375.

[4] American EEG Society. Guidelines for standard electrode position nomenclature. J Clin Neurophysiol 1991; 8:200-2.

[5] Morris HH, Luders HL, Lesser RP, Dinner DS, Wyllie E. Value of closely spaced electrodes in the localization of epileptiform foci: a study of 26 patients with complex partial seizures. Electroencephalogr Clin Neurophysiol 1987; 63:107-11.

[6] Rovit RL, Gloor P, Rassmussen T. Sphenoidal electrodes in the electrographic study of patients with temporal lobe epilepsy. J Neurosurg 1961; 18:151-158.

[7] Krauss GL, Lesser RP, Fisher RS, Arroyo S. Anterior "cheek" electrodes are comparable to sphenoidal electrodes for the identification of ictal activity. Electroencephalogr Clin Neurophysiol 1992; 83:333-338.

[8] Gorney DS. The practical guide to digital EEG. Am J EEG Technol 1992; 32:260-289.

[9] Forster O. Hyperventilations epilepsie Dtsch Z. Nervenheilkd 1924; 83:347-356.

[10] Miley CE, Forster FM. Activation of partial complex seizures by hyperventilation. Arch Neurol 1977; 34:371-373.

[11] Newmark ME, Penry JK. Photosensitivity and Epilepsy: A Review. 1979; New York: Raven Press.

[12] Gastaut H, Trevisan C, Naquet R. Diagnostic value of electroencephalographic abnormalities provoked by intermittent photic stimulation. Electroencephalogr Clin Neurophysiol 1958; 10:194-195.

[13] Gilliam FG, Chiappa KH. Significance of spontaneous epileptiform abnormalities associated with a photoparoxysmal response. Neurology 1995; 45:453-456.

[14] Degan R. A study of the diagnostic value of waking and sleep EEGs after sleep deprivation in epileptic patients on anticonvulsive therapy. Electroencephalogr Clin Neurophysiol 1980; 49:577-584.

[15] Rowan J, Veldhuisen RJ, Nagelkerke NJD. Comparative evolution of sleep deprivation and sedated sleep EEGs as diagnostic aids in epilepsy. Electroencephalogr Clin Neurophysiol 1982; 54:357-364.

[16] Robin JJ, Tolan GD, Arnold JW. Ten year experience with abnormal EEGs in asymptomatic adult males. Aviation Space Env Med 1978; 49: 732-736.

[17] Zivin L, Ajmone-Marsan C. Incidence and prognostic significance of epileptiform activity in the EEG of nonepileptic subjects. Brain 1968; 91:751-778.

[18] Okubo Y, Matsuura M, Asai T, Asai K, Kato M, et al. Epileptiform EEG discharges in healthy children: prevalence, emotional and behavioral correlates and genetic influences. Epilepsia 1994; 35:832-841.

[19] Cavazzuti GB, Cappella L, Nalin A. Longitudinal study of epileptiform EEG patterns in normal children. Epilepsia 1980; 21:43-55.

[20] Ajmone-Marsan C, Zivin LS. Factors related to the occurrence of typical paroxysmal abnormalities in the EEG records of epileptic patients. Epilepsia 1970; 11:361-381.

[21] Goodin DS, Aminoff MJ. Does the interictal EEG have a role in the diagnosis of epilepsy? Lancet 1984; i:837-839.

[22] Sundaram M, Hogan T, Hiscock M, Pillay N. Factors affecting interictal spike discharges in adults with epilepsy. Electroencephalogr Clin Neurophysiol 1990; 75; 358-360.

[23] Sharbrough FW. Scalp recorded ictal patterns in focal epilepsy. J Clin Neurophysiol 1993; 10:262-267.

[24] Geiger LR, Harner RN. EEG patterns at the time of focal seizure onset. Arch Neurol 1978; 35:276-286.

[25] Kaibara M, Blume WT. The postictal electroencephalogram. Electroencephalogr Clin Neurophysiol 1988; 70:99-104.

[26] Devinsky O, Kelley K, Porter RJ, Theodore WH. Clinical and electroencephalographic features of simple partial seizures. Neurology 1988; 38:1347-1352.

[27] Bare MA, Burnstine TH, Fisher RS, Lesser RP. Electroencephalographic changes during simple partial seizures. Epilepsia 1994; 35:715-720.

[28] Gastaut H, Broughton R. Epileptic Seizures. Clinical and Electrographic Seizures: Diagnosis and Treatment. Springfield, Ill: Charles C. Thomas 1972.

[29] Young GB, Goodenough P, Jacono V, Schieven JR. Periodic lateralized epileptiform discharges (PLEDs): electrographic and clinical features. Am J EEG Technol 1988; 28: 1-13.

[30] Handforth A, Cheng JT, Mandelkern MA, Treiman DM. Markedly increased mediotemporal lobe metabolism in a case with PLEDs: further evidence that PLEDs are a manifestation of partial status epilepticus. Epilepsia 1994; 35:876-881.

[31] Reiher J, Rivest J, Grand'Maison F, Leduc CP. Periodic lateralized epileptiform discharges with transitional rhythmic discharges: association with Seizures. Electroencephalogr Clin Neurophysiol 1991; 78:12-17.

[32] Treiman DM. Electroclinical features of status epilepticus. J Clin Neurophysiol 1995; 12:343-362.

[33] Berg AT, Shinnar S. The risk of seizure recurrence following a first unprovoked seizure: a quantitative review. Neurology 1991; 41: 965-972.

[34] Shinnar S, Kang H, Berg AT, Goldensohn ES, Hauser WA, Moshe SL. EEG abnormalities in children with a first unprovoked seizure. Epilepsia 1994; 35: 471-476.

[35] Van Donselaar CA, Schimsheimer RJ, Geerts AT, Declerck AC. Value of the electroencephalogram in adults with untreated idiopathic first seizures. Arch Neurol 1992; 42: 231-287.

[36] First seizure trial group. Randomized clinical trial on the efficacy of anti-epileptic drugs in reducing the risk of relapse after a first unprovoked tonic-clonic seizure. Neurology 1993; 43: 478-483.

[37] Hopkins A, Garman A, Clarke C. The first seizure in adult life: value of clinical features, electroencephalography, and computerized tomographic scanning in prediction of seizure recurrence. Lancet 1988; 1: 721-726.

[38] Berg AT, Shinnar S. Relapse following discontinuation of antiepileptic drugs: meta analysis. Neurology 1994; 44:601-8.

[39] Nayak DS, Sajeesh P. Technical standards for digital electroencephalogram recording in epilepsy practice. Ann Indian Acad Neurol 2007;10:121-7

Chapter 3

# Nonlinear Epilepsy Forewarning by Support Vector Machines

W.S. Ashbee, L.M. Hively and J.T. McDonald

Additional information is available at the end of the chapter

http://dx.doi.org/10.5772/57438

## 1. Introduction

Epilepsy is a neurological disorder that changes the observable behavior of an individual to the point of inducing complete loss of consciousness. Pharmaceutical drugs may reduce or eliminate the problems of epilepsy, but not all people respond to pharmaceuticals favorably, and some may find the side effects undesirable. EEG-based epilepsy prediction may offer an acceptable alternative or complementary treatment to pharmaceuticals. Invasive, intra-cranial EEG provides signals that are directly from the brain, without the muscular activity that infests non-invasive, scalp EEG. However, intra-cranial EEG requires surgery, which increases risk and cost of health care, while reducing the number of people able to receive medical attention. Algorithms to predict the seizure event—the ictal state—may lead to new treatments for chronic epilepsy. Finding solutions that involve non-invasive procedures may result in treatments for the largest section of the population.

## 2. Background

Epilepsy prediction is greater than 1 minute of forewarning before there is any visible indication that a seizure will occur. The physician does not label the pre-ictal periods that precede the seizure—states that may indicate a seizure is near. Event characterization only labels the start time of the seizure. Consequently, labeled data for the pre-ictal state is non-existent, but is necessary to train a Support Vector Machine (SVM). Other researchers address this problem by assuming that the pre-ictal phase occurs immediately prior to a seizure [1]; see Figure 1 for an example.

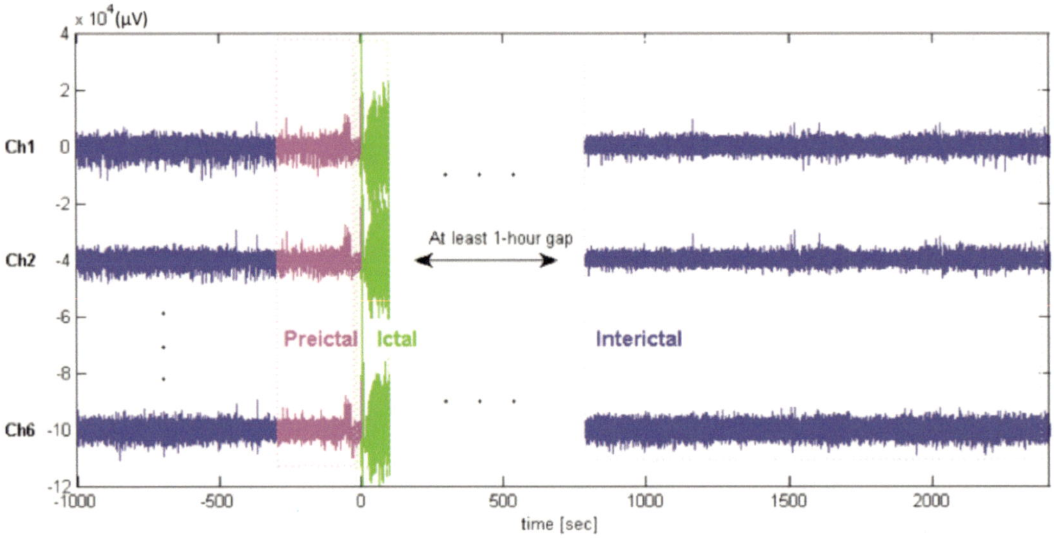


**Figure 1.** EEG with a seizure (ictal) event and potential class labels [1].

The labeling scheme of Figure 1 results in better than random predictions [1] under the assumption that the pre-ictal region immediately precedes the seizure and may be exploited for epilepsy prediction. This SVM approach provides the most obvious way to label the training and testing data without any extra information being available about the EEG.

Assuming pre-ictal dynamics occur within an hour of the seizure has the added benefit of being more likely to satisfy caregivers' requests to have forewarning within an hour of the seizure event. Netoff et al. achieve a specificity of 77% in classifying the pre-ictal region with no false positives with the above approach [1]. This level of accuracy is not high enough for a marketable prediction algorithm, but suggests that indicators of a seizure occur within an hour prior to a typical seizure. Netoff et al. use a "5 minute prediction horizon" where they label the pre-ictal region. They classify preictal as being within 5 minutes of the seizure and calculate specificities according to that labeling scheme. They assert that the short time frame makes the computational difficulty of the algorithm much more manageable than algorithms that have fewer restrictions on where the pre-ictal region is. They have a second stage of processing as well in which they look for 3 out of 5 pre-ictal indicators in a concentrated bundle in order to achieve prediction [1].

The assumption of pre-ictal indications near the event seems sound because a seizure resembles a dynamical phase transition. More specifically, the brain activity changes from some "normal" phase of brain activity into hyper-synchronous activity. The present work assumes that the brain dynamics within an hour of the seizure are approaching a phase transition, corresponding to measureable change in the scalp EEG. A simple example of a phase transition is liquid water becoming steam due to changes in pressure and temperature. However, scalp EEG exhibits nonlinear, chaotic features that are extremely difficult to predict over long periods and are extremely sensitive to initial conditions. Consequently, seizure prediction in a very complex system like the brain is very difficult. Indeed, Stacey *et al.* [2] find that no algorithm provides better-than-chance prediction of seizures in statistical tests to date.

One must also choose whether to use monopolar (single channel) or bipolar EEG (difference between two monopolar channels). Mirowski et al. assert that epilepsy can be predicted more effectively with bipolar features [3] because of changes in the brain's ability to synchronize regions during a seizure. Mirowski et al. consider their pre-ictal period to be 2 hours. They assert, "[most] current seizure prediction approaches can be summarized into (1) extracting features from EEG and (2) classifying them (and hence the patient's state) into pre-ictal or inter-ictal". They go on to enumerate more specifically on the bipolar feature set in Figure 2.

(a) Bivariate features are computed on 5s windows ($N$=1280 samples at 256Hz) of any two EEG channels $x_i$ and $x_j$.
(b) For $M$ EEG channels, one computes features on $M \times (M-1)/2$ pairs of channels (e.g. 15 pairs for $M$=6).
(c) Features are aggregated for several consecutive time frames, e.g. 12 frames (1min) or 60 frames (5min).

**Figure 2.** Approach to using bivariate features in epilepsy prediction [3].

Figure 2 enumerates a feature set from all unique channel pairs [3]. After the enumeration, they use a grid search to find appropriate parameters with their SVM with a Gaussian kernel. Mirowski et al. use intra-cranial EEG data and obtain 100% accuracy for patient-specific machine learning models. However, no single model provides 100% accuracy for all patients [3], so they choose from among a variety of algorithms to achieve high accuracy on a patient specific basis.

By contrast, the present work uses non-invasive, scalp EEG. Moreover, the present work uses a SVM to extract seizure forewarning from the entire patient population. The goal is high accuracy. The long-term objective (not addressed in the present work) is lower health-care cost by using one algorithm for all patients to analyze scalp EEG on a smartphone.

Previous work by Hively *et al.* has used bipolar, scalp EEG and found the best seizure forewarning by using electrodes in the right frontal lobe [4, 5] in the 10-20 system; see Figure 3. The present work uses this same bipolar channel. Additionally, scalp EEG from one bipolar channel facilitates a simple, ambulatory device with two electrodes, which is far more manageable than an EEG headset with many channels. Our hypothesis is that the right-frontal region acts as a filter for pre-ictal condition change—a phase transition in the brain dynamics [6] that can be induced by noise [7]. Pittau *et al.* [8] reviewed the recent technical literature on sound-induced (musicogenic) seizures, which activate the fronto-temporo-occipital area.

Conversely, soothing music (e.g., Mozart's double piano sonata K.448) decreases the intensity and frequency of epileptic seizures [9].

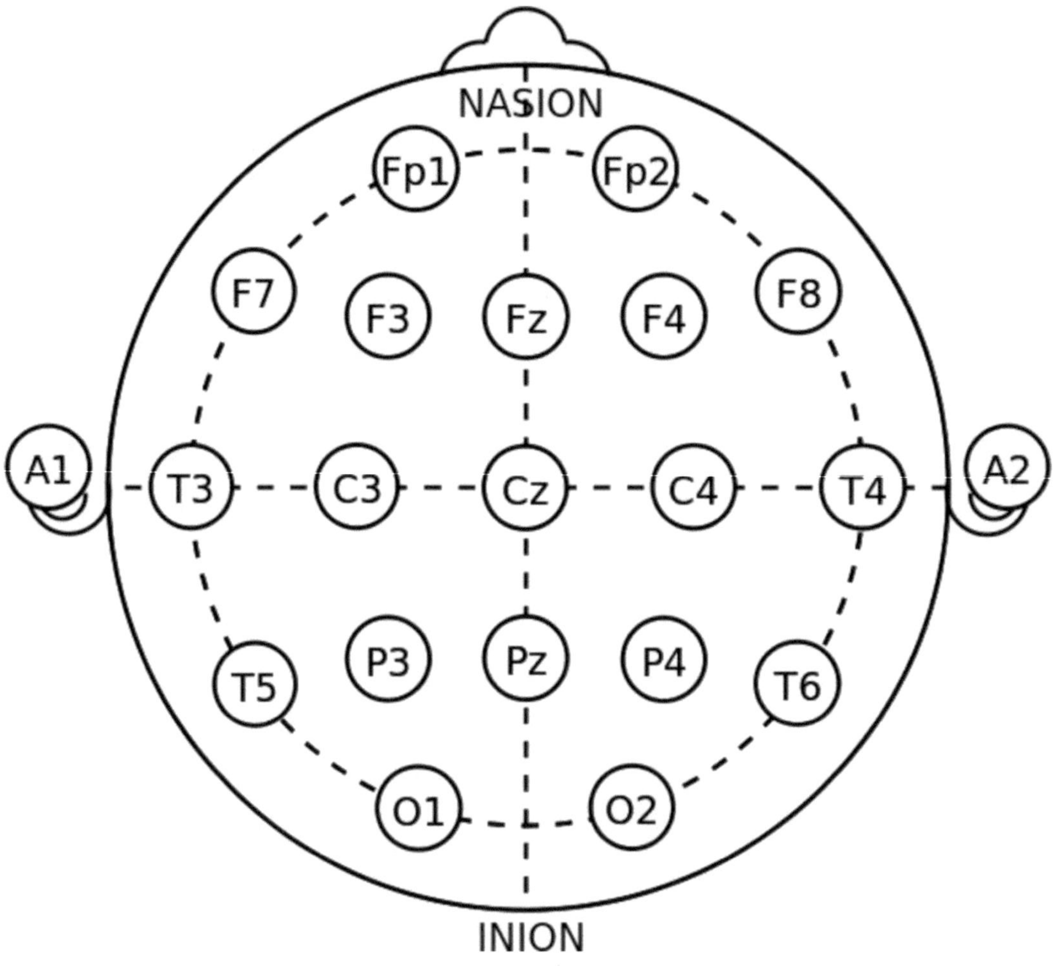


**Figure 3.** system (EEG) [10].

## 3. Phase-space analysis

We use one bipolar channel of *scalp* EEG (F8 – FP2) in the 10-20 system, as a measure of the noisy dynamics in cortical neurons over an area of roughly 6 $cm^2$. Our earlier work obtained channel-consistent forewarning across nineteen EEG channels [11]. The garbage-in-garbage-out syndrome is avoided by rejecting data of inadequate quality [12].

These data were uniformly sampled in time, $t_i$, at 250 Hz, giving $N$ time-serial points in an analysis window (cutset), $e_i = e(t_i)$. Data acquisition was under standard human-studies protocols from 41 temporal-lobe-epilepsy patients (ages from 4 to 57 years; 36 datasets from females, and 24 datasets from males). The datasets range in length from 1.4 to 8.2 hours (average = 4.4 hours). Data characterization included patient activity. Forty datasets had seizures, and twenty had no event [13].

A patented zero-phase, quadratic filter enables analysis of scalp EEG by removing electrical activity from eye blinks and other muscular artifacts, which otherwise obscure the event forewarning. This filter retains the nonlinear amplitude and phase information [14]. The filter uses a moving window of $2w + 1$ points of $e_i$-data that are fitted to a parabola in a least-squares sense, yielding $N - 2w$ points of artifact data, $f_i$. Essentially no low-frequency artifacts occur in the artifact-filtered signal, $g_i = e_i - f_i$. The value, w, is a parameter that specifies the width in points sampled of the eye blink filter; the value, N, represents the number of points in a cutset, which is represented by a graph; the value, g, is the artifact filtered set of points with eye blinks removed; the value, e, is the set of raw EEG data points; and the value, f, is the set of artifact filter points used to subtract out eye blinks.

A trade-off is required between coarseness in the data to exclude noise, and precision in the data to accurately follow the dynamics. Thus, the artifact-filtered data ($g_i$) are symbolized into $S$ discrete values, $s_i$, that are uniformly distributed between the maximum ($g_x$) and minimum ($g_n$) in the first base case cutset. Uniform symbols are generated by the form in Eq. (1).

$$0 \le s_i = \mathrm{INT}\left[S\frac{g_i - g_n}{g_x - g_n}\right] \le S - 1 \tag{1}$$

Here, INT converts a decimal number to the closest lower integer. Takens' theorem [15] gives a smooth, non-intersecting dynamical reconstruction in a sufficiently high dimensional space by a time-delay embedding. The symbolized data from Eq. (1) are converted into unique dynamical states by the Takens' time-delay-embedding vector, $y_i$:

$$y_i = \left[s_i,\ s_{i+L},\ \ldots,\ s_{i+(d-1)L}\right] \tag{2}$$

Takens' theorem allows the $y_i$-states to capture the topology (connectivity and directivity) of the underlying dynamics. The time-delay lag is $L$, which must not be too small (making $s_i$ and $s_{i+L}$ indistinguishable) or too large (making $s_i$ and $s_{i+L}$ independent by long-time unpredictability). The embedding dimension is $d$, which must be sufficiently large to capture the dynamics, but not too large to avoid over-fitting.

The states from Eq. (2) are nodes. The process flow, $y_i \rightarrow y_{i+M}$, forms state-to-state links. The nodes and links give a formal, diagrammatic construction, called a "graph." This form gives topologically-invariant measures that are independent of any unique labeling of individual nodes and links [16]. Figure 4 depicts the algorithmic steps to: extract the analysis window from the stream of EEG data; remove the artifacts from scalp EEG; symbolize the artifact-filtered data; and construct the graph nodes and links [4]. The parameter space in Figure 4 enumerates the parameters used to generate the phase space graphs.

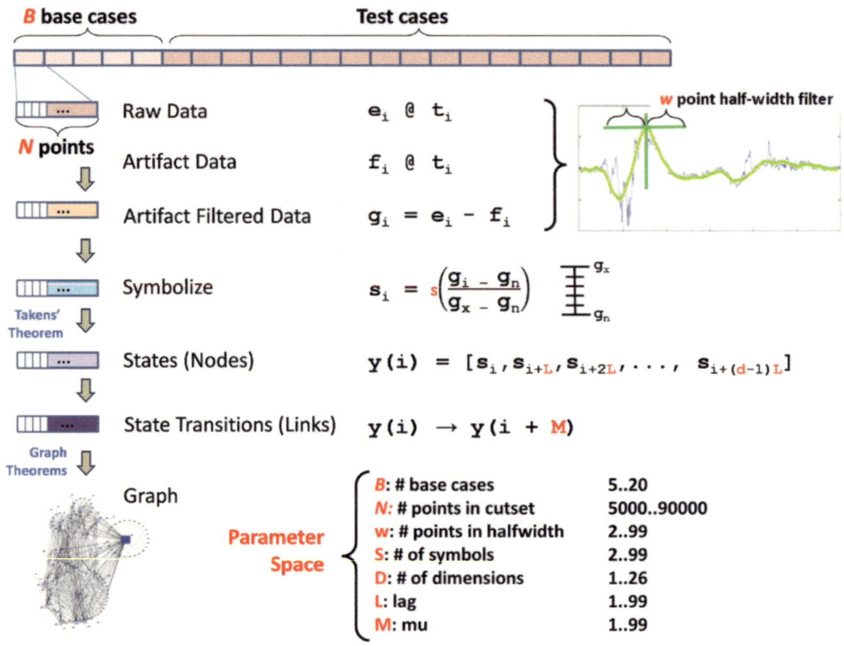


Figure 4. Nonlinear phase space construction [4].

The value, B, is the number of base cases, which establishes a normal range of activity for the patient. The value, N, is the number of sampled points that are in a cutset and graph, The value, w, as mentioned previously is the half-width of the eye-blink filter in sampled point units. The value, S, is the number of bins that the EEG is discretized into in order to create the base-s number represented by the vector y(i) in Figure 4. The value, d, is number of numerals in the base-s number or elements in the d dimensional vector, y(i). L is a time delay embedding that specifies the interval between points sampled in order to create a node. M is a second time delay embedding that specifies the interval between two connected nodes. The parameters mentioned are all used to generate the phase space graphs illustrated in Figure 4.

The dissimilarity measures involve counting unique nodes and links (those not in common between the two graphs): (1) nodes in graph A but not in B; (2) nodes in B but not in A; (3) links in A but not in B; and (4) links in B but not in A. Nodes and links in common between graphs do not indicate change and are not useful. These measures sum the absolute value of differences, which is better than traditional measures that use a difference of averages. Each measure is normalized to the number of nodes (links) in A (for A not in B) or in B (for B not in A). This feature vector, $V$, is used to classify the EEG as pre-ictal or inter-ictal. The analysis obtains a vector of mean dissimilarities, $V$, and matching standard deviations, $\sigma$, by comparison among the $B(B-1)/2$ combinations of the $B$ base-case graphs, as shown in Fig. 4. Subsequent test-case graphs are then compared to each of the $B$ base-case graphs to get an average dissimilarity vector, $v$. Our previous approach to obtain forewarning was several successive

http://dx.doi.org/10.5772/57438

instances ($K$) above a threshold ($U_T$) for each of $J$ features, $U\left(V\right)=\frac{|v-\bar{V}|}{\sigma}$. This normalization allows regions of the feature space to be found that forewarn for many patients. Because the graphs are diffeomorphic to the underlying dynamics (from Takens' theorem), changes in the scalp EEG are captured by changes in the graph measures.

The present work uses a SVM approach to obtain forewarning from the normalized dissimilarity measures—namely, we find nonlinear regions in the feature space using a SVM. Figure 5 shows the calculation of the dissimilarity measures [4]. The frequency of nodes and links is not used because Takens' theorem guarantees topology, but not density—meaning Takens' theorem doesn't guarantee useful information in the repetition of nodes or links.

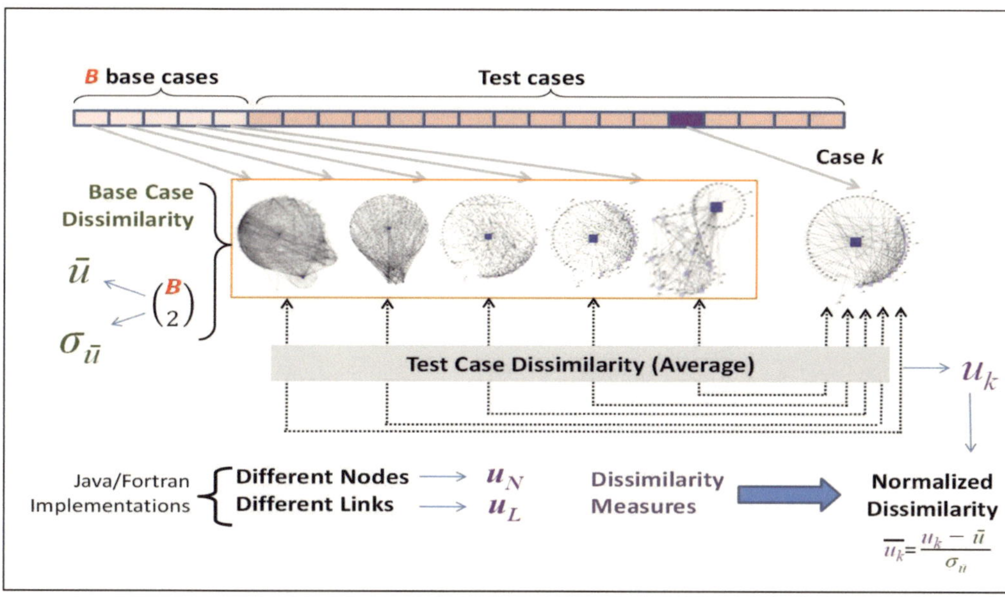


**Figure 5.** Normalized Graph dissimilarity measures, based on [4].

The dissimilarity measures in Figure 5 capture topology changes between two graphs. While node and link differences are basic graph measures, they quantify the hypothesis in a very simple and general way. Less commonality of nodes and links between two graphs produces larger dissimilarity measures, which are used to capture changes in topology. Topology change is a necessary, but not sufficient condition for a phase transition [17]. Our results show that changes in topology over extended periods indicate a higher likelihood of observing a phase transition as an indicator of an impending seizure. Additionally, the four graph dissimilarity measures from nodes and links rely on two concepts from set theory and Venn diagrams. Node dissimilarity and link dissimilarity are broken into two measures of dissimilarity each. Comparing two graphs (A and B) results in differences in nodes, as well as links. The dissimilarity measures are used as SVM features (for a total of 4 features in the Stage-1 SVM described below), and include the nodes in graph A that are absent from graph B, links in graph A that are absent from B, nodes in graph B has that are absent from graph A, and links in graph B that are absent from graph A. All four dissimilarity measures are normalized and vary with cutset. Figure 6 shows these dissimilarity measures varying with time and how each cutset results in features and labels ("+" for pre-ictal, and "–" for inter-ictal).

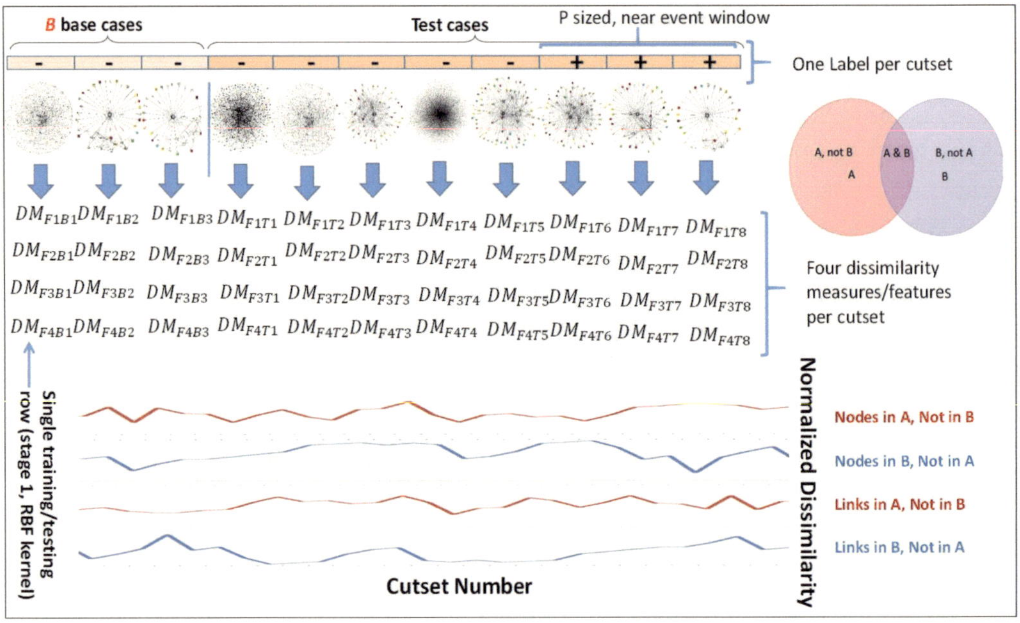


**Figure 6.** Cutset labels (+/-) and features (four dissimilarity measures).

Analysis of graph dissimilarity measures by a SVM allows quantification of the change in topology over time by determining how dissimilar the graphs must be to predict an epileptic event. The details of the forewarning algorithm are in Section 5—with a brief overview of Support Vector Machines in Section 4.

## 4. SVM with RBF kernels

SVMs are one of the most commonly used supervised learning tools. The SVM approach was originally designed as a two-class (binary) classifier, but has been expanded to single and multiple classes. A SVM without a kernel function performs linear classification by finding a hyper-plane in the feature space that best maximizes a margin of separation between two classes with a given list of features.

SVM kernels define the similarity between two points in the feature space. For example, with a radial-basis-function (RBF) kernel, two points are said to be similar when they are proximate to one another in the feature space. The RBF kernel function for two points in a feature space evaluates to 1 when the distance between the two points approaches zero. The RBF, Gaussian kernel function evaluates to 0 as the distance between the two points becomes very large. The region where the kernel function evaluates to zero is parameterized by the value of gamma ($\gamma$), which is inversely proportional to the width of a multi-dimensional Gaussian function. SVMs with RBF kernels transform the decision boundary from a hyper-plane into much more amorphous decision boundary in the feature space. Figure 7 shows the difference in boundaries found by linear and RBF kernels for a representative SVM example in two dimensions.

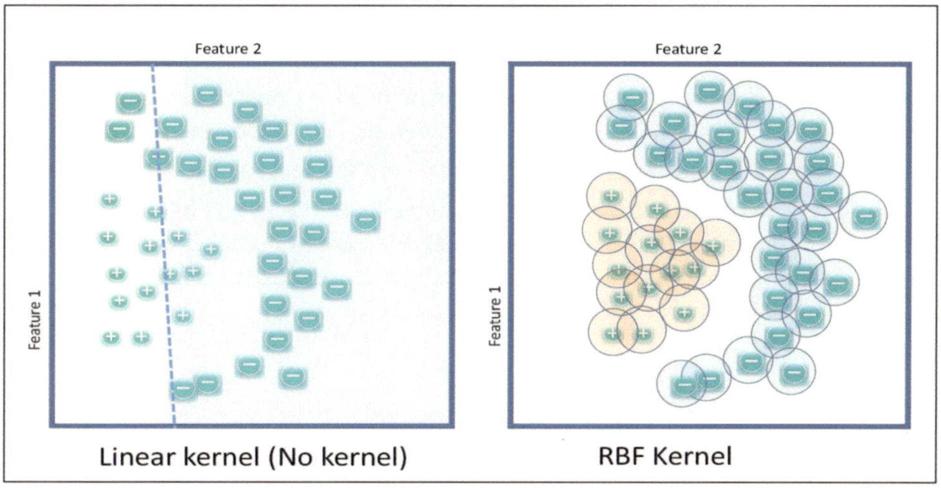


**Figure 7.** Types of boundaries found with two main SVM kernel types.

Each point in Figure 7 is equivalent to one instance of a class. Positive class values are denoted by positive signs and negative class values are denoted by negative signs. The Cartesian dimensions are the feature values—such as a dissimilarity measure. More than two features (two dimensions) can be used with a SVM, but it is more difficult to visualize when more than 3 dimensions (features) are involved. The main requirement of a RBF kernel is that the training set has a representative sample of the data that will be observed in the future and enough features to make distinctions between classes. Additionally, the range and scale of each feature has a large effect on the value of $\gamma$, the results, and the accuracy of predictions. Given a SVM model, future points are likely to be labeled as the class, to which they are most similar in the feature space. Similarity is defined by the kernel function as closeness between points of one class in the feature space. In essence, points from the training set are stored along with a weight associated with that point. A weighted sum of inner products is computed to evaluate how similar a new point is to all of the training data. The SVM training phase minimizes a cost function of several parameters, one of which is a vector, $\theta$, of weights. Eq. (3) [18] gives the SVM cost function to be minimized with a kernel.

$$\min_{\theta} C\sum_{i=1}^{m}\left(y^{(i)}\mathrm{cost}_1\left(\theta^T f^{(i)}\right) + \left(1 - y^{(i)}\right)\mathrm{cost}_0\left(\theta^T f^{(i)}\right)\right) + \frac{1}{2}\|\theta\|^2 \tag{3}$$

Here, $y^{(i)}$ is the class label of a training point i (positive or negative one); $f^{(i)}$ is the feature vector of the point $(i)$, which compares a single point with the other points in the training set. Cost is an input parameter to the SVM and is a penalty for being in one class or the other.

Once the vector $\theta$ is found through the minimization, it can be used to determine the classes of new points. The method of determining the label of new points is shown in Eq. (4) [18].

$$\text{class label of new point} = \begin{cases} \textit{Predict class } 1 \textit{ when } \Theta^{T}f^{(i)} > 0 \\ \textit{Predict class } -1 \textit{ when } \Theta^{T}f^{(i)} < 0 \end{cases} \tag{4}$$

Once the vector $\theta$ is obtained from training, it can be used to determine whether a new point in the feature space is of one class or another as given by Eq. (4). For each new point of index $i$ in a test set, a vector $f^{(i)}$ is computed. The vector $f^{(i)}$ is a function of the kernel and the points in the training set. Each row in the vector $f^{(i)}$ is the value of the kernel function when the test set point and the training set points are inputs. The Gaussian kernel function in Eq. (5) will evaluate to 1 for points that are close together in the feature space and to zero when they are not. Eq. (5) gives the kernel function comparing a test set point $x_i$ to a landmark training set point, $l^{(k)}$ [18].

$$f_k^{(i)} = \exp\left(-\gamma \| x_i - l^{(k)} \|^2\right) \tag{5}$$

Without a kernel, the vector $\theta$ has n+1 dimensions for n features. With a kernel, the cardinality of $\theta$ and $f^{(i)}$ is m+1 for m training points. Often, this comparison to known points in the training set with a kernel function is referred to as mapping the feature space into a higher dimensional space. Indeed, the dimensionality of $f^{i}$ and $\theta$ increases from approximately the number of features without a kernel to roughly the number of training points with a kernel. The kernel results in linearly separable classes in the higher dimensional space, where the classes are not linearly separable in the feature space. The above description applies to binary classification with a RBF kernel and is intended to give the reader intuition on the types of boundaries that are found in the feature space as illustrated in Figure 7.

LIBSVM makes the details of the linear algebra of the training and testing transparent to the user and is easily used with the intuitions given in this Section [19]. The effect of a RBF kernel is that points proximate to one another will be labeled as belonging to the same class. Multi-class classification is treated as several binary classifications and is beyond the scope of the present work.

## 5. SVM forewarning algorithm using graph dissimilarity features

Labeling training data as pre-ictal or inter-ictal requires assumptions that must be sound. Current epilepsy prediction algorithms offer guidance about acceptable assumptions. The goal

http://dx.doi.org/10.5772/57438

is enough forewarning to stop or mitigate an event. Patients and caregivers [20] suggested 1-6 hours for safety, planning the day, and "driving myself to the hospital." Non-parent caregivers preferred 25 minutes to 1 hour for travel to the patient's location. Others gave 3-5 minutes, because longer forewarning was seen as more stressful to the patient. These requirements—as well as previous research indicating that these constraints are a reasonable request—led to the labeling scheme used for pre-ictal indications for a SVM. For epileptic event data sets, the pre-ictal region is labeled as being anywhere from 3.3 minutes to 70 minutes before the seizure. Each epileptic patient is labeled as being pre-ictal for the same length of time prior to the seizure. Each plus and minus sign in Table 1 represents a 3.3 minute window (consistent with a cutset length of 49716 points, sampled at 250 Hz). The number of pluses is determined by a parameter (p) that is varied during cross validation. Figure 6 shows how the signs in Table 1 relate to graphs, features, cutsets, and dissimilarity measures.

| Patient data type | 3.3 minute labeling | 70 minute labeling |
|---|---|---|
| Event data set | ----------------------------------+ | -------------------------+++++++++++++++++++++ |
| Non-event data set | ----------------------------------- | ---------------------------------------------------------- |
| Value of p | 1 | 21 |

**Table 1.** Effect of variable p (number of + values) on Stage-1 pre-indication labeling of training data.

The input labeling (e.g., Table 1) assumes only approximate correctness and uses class weights to vary the correctness likelihood. The SVM methodology is implemented in three Stages with 10-fold cross validation. Stage-1 constructs a classifier that can label the pre-ictal state indicators. Stage-2 determines how long a patient must exhibit pre-ictal indicators in order to be certain of a seizure. Stages 1 and 2 establish cross validation accuracy and error. The SVM forewarning algorithm and previous voting method algorithm [4] both imply that patients must be in abnormal states a higher portion of the time before they are likely to have a seizure. Datasets without seizures can have infrequent abnormal states as well. Stage-3 obtains two models that can be used for seizure prediction in an ambulatory setting. Cross validation results in k different classifiers that leave out disjoint sets of data to establish an off-training-set error (OTS error) to avoid overconfidence in accuracy. However, k slightly handicapped classifiers result in either less accuracy than is possible or more complexity in creating an ensemble. Stage-3 avoids this unnecessary choice by performing cross validation to create a final SVM model that includes all of the available data. Accuracy and error rates are statistically stronger, when they are reported from cross validation. The statistical claims are less robust, when one trains and tests on the same data. SVM with a RBF kernel is particularly susceptible to over-fitting—implying the need for cross validation. Figure 8 shows an outline of this three-Stage algorithm.

I. Stage-1
   a. Obtain 4 dissimilarity measures from phase space analysis (for each cutset in a patient data set). These dissimilarity measures become the features of the RBF kernel;
   b. Labels
      i. Labels for +1: p cutsets immediately before an event;
      ii. Labels for -1: all other cutsets. See Table 1;
   c. Divide datasets into 10 sets of patients: each set contains 4 event patients and 2 non-event patients;
   d. Train RBF Model on 9 sets of patients – obtain SVM model (see Figure 9);
   e. Predict on the remaining, 10th set with RBF Model from (1d) ;
   f. Scan the results from (1e) for max # of contiguous +1. See Table 4-5.
   g. Repeat 1d-1f (and save results): 10 predicted sets. See Table 4-5.

II. Stage-2
   a. Label each patient's feature set from (1g)
      i. Event data sets are labeled as +1;
      ii. Non-event data sets are labeled as -1;
   b. Train on 9 sets of max contiguous pre-ictal indicators from (1g): linear kernel (see Table 5);
   c. Predict on 1 set of max contiguous indicators from (1g): via linear model from (2b);
   d. Get false positive rate (FP) and false negative rate (FN) from (2c);
   e. Get $D_i = \sqrt{\left(\frac{FP}{2}\right)^2 + \left(\frac{FN}{4}\right)^2}$ (We use stratified cross validation);
   f. Repeat (2b) – (2e) 10 times;
   g. Get the average over $D_i$ (average cross validation OTS error rate);

III. Stage-3
   a. Use D(average)=(Average prediction distance) from (2g);
   b. If D(average) < 0.7 create models to use in future via 3c-g;
   c. Use results (see Table 4-5) from all 60 data sets from (1g);
   d. Retrain linear model (similar to Stage-2, but with all 60 patients' max successive indicators) to get the number of successive occurrences to trigger forewarning (see Table 5);
   e. Use data from (1b) to retrain RBF model on all 60 patients' cutsets (a total of 4244 cutsets) to obtain the RBF model (see Figure 10);
   f. Use RBF Model (3e) result to predict +/- on data from (1b); see Table 7;
   g. Use linear model from (3d) to do event forewarning on all 60 patients (see Table 4-5, Table 7, Figure 9, Figure 10) and get $D_{final\ models}$ (represents an optimistic over-fit);

**Figure 8.** Steps in the three stages for cross validation and final model construction.

Figure 9 shows how Stages 1 and 2 flow together. Stage-3 involves training the RBF Model on all of the Stage-1 cutsets (4244 rows, instead of approximately 90% of it) and training the linear model on the all of the Stage-2 results (60 rows, instead of 90% of it). Then, one predicts on the training data to verify that the model is working as expected to produce $D_{final\ models}$.

Figure 9 shows that event datasets are labeled in Stage-1 as pre-ictal (+) in a window of p cutsets prior to the seizure and inter-ictal (-) outside of this window. All cutsets in non-seizure datasets are labeled as inter-ictal (-). A cost sensitive SVM is used to account for the uncertainty in the pre-ictal and inter-ictal labeling. The motive for this labeling scheme is the caregiver's desire to have forewarning within an hour of the event. Indicators are assumed to be near the event, and the time window is varied by the parameter (p) that is tested during cross validation. The assumption behind the design choice is that the pre-ictal state is a rare occurrence. Because

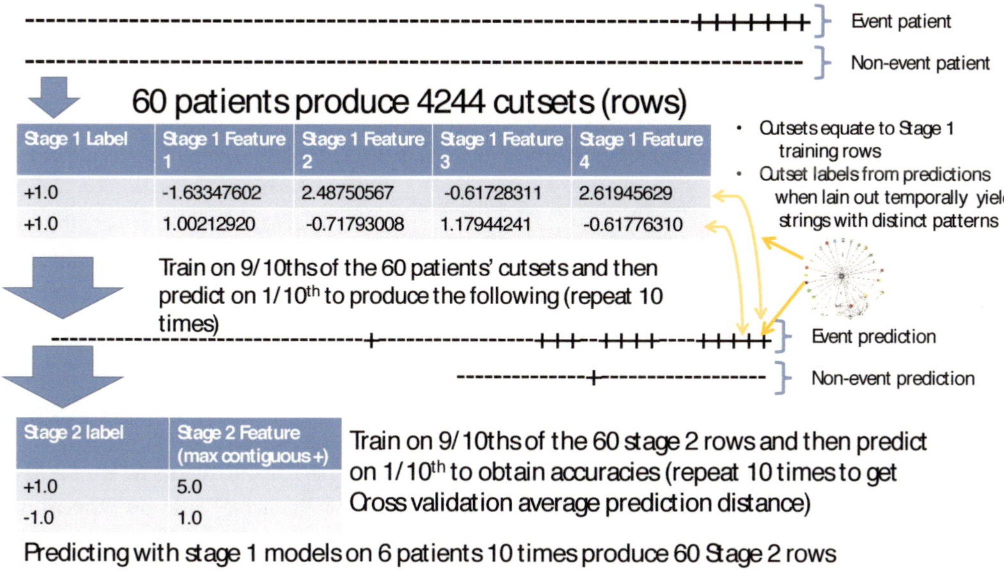


**Figure 9.** Stage-1 and 2 flow (each + or – represents a 3.3 minute cutset).

there may be similar points in the feature space outside the hour time-frame being labeled as pre-ictal and inter-ictal, the variable for class weights are varied via a Monte-Carlo search over the parameter space during cross validation to determine how to tie-break. Other parameters are also varied randomly during the Monte-Carlo search, as shown in Table 2. To compensate for labeling uncertainty, the cost sensitive SVM adjusts a weight on the class labels to indicate how certain the labeling scheme is for the pre-ictal and the inter-ictal classes. The labeling scheme combined with the features, training set, class weights, and gamma creates regions in the feature space that will be associated with one class or another. Additionally, we use stratified cross validation—maintaining a ratio of 4 event patients and 2 non-event patients in each strata of cross validation. Cross validation is performed on 90% of the patients—54 patients in each training set and 6 patients in each test set—having varying numbers of cutsets due to having varying length observations. This process is repeated 10 times with disjoint sets of patients in each test set.

Successive, contiguous occurrences of pre-ictal indicators trigger an alert (prediction of an event). Non-event datasets have more inter-ictal indicators labeled with negative symbols, while event patients have fairly dense pre-ictal indicators—labeled with plus symbols. A single pre-ictal indicator is usually not enough to make accurate predictions.

The accuracy of the assumptions is reflected in the success rate of the predictions during cross validation. The parameters that appear to be uncertain are left as search variables, recognizing that more free parameters create a more computationally complex search. Too many variables result in a computational explosion in the CPU time to explore the search space. Each point in the parameter space corresponds deterministically with a cross-validation error rate. Assumptions about the certainty of the training or testing example's labels are represented by class weights in a cost sensitive SVM. Table 2 shows the variables for the SVMs that were searched during the research for this paper.

| $\gamma$ | $c_{rbf}$ | $weight_+$ | $weight_-$ | $c_{linear}$ | p |
|---|---|---|---|---|---|
| Sets the radius of the contribution of a single point to the decision boundary | Adjusts the pliability of the Stage-1 decision boundary given additional points | Weighs how powerfully the + class influences the decision boundary | Weighs how powerfully the – class influences the decision boundary | Adjusts the pliability of the Stage-2 decision boundary given additional points | Number of cutsets prior to the event that are labeled as being in the positive class. |

**Table 2.** Parameters for the SVM portion of analysis.

Figure 4 and Table 3 lists other variables that might be searched in addition to those of the SVM. Those parameters were fixed in the present analysis because Takens' theorem is sufficiently powerful to show significant changes in topology with many sets of parameters when those topological changes are normalized properly. The dissimilarity measures for a patient reflect relative differences in graph topology. The parameter values for the phase-space graph generation are shown in the Table 3. Recall that these parameters were described near and are contained in Figure 4. Throughout the entire analysis, the parameters to generate the phase space graphs were kept fixed. These values were found in our prior work [4] to give good forewarning using ensemble voting methods mentioned in the background section.

| B | D | L | M | N | S | W |
|---|---|---|---|---|---|---|
| 12 | 7 | 56 | 77 | 49716 | 3 | 29 |

**Table 3.** Parameters used in generation of phase space graphs [4].

Statistical validation of forewarning requires measures of success. One measure is the number of true positives (TP) for known event datasets (Ev), to yield the true positive rate (sensitivity) of TP/Ev. A second measure is the number of true negatives (TN) for known non-event datasets (NEv). The true negative rate is TN/NEv (specificity). The goal is a sensitivity and specificity of unity. Consequently, minimizing the distance from ideal ($D$ = prediction distance) is an appropriate objective function for any event type:

$$D = \sqrt{\left[1 - \left(\frac{TP}{Ev}\right)\right]^2 + \left[1 - \left(\frac{TN}{NEv}\right)\right]^2} \quad (6)$$

Eq. (6) is the objective function to be minimized for the OTS error rate and over-fit error rate. The OTS error rate is found from 10-fold cross validation from Stages 1 and 2. The over-fit error rate verifies that the final models in Stage-3 can correctly predict the training examples. Excessive false positives (inverse of a true negative) will cause real alarms to be ignored and needlessly expend caregiver resources. False negatives (inverse of a true positive) provide no

http://dx.doi.org/10.5772/57438

forewarning of seizure events. A Monte-Carlo search is used over variables in Table 2 because the prediction distance has very irregular, fractal behavior—with sparse parameters generating good predictions and the gradients of the parameter space being highly irregular with many local maxima and minima [4].

The algorithm for forewarning is done in two Stages of processing after producing diffeomorphic graphs and their dissimilarity measures for each graph. Stage-1 uses a cost sensitive SVM type (CSVC) from LIBSVM [19] with a RBF kernel. For each iteration of the cross validation, the algorithm labels event data sets as having pre-ictal indicators within a one-hour window prior to the seizure; see Table 1. The length of this window (p cutsets) is part of the Monte Carlo search. All other values—non-event data and event data far away from the seizure (outside the variable window)—are labeled as inter-ictal indicators. The analysis trains on k-1 sets, then predicts on a single, left-out set; this analysis is repeated k times in a k-fold cross validation (with k being 10). Table 4 shows a small sample of Stage-1 predictions for two patient outputs, but in actuality there are 60 sets of predictions—like Table 7 in the results section.

| Common prediction output | Example predictions from Stage-1 | Maximum Successive occurrences of + |
|---|---|---|
| Event Data set | ------------------+------------------+-----++-++----------+++E 53 | 3 |
| Non Event Data set | -------------------------------------------------NE no prediction | 0 |

**Table 4.** Representative predictions from Stage-1 (E=event; NE=no event).

After making the six predictions on six patients for one of ten cross validation runs, one creates a new set of cross validation folds (representing sets of patients) for the Stage-2 analysis out of the Stage-1's predictions on the omitted set (similar to the middle column of Table 4). For the Stage-2 single feature, one scans each of the predictions made by the Stage-1 for the maximum number of contiguous occurrences of pre-ictal indicators. Specifically, the third column of Table 4 is obtained from scanning the second column of Table 4. Stage-2 of the algorithm has training and testing files that follow the format of the second and third columns of Table 5.

| Label meaning | Label | Single Feature (Maximum Successive occurrences of +) |
|---|---|---|
| Event Data set | +1.0 | 3 |
| Non Event Data set | -1.0 | 0 |

**Table 5.** Training/Testing set format for Stage-2.

The analysis labels the Stage-2 training and testing values as either event or non-event data sets. In general, there are fewer pre-ictal indicators (+ in Table 4) in the non-event data sets on successful cross validation runs. One determines the cross validation average prediction distance by training on values of maximum contiguous, pre-ictal indicators from k-1 subsets at the Stage-2, making k predictions on the omitted subset, and then taking the average of the prediction distances.

Stage-3 obtains the cross-validation prediction distance via two models that predict on the basis of all of the data available instead of 90% of it. Specifically, Stage-3 takes the 4244 cutsets labeled in Stage-1 and the sixty predictions from Stage-2 to create two SVM models for predicting on future patients. Stage-3 involves retraining both the RBF model (from Stage-1) and linear model (from Stage-2). Figure 10 illustrates the flow in Stage-3 after optimal parameters have been discovered via cross validation. Once the two models are obtained, they are used to predict on the original data sets to verify the model's validity.

| Stage 1 Label | Stage 1 Feature 1 | Stage 1 Feature 2 | Stage 1 Feature 3 | Stage 1 Feature 4 |
|---|---|---|---|---|
| +1.0 | -1.63347602 | 2.48750567 | -0.61728311 | 2.61945629 |
| -1.0 | 1.00212920 | -0.71793008 | 1.17944241 | -0.61776310 |

Retrain the RBF model on all 4244 rows.

| Stage 2 label | Stage 2 Feature (max contiguous +) |
|---|---|
| +1.0 | 5.0 |
| -1.0 | 1.0 |

Retrain the linear model on all 60 patient predictions (10 sets of 6 predictions from stage 1, scanned for max contiguous + features)

- Use the RBF model to predict on all 60 patients' 4244 cutsets (**Stage 1 table**)
- obtain Max contiguous + for each patients' prediction from stage 1 (**Stage 2 table**)
- Test the linear model to obtain $D_{final\ model}$

**Figure 10.** Stage-3 process (builds on Stages 1 and 2).

## 6. Representative results

From Eq. (6), the scenario of a classifier never getting the answer correct is $D = \sqrt{(1)^2 + (1)^2} \approx 1.41$. A random number generator that is guessing each class with equiprobability will have a prediction distance over time, $D = \sqrt{\left(\frac{1}{2}\right)^2 + \left(\frac{1}{2}\right)^2} \approx 0.7$. A perfectly ideal classifier will have an average prediction distance (OTS error rate) of 0 during cross validation. A Monte-Carlo search over the parameters for the SVM attempts to find the set of parameters and corresponding SVM models that minimize the prediction distance at Stage-2 for classification of each dataset as an event or a non-event. Cross validation on Stage-2 leads to k prediction distances that are averaged. The average is minimized and corresponding "ideal" parameters for the SVM are found. Once the Monte-Carlo search has found parameters associated with an acceptable OTS error rate, one retrains the Stage-1 model on all of the data.

http://dx.doi.org/10.5772/57438

One then retrains the Stage-2 model on all of the predictions from Stage-1. The retraining process results in two SVM models—one for Stage-1 and one for Stage-2. With these models, one can make predictions on new data.

Table 6 shows representative SVM parameter values (from the Monte Carlo search) and results. $N_{OCC}$ is the number of contiguous, pre-ictal indicators (+ labels) that must be present before forewarning occurs and is found by training the Stage-2 SVM model (and Stage 3g). A value of Nocc of 2 implies that dynamics over a 6.6 minute period must be observed to be abnormal for prediction. Nocc of 1 implies that dynamics over a 3.3 minute period must be observed to be abnormal to have forewarning. D(AVG) is the average OTS error rate during cross validation runs; D(AVG)<0.7 indicates that the algorithm is performing more accurately than random guessing and simple heuristics. D(final) is a value verifying that the final models in Stage-3 have the capability of accurately classifying the training data. D(final) represents an over-fit, but is valuable in narrowing the Monte-Carlo search and determining whether the algorithm has any merit.

Table 6 shows that the best cross validation accuracy with an average prediction distance of 0.287 and a final model prediction distance of 0.056. Table 6 shows additional representative cross validation averages—D(Avg)—and final model prediction distances—D(final). Hundreds of runs resulted in cross validation prediction distances of <0.5. The best cross validation accuracy of.287 achieved thus far also has a fairly decent over-fit error rate of 0.056—which has a value of the objective function, $D=\sqrt{\left(\frac{1}{20}\right)^2+\left(\frac{1}{40}\right)^2}\approx.056$, corresponding to one false positive and one false negative in the final model's ability to predict the original training data. Although, due to the fact that the number of events and non-events was unbalanced, the probability of having a false negative was twice as likely as false positive, which is still desirable.

| Ex# | $\gamma$ | $c_{rbf}$ | $weight_+$ | $weight_-$ | $N_{OCC}$ | $c_{linear}$ | D(AVG) | D(final) | size of + label window |
|---|---|---|---|---|---|---|---|---|---|
| 1 | 3.929 | 9.732 | 19.160 | 24.860 | 1 | 3.723 | 0.287 | 0.056 | 21 |
| 2 | 2.640 | 87.825 | 32.867 | 81.832 | 1 | 2.945 | 0.342 | 0.025 | 22 |
| 3 | 2.860 | 10.022 | 85.200 | 50.843 | 2 | 8.033 | 0.374 | 0.075 | 18 |
| 4 | 0.964 | 85.719 | 71.740 | 33.237 | 2 | 9.259 | 0.396 | 0.125 | 19 |
| 5 | 7.709 | 2.587 | 28.705 | 26.903 | 2 | 2.347 | 0.413 | 0.050 | 20 |
| 6 | 2.207 | 75.920 | 52.493 | 36.102 | 2 | 8.994 | 0.438 | 0.125 | 18 |
| 7 | 6.783 | 18.035 | 49.974 | 30.441 | 2 | 6.985 | 0.456 | 0 | 21 |

**Table 6.** Summary of typical best results to date.

Example 7 in Table 6 has an average cross validation accuracy of.456 with D(final)=0 (perfect prediction). Cross validation average accuracy or error rate is the more valid statistical claim. The best cross validation accuracy represented in Table 6 is in the same realm of accuracy as Netoff et al.'s intracranial methodology. Recall that Netoff et al. claim a specificity of 77.8% with no false positives, which is approximately a prediction distance of approximately.22 ($D_{Netoff} \approx \sqrt{(1 - .78)^2 + (0)^2} \approx .22$) [1]. Our best cross validation accuracy is of the same order of magnitude. Given that we are using scalp EEG, and Netoff et al. are using intra-cranial EEG, this is a statistically significant result that cannot be said to be an over fit. Mirowski et al. claim 100% accuracy, which would be a prediction distance of zero ($D_{Mirowski} = 0$). However, Mirowski et al. are making patient specific machine learning models that are tailored to individual patients. We are creating a novel algorithm that can be applied to a group of patients with non-invasive EEG. Very little research is being done to advance non-invasive EEG prediction algorithms in this way. Furthermore, comparing cross validation accuracies and error rates dispels any arguments of overconfidence due to over-fitting.

Figure 11 shows a plot of forewarning times (typically less than 1.5h) for the final-model of Example 7. The number of successive contiguous indicators to trigger forewarning was found to be 2 successive + values. Table 7 shows the Stage-1 predictions (Stage-3g of Figure 8 for all 60 patients) to produce the distribution of forewarning times in Figure 11. See Table 7 for the cutset indications (+ or -) that correspond to the parameters in Example 7 from Table 6.

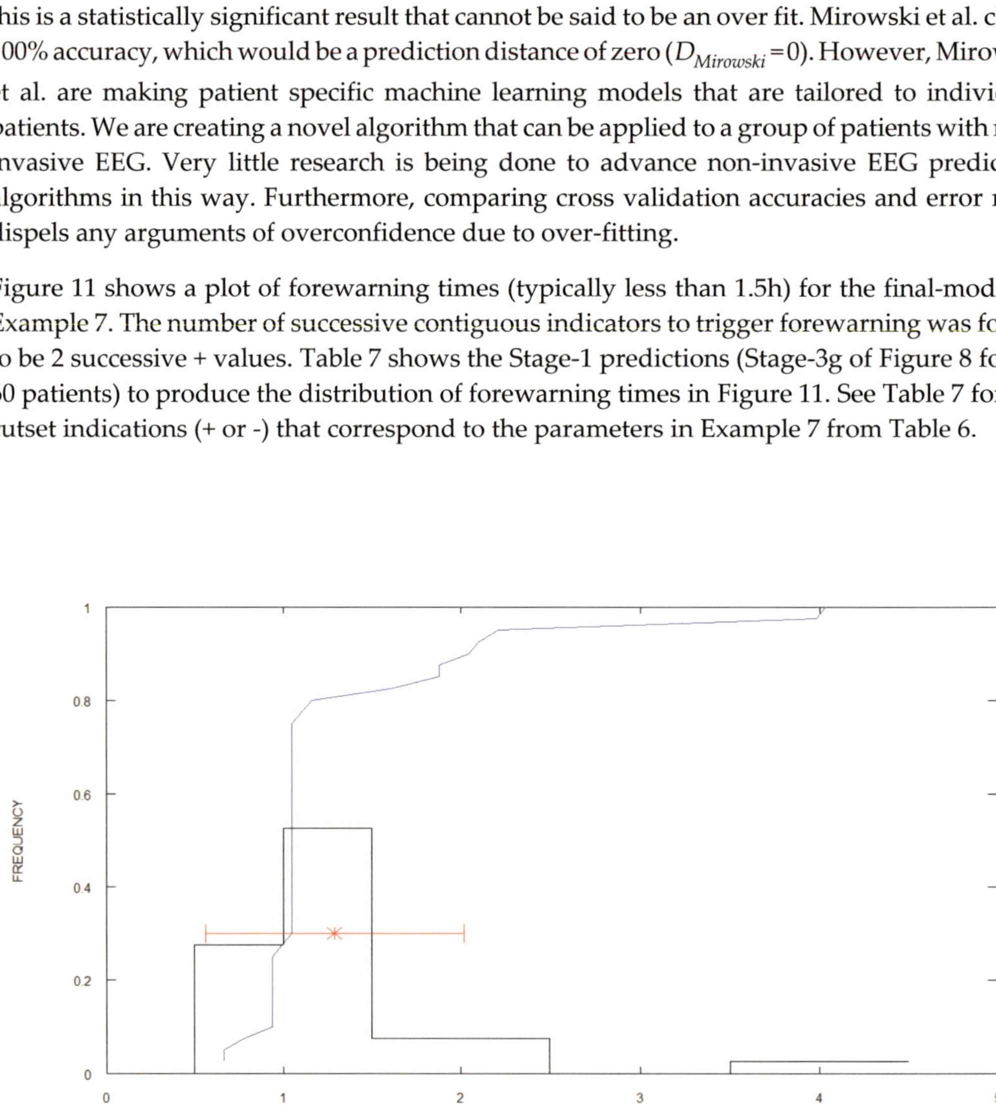


**Figure 11.** Distribution of forewarning times for Example #7 in Table 6.

| Patient | --------------------------------NE no prediction |
|---|---|
| 1 | --------------------------------NE no prediction |
| 2 | --------------------------------------NE no prediction |
| 3 | -------------------------------------------------------------NE no prediction |
| 4 | -+-------------+---------------------------------------------------------------+---------------------------NE no prediction |
| 5 | -------------------------------------------------------------------------------------------------NE no prediction |
| 6 | ------------------------------NE no prediction |
| 7 | --------------------------------NE no prediction |
| 8 | --------------------------------NE no prediction |
| 9 | ------------------------------NE no prediction |
| 10 | ----------------------NE no prediction |
| 11 | ---------------------------------+----------------------NE no prediction |
| 12 | --------------------------------------+---------------+-------------------------+----------NE no prediction |
| 13 | -------+----------+----+----NE no prediction |
| 14 | ------------------------------------------------------------------------------------------------------------------------NE no prediction |
| 15 | ----------------------NE no prediction |
| 16 | --------------------+-------------------------------+-------------------------------------NE no prediction |
| 17 | ----------------+----------------------------------------------------------------------------------------NE no prediction |
| 18 | --------------+---------------------------------------------NE no prediction |
| 19 | ---------+---------------------+-------------------------------------------------+---------------NE no prediction |
| 20 | -------------------------+------------------------------------------------------NE no prediction |
| 21 | --+----------------+---------------------------++++++++++++++++++++E 62 |
| 22 | --------------------------------------+++---++--+-+++-+--++E 62 |
| 23 | -------------------------+----------------------------++++++++++++++++++++E 62 |
| 24 | --------------------------+-++++++++++-++-+++++E 56 |
| 25 | -----------------------+---------------+--+---++-++++++++++-++E 46 |
| 26 | --------------+--------------+-+-+-++-++++++++++++++++E 56 |
| 27 | ----------------------------------+-++--++++++-++--+-++E 56 |
| 28 | ----------------------+-------+---------+----------------++--++++-+-+++-+-++++E 62 |
| 29 | --------------------+-------+-------+----++-+++++++++++E 39 |
| 30 | ------------------------------+--+-++++++++++++++++++E 56 |
| 31 | ------------------------+----------------------+----------------+++++++-++++++++-+-++E 62 |
| 32 | --------------+---+-+---------------------+--------------------+++++++++++---+++++++E 62 |
| 33 | -------------------------+-+-+++++++++-++++-++-+--+E 66 |
| 34 | ------------------------------+++++++++++++--++++++E 62 |
| 35 | -----------------------------------------------------------------------------------------++++--+-++-+++E 39 |
| 36 | ------------+-------------------------------+++-++++++-----+--+--E 62 |
| 37 | --------------------------------------------------------------------------++++++++++++++++++++--E 62 |
| 38 | -------------------------+----------------++--------------+-----+++++-----------------------++++++++-++++++++++-+E 238 |
| 39 | -------------------------------------+-----------------------------++++-+++++++++++++++-E 62 |
| 40 | -----------++----------------+----------------+---+------------+++++++++++++++++++++-E 241 |
| 41 | -------------------------+++-+++++++++-+++-+++E 62 |
| 42 | ---------------------------------++++++-+++++++-+---++++E 69 |
| 43 | ----------------+--------++++++++++++++++++++++E 62 |
| 44 | -----------------------------------+---------------------------------------++-++++++++-++++++++E 62 |
| 45 | -----------------------------------------++-++-+++++++-+--++++E 62 |
| 46 | -----------------------+---+-++-+-++-+-+-++-+++-E 56 |
| 47 | ------------------+++----+----+++++++++++++++-+--+++++++++E 125 |
| 48 | --------------------------------------------------+------+---------------------+-----+-------++----++----++-----E 56 |
| 49 | -------------------------------------------------++++++++++++++++++++++E 62 |
| 50 | ------------++--+++--+-+-----++++++++-++-+-+++++++++E 122 |
| 51 | --------------------------+-------------------------+++---++++-+++-+--++E 59 |
| 52 | -------------------+---------+--------+--+--+-++-++-++-++++++----E 56 |
| 53 | --------++------------------++-++++++-++++++++++-E 132 |
| 54 | ------------------------+-----------------------+-++++--+-----+-+++-++++++++++++++++++E 112 |
| 55 | ---------------+------------------------------+++++-++++++++++++++-E 62 |
| 56 | ------------+-----+-----++----++++-----+++++++++-+-+++++++++E 112 |
| 57 | ------------------------------------------------------------------++++++++++++++++++++++E 62 |
| 58 | ----------------------------------------++++++++++++++++++++-E 62 |
| 59 | -----------------------++------+++++++++++++++++--+++E 96 |
| 60 | -----------------------------------------------------------------++++++++++++++++++++++E 62 |

**Table 7.** Stage-3g predictions on all 60 patients for Example 7 in Table 6 (E=event; NE=no event; number following "E"=forewarning time in minutes).

In Figure 11, the solid black line is the occurrence frequency (arbitrary units) in half-hour bins. The blue line is the cumulative distribution of forewarning versus time. The red H-bar with the star in the middle indicates the mean value of the forewarning times (approximately 1 hour) and the sample standard deviation. The result in Example 7 of Table 6 is better than random guessing or biased heuristics with D(final)=0, despite poorer cross validation accuracy than other examples. This example shows most of the forewarning times of ≤1.5 hours with a statistically significant accuracy. One can visually make a prediction by scanning Table 7 from left to right and looking for 2 contiguous plus values. When 2 values are found, the seizure is highly likely to occur. One may be tempted to reduce the forewarn time by increasing the value of positive values that trigger a forewarning, but that would likely result in bad OTS error, which is why it is not the value found by the second and third stage SVMs.

## 7. Discussion

Ideally, we would like to achieve an average cross validation OTS prediction distance of zero, final model prediction distance of zero, and all forewarning times <1 hour. In order to achieve this goal, additional features will need to be explored that exploit topology and the distance metric that Takens' theorem guarantees. Some modifications for Stage-2 improve the results (e.g., the choice of p cutsets prior to the event as a variable, and use of a RBF kernel instead of a linear one). More search parameters in Stage-1 (e.g., those in Table 3) should lead to better results with enough CPU time. Additional graph dissimilarity measures may be helpful. We have discovered more features for Stage-1 that may be of use. More data is needed for a robust statistical validation of the model.

The choice of optimal features is very difficult. Theorems guide the choice of parameters, features, and the algorithm. Some combination of theorem-based feature selection and occasional intuition derived from experimentation is the only way to keep the cost of the research initiative practical. Feature selection is one of many hard problems involved in epilepsy prediction. When one adds features, one often needs more data to make meaningful statistical assertions. Other important choices involve the type of kernel and the thresholding strategy. Linear kernels and thresholding strategies may perform well while Radial Basis Function (RBF) kernels perform poorly and vice versa. Our previous work [4] used a voting method that performed well. There is no guarantee that a set of features will behave similarly with different kernels and strategies to determine the threshold. Other measurement functions are possible under Takens' theorem to create the phase-space states. Use of a single-class or multi-class SVM could also prove fruitful.

The results in Tables 6-7 and Figure 11 are encouraging, despite several limitations, which are discussed next. (1) We analyzed 60 datasets, 40 with epileptic events and 20 without events. Much more data (hundreds of datasets) are needed for strong statistical validation. (2) These data are from controlled clinical settings, rather than an uncontrolled (real-world) environment. (3) The results depend on careful adjustment of training parameters. (4) Only physician-selected portions of the EEG are available, rather than the full monitoring period. (5) The present approach uses retrospective analysis of archival data on a desktop computer. Real-world forewarning requires analyst-independent, prospective analysis of real-time data on a portable device. (6) The results give forewarning times of 4 hours or less. A time-to-event estimate is needed. (7) All EEG involved temporal lobe epilepsy; other kinds of epilepsy need to be included. (8) A prospective analysis of long-term continuous data is the acid test for any predictive approach. Prospective data were unavailable for the present analysis. Clearly, much work remains to address these issues.

## 8. Conclusions

The present work uses Support Vector Machine analysis to extend earlier work by Hively *et al.* [4] for forewarning of epileptic seizures. The previous work obtained a prediction distance of 0.0559 and a maximum forewarning time of 5.1 hours. The present analysis divides the continuous data stream from one bipolar channel of scalp EEG into contiguous, non-overlapping windows; removes the muscular artifacts with a novel zero-phase quadratic filter; converts the artifact-filtered data into discrete symbols; applies the time-delay-embedding (Takens') theorem to create unique phase-space states that capture the brain dynamics; forms a graph from the nodes (phase-space states) and links (dynamical state-to-state transitions); extracts dissimilarity measures by pair-wise comparison of graphs (e.g., nodes in graph A that are not in B); uses these dissimilarity measures as features for a novel Support Vector Machine to classify the data as forewarning of a seizure event or not (i.e., not characteristic of the baseline, but characteristic of data near the event). The present work obtains a prediction distance as small as zero (sensitivity =1, specificity =1) with most of the forewarning times ≤1.5 hours. The best off-training-set error rate was.287 using 10-fold cross validation.

Our non-invasive (scalp) EEG analysis resulted in cross validation error rates comparable to other invasive EEG approaches. Additional accuracy could be obtained by applying this methodology to specific patients on a per patient basis for custom EEG models if the data were available. Modifications could conceivably be made to the algorithm to improve the computational feasibility of per patient machine learning models. A research team could allow patients to start with group-based models that are less accurate while the patients collect and upload ambulatory data from their devices as they use them in real-world settings. Furthermore, businesses could be compensated for creating patient specific models from patients' ambulatory data. Additionally, our algorithms have other applications as well, such as failure forewarning in machines [21] and bridges [22].

## Acknowledgements

This manuscript has been authored by UT-Battelle, LLC, under contract DE-AC05-00OR22725 with the U.S. Department of Energy. The United States Government retains and the publisher, by accepting the article for publication, acknowledges that the United States Government

## Author details

W.S. Ashbee[1], L.M. Hively[2] and J.T. McDonald[3]

1 School of Computing at University of South Alabama, USA

2 Computational Sciences and Engineering Division at Oak Ridge National Lab, USA

3 School of Computing at University of South Alabama, USA

## References

[1] Netoff T, Park Y, Parhi K. Seizure prediction using cost-sensitive support vector machine. Engineering in Medicine and Biology Society (2009) EMBC 2009 Annual International Conference of the IEEE, pp. 3322–5.

[2] Stacey W, Le Van Quyen M, Mormann F, Schulze-Bonhage A, "What is the present-day EEG evidence for a preictal state?" *Epilepsy Res.* (2011);97(3):243–51.

[3] Mirowski PW, LeCun Y, Madhavan D, Kuzniecky R. Comparing SVM and convolutional networks for epileptic seizure prediction from intracranial EEG. Machine Learning for Signal Processing (2008) IEEE Workshop on MLSP, pp. 244–9.

[4] Hively LM, McDonald JT, Munro NB, Cornelius E, "Forewarning of Epileptic Events from Scalp EEG," peer-reviewed proceedings paper for *Biomedical Science and Engineering Conference at ORNL* (May 2013).

[5] Hively LM, Protopopescu VA, Munro NB, "Enhancements in epilepsy forewarning via phase-space dissimilarity," *J Clin Neurophysiol.* (2005);22(6):402–9.

[6] Percha B, Dzakpasu R, Żochowski M, and Parent J, "Transition from local to global phase synchrony in small world neural network and its possible implications for epilepsy," *Phys. Rev. E* (2005);72, paper #031909

[7] Van den Broeck C, Parrondo JMR, and Toral R, "Noise-induced nonequilibrium phase transition," *Phys. Rev. Lett.* (1994);73, 3395-3398

[8] Pittau F, Tinuper P, Bisulli F, Naldi I, Cortelli P, Bisulli A, et al., "Videopolygraphic and functional {MRI} study of musicogenic epilepsy. A case report and literature review," *Epilepsy & Behavior* (2008);13(4):685 – 692.

[9] Jenkins JS. "The Mozart effect," *J. Royal Soc. Med.* (2001); 94:170-172

[10] "10-20 System (EEG)."*Wikipedia*. Wikimedia Foundation, 22 July 2013. Web. 16 Aug. 2013.

[11] Hively LM, Protopopescu VA, "Channel-consistent forewarning of epileptic events from scalp EEG," *IEEE Trans Biomed Eng*. (2003);50(5):584–93.

[12] Hively LM, "Prognostication of Helicopter Failure," *ORNL/TM-2009-244*, Oak Ridge National Laboratory, Oak Ridge, TN (2009).

[13] Protopopescu VA, Hively LM, Gailey PC, "Epileptic event forewarning from scalp EEG," *J Clin Neurophysiol*. (2001);18(3):223–45.

[14] Hively LM, *et al.*, "Nonlinear Analysis of EEG for Epileptic Seizures," *ORNL/TM-12961*, Oak Ridge National Laboratory, Oak Ridge, TN (1995).

[15] Takens F, "Detecting strange attractors in turbulence," In: Rand D, Young L-S, editors. *Dynamical Systems and Turbulence*, Warwick 1980 [Internet]. Springer Berlin Heidelberg; 1981. p. 366–81.

[16] Bondy JA, Murty USR, *Graph Theory*, Springer (2008).

[17] Franzosi R, Pettini M, "Topology and phase transitions II. Theorem on a necessary relation," *Nuclear Physics B* (2007);782(3):219 – 240.

[18] Ng A. "Machine Learning: SVM Kernels." Coursera. Stanford University, n.d. Web. 10 Oct. 2013.

[19] Chang C-C, Lin C-J, "LIBSVM: A library for support vector machines" *ACM Trans. Intell. Syst. Technol.*( 2011);2(3):27:1–27.

[20] Arthurs S, Zaveri HP, Frei MG, Osorio I, "Patient and caregiver perspectives on seizure prediction," *Epilepsy Behav*. (2010);19(3):474–7.

[21] Protopopescu V, Hively LM, "Phase-space dissimilarity measures of nonlinear dynamics: Industrial and biomedical applications," *Recent Res. Dev. Physics* (2005);6, 649-688.

[22] Bubacz JA, Chmielewski HT, Pape AE, Depersio AJ, Hively LM, Abercrombie RK, "Phase Space Dissimilarity Measures for Structural Health Monitoring," (2011) *ORNL/TM-2011/260* (Oak Ridge National Laboratory).

Chapter 4

# Juvenile Myoclonic Epilepsy: An Update

Boulenouar Mesraoua, Dirk Deleu, Hassan Al Hail,
Gayane Melikyan and Heinz Gregor Wieser

Additional information is available at the end of the chapter

http://dx.doi.org/10.5772/57431

## 1. Introduction

We review the most important electroclinical aspects and possible subsyndromes of Juvenile myoclonic epilepsy (JME), as well as its genetic background, its pathophysiological and neuroimaging correlates, and treatment. JME is among the most common types of genetic epilepsies. The prevalence of JME in large cohorts has been estimated to be 5% to 10% of all epilepsies and around 18% of idiopathic generalized epilepsies but may be lower in some settings. There is a marked female predominance. Today JME is a widely recognized electro-clinical idiopathic generalized epilepsy syndrome. Onset is around the time of puberty. The most typical ictal phenomenon is bilateral myoclonia without loss of consciousness. Most patients also present with generalized tonic-clonic seizures (GTCS), and some with absence seizures. The typical circumstance at diagnosis is a first GTCS episode, after the patient has had myoclonia in the morning. Typically seizure episodes occur after awakening from a sleep period or in the evening relaxation period and are facilitated by sleep deprivation and sudden arousal. Diagnosis of JME can be made with the history of myoclonus plus a single GTCS plus generalized polyspike-waves or fast spike-waves on the EEG. The prevalence rate of photo-sensitivity (photoparoxysmal EEG response) in patients with JME ranges from 8 to 90%. Hyperventilation can induce absence seizures in patients with JME, while cognitive tasks are efficient in precipitating myoclonic seizures. Most patients have a good prognosis when treated with appropriate drugs, but with a well-known tendency to relapse after withdrawal. However, around 17% are able to discontinue medication and remain seizure-free thereafter. There is a small but still considerable subgroup of JME patients whose seizures are difficult to treat. Recent findings suggest that patients with JME have worse social adjustment in relevant aspects of their lives, works and familiar relationship. Differential diagnoses include the adolescent-onset progressive myoclonus epilepsies, or other forms of idiopathic generalized epilepsies of adolescence.

## 2. History of juvenile myoclonic epilepsy

Juvenile Myoclonic Epilepsy (JME) has been recognized by early distinguished physicians as Theodore Herpin in 1867 [1] and Robot in 1899 [2]. However, it was not until 1957 that Janz and Christian gave the first and precise description of JME in 47 German patients [3]. Later on, Castells and Mendilaharsu described JME in 70 Uruguayans patients [4]. Following that, Delgado-Escueta and Enrile-Bacsal reported 43 cases of uncontrolled convulsive seizures because the syndrome of JME was not recognized in those cases. [5]. Since then JME was reported in different ethnic groups around the world (Asia, Europe, North and Latin America, Oceania and Africa).

## 3. Classification and definition of JME

The ILAE Commission proposed and defined JME as a distinct syndrome of IGE: "Juvenile myoclonic epilepsy (impulsive petit mal) appears around puberty and is characterized by seizures with bilateral, single or repetitive, arrhythmic, irregular myoclonic jerks predominantly in the arms. Jerks may cause some patients to fall suddenly. No disturbance of consciousness is noticeable. The disorder may be inherited, and sex distribution is equal (but see Introduction!). Often, there are GTCS and less often infrequent absences. The seizures usually occur shortly after awakening and are often precipitated by sleep deprivation. Interictal and ictal EEG have rapid, generalized, often irregular spike-waves and polyspike-waves; there is no close phase correlation between EEG spikes and jerks. Frequently, the patients are photosensitive. Response to appropriate drugs is good" [6].

## 4. Incidence, prevalence, and sex ratio of JME

JME is the most common form of genetic/idiopathic generalized epilepsy (IGE): The IGEs comprises 40% of epilepsies in the US, 20% in Mexico, 8% in Central America [7] and is much higher (between 68–82.6%) in the Arab countries [8]. JME is responsible for 6 to 12 % up to 30% of all epilepsies in hospital and clinics records [3, 9-11] and for 3% according to a door-to-door population survey [12]. The incidence of JME varies from 0.5 to 6.3%/100.000. Based on 1% of population risk for epilepsy by age 20 [13], the risk of JME in the general population would be 1 per 1000 to 2000. The prevalence of JME has been estimated to be 5% to 10% of all epilepsies and 18% of IGEs [14, 15]. Gender differences are evident for JME with a marked female predominance [15, 16].

## 5. Age of onset

As shown by Janz [17], one of its first investigators, JME has an age-related onset [Figure 1]. The seizures appear between 8 and 26 years with a mean age of onset at 14.2 years; the majority of seizures occur between 12 and 18 years [17, 18]. Absence seizures (AS) are reported in 33.3% to 66.7%, myoclonic jerks (MJ) in 97%, and GTCS in 78.8% of the patients [14, 19]. In one study, AS antedated other types of seizures in all patients [20]. An earlier or later age of onset has also been reported [20-23]. Female preponderance for photosensitivity may explain an early onset of JME in female patients with photosensitivity [24, 25].

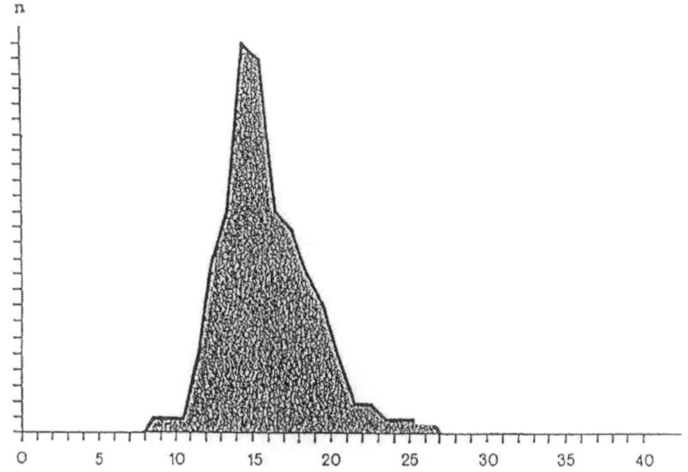


**Figure 1.** Age of onset of 139 patients with myoclonic seizures (i.e., beginning cases of juvenile myoclonic epilepsy). From Janz D et al. In: Epilepsy, J. Engel Jr and T A Pedley (ed), 1998. p.2391

## 6. Clinical presentation

JME is probably the most common and the most characteristic form of the IGE of adolescent-onset group. It is characterized by:

### 6.1. Myoclonic Jerks (MJ)

MJ occur prominently and spontaneously in the morning, after awakening; they are sudden, short-lasting, irregular, frequently symmetric; they may be self- limited (isolated) or may occur in clusters; if prolonged, these clusters may lead to a convulsive tonic-clonic seizure. MJ involve prominently the upper limbs; however the lower limbs and trunk are often not spared. Both distal and proximal jerky movements do occur; flexion of both forearms, flexion of both arms, flexion and abduction of the thighs and extension of the back, are typical. Patients may throw things out of their hands when the MJ involve or are restricted to the fingers; occasionally, the jerks are so intense that the patient falls to the ground (myoclonic-astatic seizures). Several authors reported some degree of asymmetry in MJ as well as focal features in the EEG, leading to the false diagnosis of focal epilepsy [26-28]. This issue will be discussed in the paragraph below (*Electroencephalography*)

Myoclonic status epilepticus (MSE) is not so rare. In MSE, consciousness may be intact. Drug withdrawal, sleep deprivation and alcohol intake are the main causes [29, 30].

### 6.2. Generalized Tonic Clonic Seizures (GTCS)

Frequently, in the outpatient clinic or emergency unit, a young patient is examined because he/she was a victim of a generalized seizure upon awakening. Often, when asked, the family reports that the GTCS followed repeated, severe MJ (generalized clonic-tonic-clonic seizure type) [5].

### 6.3. Absence Seizures (AS)

As indicated above, AS are generally reported in one-third of patients with JME [31, 32]. However, their frequency might be much higher (66.7%) [19]. There is general agreement that AS associated with JME are mild and short, when compared to the childhood absences and absences of *Juvenile Absence Epilepsy*. They are less severe with age and are often unnoticed by the patient [33].

In a recent prospective study, with long-term follow up of 257 patients with JME, Martinez-Juarez IE and coworkers encountered four JME groups: (1) Classic JME (72%), (2) Childhood Absence Epilepsy (CAE) evolving to JME (18%), (3) JME with Adolescent Absence (7%), and (4) JME with Astatic Seizures (3%). There was a female preponderance in the second group (CAE evolving to JME); the authors concluded that all 4 subtypes are chronic and probably lifelong [34].

### 6.4. Precipitation of seizures

As reported above, occurrence of MJ in the early morning is one of the hallmarks of JME. MJ and GTCS are induced by sleep deprivation, fatigue and excessive alcohol intake [35]. Sleep deprivation is understood as falling asleep late at night and getting up or awaken early in the morning (short sleep).

Seizure-provoking factors in JME are numerous; among them are: stress, fatigue, fever, sleep, flashing sunlight, music, reading, thinking, and excessive alcohol intake. In JME, photosensitivity (photoparoxysmal EEG response), is age-related and it varies considerably [35]. However, only a small number of patients experiences seizures by photic stimulation in daily life. In patients with JME, absence seizures are induced by hyperventilation, while myoclonic seizures are provoked by cognitive tasks [35, 36].

## 7. Electroencephalography

JME is widely underdiagnosed despite a characteristic clinical picture and a distinct EEG profile. If a patient, who is suspected clinically to suffer from JME, has a normal EEG, a sleep EEG and an EEG on awakening should follow.

The background activity is usually normal; some authors report theta slowing during poor seizure control [37], others found an increase in *Absolute Power* of delta, alpha and beta bands, more evident in frontoparietal regions in patients with JME [38].

### 7.1. Interictal EEG

Interictal EEG shows diffuse or generalized spike-wave (SW) and polyspike-wave (PSW) discharges at 3-6 Hz [Figure 2].

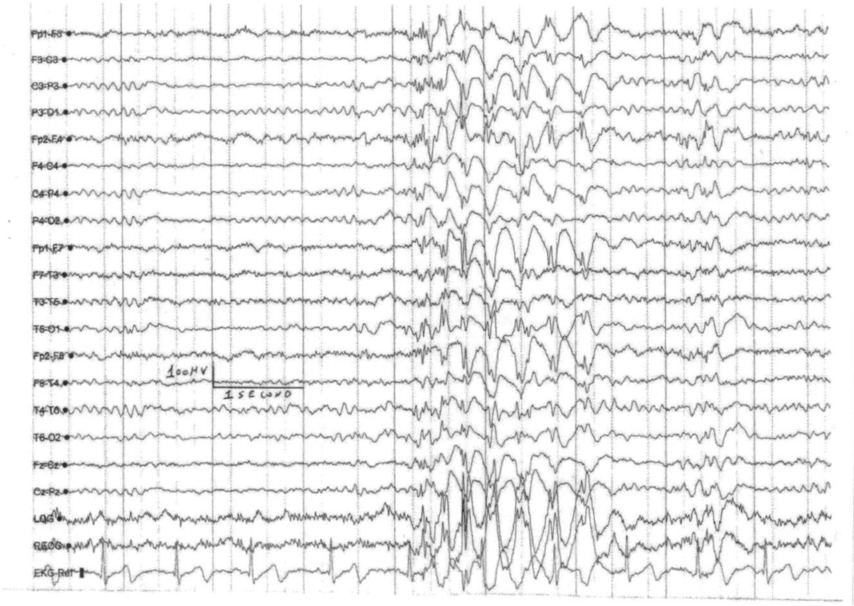

**Figure 2.** Thirteen year-old girl with clinical JME since the age of 9 years showing interictal polyspike wave discharges not associated with clinical manifestations.

Localization related EEG abnormalities are found in 16.9-57.1 % of patients [14, 27, 28]. These focal abnormalities include unilateral discharges, paroxysms with unilateral onset, and frequently discharges with above 50% voltage asymmetries. In general these EEG changes are predominantly seen at sleep onset and after provoked awakening [26].

Photosensitivity (photoparoxysmal EEG response), with or without MJ, in patients with JME varies from 8 to 90% and is more frequent in females and adolescents [35] [Figure3]

As mentioned above, in patients with JME, absence seizures can be induced by hyperventilation [Figure 4],

while cognitive tasks can precipitate myoclonic seizures.

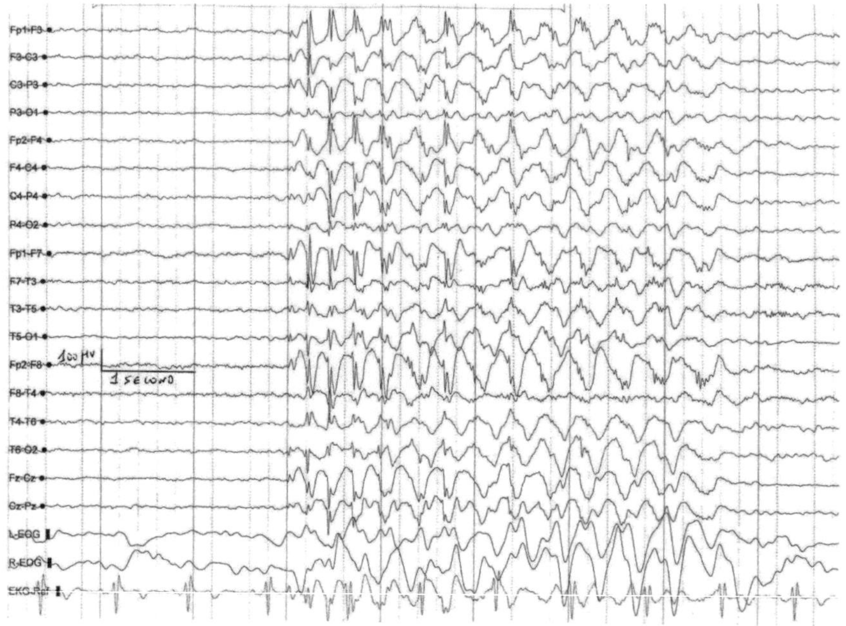


**Figure 3.** Photoparoxysmal response in a 20 year old female patient with JME since the age of 11 years accompanied by jerky movements of both upper extremities.

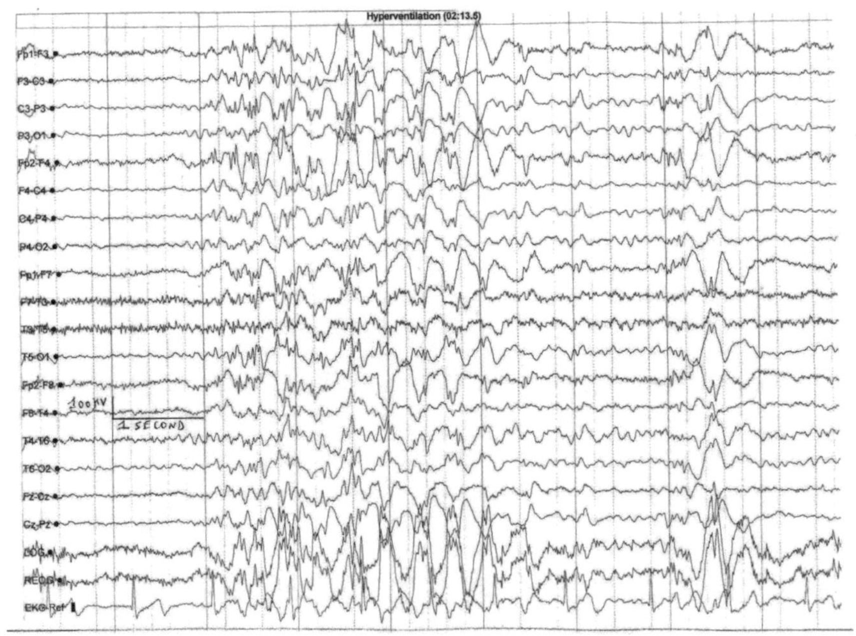


**Figure 4.** Same patient as in Figure 3 showing generalized polyspike-wave complexes accompanied by mild impairment of cognition during hyperventilation.

### 7.2. Ictal EEG

The characteristic ictal EEG manifestations of a MJ are a generalized burst of multiple spikes of short duration (0.5-2s). Frequently, however, the spikes are followed by slow waves with poorly structured spike-wave sequence [Figure 5].

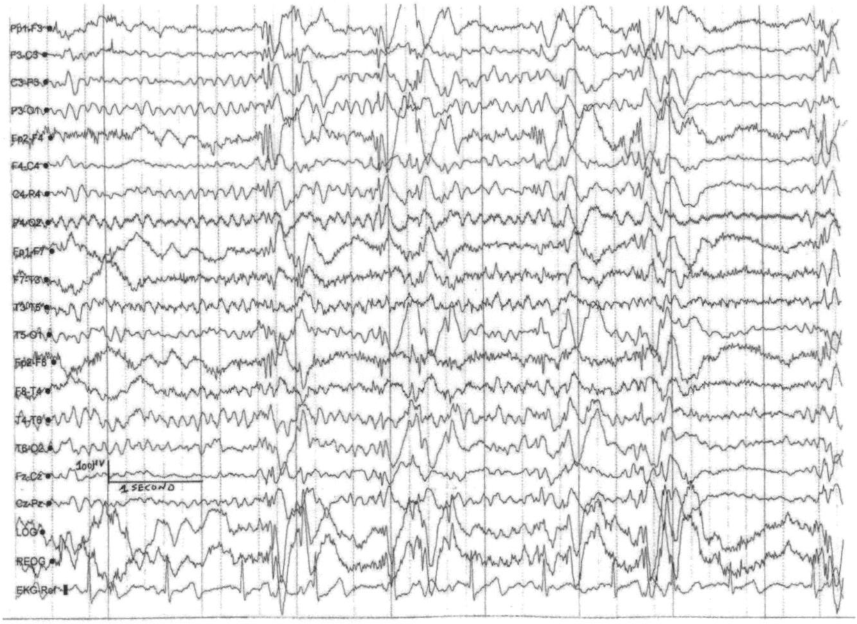

**Figure 5.** Seventeen year-old male with JME suffering from mild MJ of the upper extremities accompanied by bursts of brief polyspike wave sequences.

Ictal EEG discharges of absences consist of multiple spikes usually preceding or superimposed on a slow wave. These discharges often show a characteristic fragmentation and last from 1-4 s.

## 8. Genetics of JME

JME is the most common cause of hereditary grand mal seizures in people with epilepsy in the population at large [5, 17]. It has both Mendelian inheritance and complex genetic inheritance [12, 39]. 49% of JME families have clinical and EEG traits suggesting an autosomal dominant inherited disease. Variants of JME genes, with small to modest effects, contribute to risk/susceptibility in the remaining 51% [5, 12, 39]. Linkage disequilibrium is understood as the occurrence, in a specific population, of both DNA markers (DNA microsatellites or SNPs, single nucleotide polymorphisms) and a JME mutation at a higher frequency than would be predicted by random chance. With the passage of time, linkage disequilibrium decays through recombinations and transmissions into thousands of generations resulting in the fact that the epilepsy allele will have smaller and smaller genetic effects and will require other epilepsy alleles or environment to produce the epilepsy phenotype [40]. However, linkage disequilibrium is strongest and covers the widest region of a chromosome when the epilepsy allele is of recent origin, and has large genetic effects, e.g., Mendelian dominant or recessive effects.

| Chromosome locus | HGNC gene symbol[a] or phenotype | Gene/encoded protein | Country/ethnic group/ancestral origin | Mode of inheritance |
|---|---|---|---|---|
| 1p36.22 | Absence and JME | *Suggestive linkage* | Europe (Germany, Netherlands, UK, France, Italy, and Greece) | AD/oligogenic: meta analysis of 235 absence families and 118 JME families |
| 1p36.33 | **EJM7** | **GABRD**[b] | Australian families incl GEFS+ | AD |
| 2q23.3 | **EJM6** | **CACNB4**[b] | One family (father and son with absences and tonic-clonic) from Germany; one woman with JME and a daughter with 3-Hz spike waves | AD |
| 2q34 | JME | Unknown | Europe (Germany, Netherlands, UK, France, Italy, and Greece) | AD/oligogenic: meta analysis of 118 JME families |
| 2q23.3 and 2q24.1 | JME | Unknown | Tunisia – one large consanguineous family | AR |
| 2q33–36 | **EJM9** | Unknown | India – one large 4 generation family | AD |
| 3p14.2 | JME/absence | *Suggestive linkage* | Europe(Germany, Netherlands, UK, France, Italy, and Greece) | AD/oligogenic: meta analysis of 235 absence families and 118 JME families |
| 3q21.1 | JME | **CASR** (*calcium sensor receptor gene*)[b] | One large Indian family; 5 of 96 unrelated JME patients from Bangalore, India | AD |
| 5q12–14 | **EJM4** | Unknown | India | AD |
| 5q35.3 | JME with PPR (photosensitivity) | Unknown | Netherlands | AD |
| 5q34 | **EJM5** | **GABRA1**[b] | French Canadian (Quebec, Canada) – one family; one absence patient from Germany; one absence patient from France | AD |
| 5q34 | (Absence and JME) | Unknown | Europe – Germany, Netherlands, UK, France, Italy, and Greece | AD/oligogenic: meta analysis of 235 absence families and 118 JME families |
| 6p12 | EJM1 | **Myoclonin1/EFHC1**[b] | Hispanics [LA, California]<br>Hispanics [Belize]<br>Hispanics [Mexico[c]]<br>Hispanics [Honduras and Mexico]<br>Japan<br>Italy and Brazil<br>Tennessee<br>Austria | AD |
| 6p12 | JME | Unknown | Netherlands | AD |
| 6p21.3 | **EJM3** | **BRD2**[c] (bromodomain-containing gene) | European descent [Los Angeles, California]<br>European descent [NY[c]];<br>Germany | AD/oligogenic |
| 6p20 | JME | – | Germany | AD |
| 6p21.2 | JME | – | Germany | AR |
| 6q24 | JME | – | Saudi Arabia | AR |
| 7q32 | Photoparoxysmal response 3 (PPR3), with or without myoclonic epilepsy | Unknown | Netherlands<br>16 multiplex families | AD |
| 8q21.13 | JME with PPR (photosensitivity) | JRK, glutamate receptor GRIN2A are suspected | Netherlands | AD |
| 10q25–q26 | JME with PPR | – | India (New Delhi), 5 families | AD |
| 13q13 | JME with PPR (photosensitivity) | – | Netherlands | AD |
| 13q31.3 | CAE | – | Europe and Australia (Germany, Netherlands, UK, France, Italy and Greece) | AR/meta analysis of 235 absence families |
| 13q31 | JME with PPR (photosensitivity) | | Netherlands, United Kingdom, Denmark, France, Greece, Portugal, Sweden | AD |
| 15q14 | | **Cx36**[c] (connexin36) or a gap junction protein2 or GJP2 | European descent (New York) | AR |
| 15q14 | **EJM2** | CHRNA7 is candidate gene | UK, Sweden families | AR |
| 18q21 | JME | **ME2**[c] (malic enzyme2) | European descent (New York) | Association studies with JME |
| 19q13 | Generalized epilepsy with febrile seizures plus, including JME (GEFS+1) | **SCN1B** | Europe and Australia (Germany, Netherlands, UK, France, Italy, and Greece) | Meta analysis of 235 absence and 118 JME families<br>AR |
| Xp11.4–11.3 | JME | **EFHC2** | Germany, 81 patients | Association study with JME |

[a]Human Genome Nomenclature Committee gene symbol in bold letters.

[b]Mutation segregate with epilepsy affected members across 2 to 4 generation families or in singletons.

[c]SNP-associated variants of BRD2, Cx36 and ME2; AD, autosomal dominant; AR, autosomal recessive; JME, juvenile myoclonic epilepsy; and pCAE, pyknoleptic childhood absence epilepsy.

**Table 1.** Juvenile myoclonic epilepsy genes and chromosome loci. (Modified from Delgado-Escueta AV,2004,2007)

5 Mendelian JME genes have been reported (http://omim.org and http://www.ncbi.nlm.nih.gov/omim/). These are as follows: CACNB4 (calcium channel beta4 subunit) [41], CASR (calcium channel sensory receptor) [42], GABRA1 (GABA receptor alpha one subunit) [43], GABRD (GABA receptor delta subunit) [44], and Myoclonin1/EFHC1 (myoclonin1/one EF-hand containing gene) [45] [Table 1].

Also, three SNP susceptibility alleles of putative JME genes that contribute to the complex genetics of JME have been reported [12, 39, 46]: bromodomain-containing 2 (BRD2) [47], connexin 36 (Cx-36) [48], and malic enzyme2 (ME2) [49]. Additionally, more than 22 chromosome loci linked to JME have been described [Table 1] [46].

Familial segregation or association with disease identifies putative JME disease genes. In autosomal dominant JME, a candidate epilepsy gene should show at least one variant per affected individual; each candidate epilepsy gene should show homozygous mutations or compound heterozygous mutations in autosomal recessive JME [50-52]. Contributions of de novo mutations in the epilepsies have been demonstrated through studies of copy number variations (CNVs) which in fact contribute to genetic generalized epilepsies with complex inheritance, including JME. Consequently, pathogenic de novo mutations could be identified in JME patients [51, 52].

It has been shown that mutations in Cchb4, the mouse homologue of human CACNB4 or mutations in GABRA1 are sufficient to produce the absence phenotype while mutations in Myoclonin1/EFHC1 or BRD2 are sufficient to produce a convulsive phenotype (myoclonic or clonic or tonic-clonic seizures).[41, 53-55].

The identification of epilepsy alleles that cause JME could lead to new AED discoveries, to early diagnosis and curative treatment of JME.

## 9. Neuropsychological and behavioral studies in JME

There is an increasing interest in the behavioral and neuropsychological aspects of JME patients. Several studies have suggested specific cognitive deficits that explain some of the clinical and special behavioral findings in patients with JME. They also reported an increased incidence in psychiatric comorbidity in patients with JME. Pung and coworkers reported a circadian dysrhythmia in patients with JME which might explain the poor social outcome observed in those patients [56]. Several recent neuropsychological studies suggest that JME has a specific cognitive profile, with some deficiencies in areas related to the frontal lobes. These studies report impairment in word fluency and interference as well as dysfunctional planning abilities [57-59]. JME patients often show intolerance towards multiple tasks under time pressure. This might explain the occurrence of seizures in some of these patients during cognitive tasks. Also, patients with JME often fail to adhere to treatment plans. This might be linked to impairment in prospective memory. Interestingly Wandschneider and coworkers found this impairment also in their siblings indicating that it might be genetically determined [60].

## 10. Neuroimaging findings in JME

A particular personality profile is associated with JME. Behavioral studies suggest a possible frontal lobe dysfunction [57]. Modern neuroimaging techniques have proven to be very useful in understanding the underlying mechanisms of JME. A PET study using H215O to measure cerebral blood flow in patients with IGE and a history of absence seizures showed that there was a significant focal increase in thalamic blood flow during absence seizures. This result suggests that the thalamus plays a key role in the pathogenesis of typical absence seizures [61]. In another study using 18F-FDG PET and a visual working memory paradigm in nine JME patients and 14 controls, in which pairs of abstract images were presented and subjects had to indicate (by pressing a button) whether the images were matching or not, Swartz and colleagues showed that JME patients' performance was impaired during the working memory condition. The authors concluded that dysfunction in thalamo-fronto-cortical networks might account for poor working memory performance in JME patients. The decreased uptake of 18F-FDG in the ventral premotor cortex, the caudate, the dorsolateral prefrontal cortex bilaterally, and the left premotor area, was in favor of a widespread frontal impairment [62]. Using PET and the radioligand 11C-WAY-100635, Meschaks and colleagues observed reduced WAY-100635 binding potential in the dorsolateral prefrontal cortex, the raphe nuclei, and the hippocampus, but not in motor cortex. The observed reductions in serotonin 1A receptor binding suggest that the serotonin system is affected in JME, and also that serotonergic processes are involved in the pathophysiology of myoclonus in JME [63]. In another PET study, Ciumas and coworkers compared JME patients with patients suffering from generalized tonic-clonic seizures (GTCS): alterations in the dopamine system were found in both GTCS and JME [64].

One study using Magnetic Resonance Spectroscopy (proton MRS) found that N-acetyl aspartate (NAA) levels are reduced in the thalami of JME patients. This finding supports the idea that thalamic dysfunction is part of the underlying mechanism of epileptogenesis in JME [65]. Moreover, other interesting studies using 1H-MRS demonstrated a significantly reduced prefrontal concentration of NAA in JME patients compared with controls [66, 67]. This finding seems to be specific to JME, compared with other forms of IGE [67].

Using Functional MRI (fMRI), Vollmar et al. [68] investigated 30 JME patients with a challenging working memory fMRI paradigm. The authors found an increased functional connectivity within the frontal and parietal lobes, between the motor system and areas of higher cognitive functions. They correlated those findings with the well-known fact that cognitive tasks can precipitate MJ in some JME patients [69].

Quantitative MRI has been used to demonstrate subtle but widespread cerebral structural changes in patients with IGE, particularly in patients with JME [70]. In a related study, Woermann and coworkers have shown that patients with JME have an increase in cortical gray matter in the mesial frontal lobes compared with healthy subjects [71].

Using T1weighted MRI and diffusion tensor imaging (DTI) O'Muircheartaigh et al. found a decreased mesial frontal gray matter volume and a reduced fractional anisotropy (FA) in the underlying white matter tracts. These findings may represent the anatomical basis for the reported neuropsychological and psychiatric changes seen in patients with JME [72].

## 11. Animal model of JME

Genetic Absence Epilepsy Rats from Strasbourg (GAERS) is a well-established genetic model of absence epilepsy [73]; however, the baboon represents a more advanced non-human primate model of epilepsy, specifically of IGE [74]. It offers a natural model of photosensitive epilepsy with myoclonic and generalized tonic clonic seizures occurring spontaneously or provoked by intermittent photic stimulation [74]. In the baboon, seizures occur spontaneously or are triggered by ketamine or under other circumstances, such for example,fighting among baboons; also, baboons with seizures have normal brain anatomy [75].

In a recent study by Szabó and coworkers [76] involving a pedigree baboon colony, seizures were defined as generalized myoclonic or tonic-clonic; characteristically two thirds of the seizures occurred in the morning. Also, seizure onset occurred in adolescence (age, 5 y), the prevalence of recurrent seizures in this pedigree was 15%. Contrary to human recent findings, seizures in the baboon were more prevalent in male baboons, with a tendency of an early onset and more frequent seizures compared with female baboons. Electroencephalographically, on the baboon scalp, interictal epileptic discharges present as generalized spike-and-wave discharges of 4-6 Hz frequency. All the above clinical and EEG features in the baboon suggest similarities to juvenile myoclonic epilepsy in humans. The baboon also represents an excellent model for testing the efficacy and electrophysiological mechanisms of action of future AEDs for IGEs [77, 78].

## 12. Management of JME

### 12.1. AEDs treatment

According to the international League Against Epilepsy, "In JME, response to appropriate drugs is good" [6]. However, 15 % of patients with JME might be drug resistant [79]. Today, with the available new AEDs the rate of drug resistance might be lower. The choice of AEDs is based on clinical experience and the available studies and trials. Several AEDs can be used with success in patients with JME. However, it is important to know, that some AEDs can aggravate myoclonic jerks. Valproate is still considered the first-line treatment in JME in male and females without childbearing potential. The dosage in adults varies from 1000mg to 2000mg/day. The control rate varies from 84.5% to 90% in different studies [5,,80-82].

Several studies have shown the efficacy of Lamotrigine (LTG) in the treatment of JME [83,84]. LTG is useful in younger women because of the potential teratogenicity of VPA [85], in patients with migraine (with aura) [86] and in patients with bipolar depression [87]. However, LTG appears to be less effective than VPA [82] and has the potential to exacerbate seizures in IGE and can aggravate MJ or GTCS [88]. The same authors [88] reported de novo appearance of MJ in IGE in five women among 93 patients treated with LTG (5.4%) with a phenotype close to JME at the time of aggravation.

Levetiracetam (LEV) is highly effective in controlling seizures in JME as shown by the studies by Berkovic and co-workers [89], Noachtar and colleagues [90] and Rosenfeld and colleagues [91]. These randomized, double-blind, placebo-controlled studies showed a responder rate of 61% in patients with JME, with 20.8% of them becoming seizure-free. LEV should be one of the options in the treatment of JME [92] as a first line or add on, particularly in women of childbearing potential. However, LEV also may exaggerate myoclonus [93].

If tolerated, Topiramate (TPM) can be useful in the treatment of JME particularly in overweight patients and in patients with associated migraine. Several authors have shown its efficacy as an add-on therapy in JME [94-96]. TPM was even reported as slightly more efficacious then VPA in a study by Levisohn and Holland in 2007 [97]. TPM may produce neuropsychiatric side effects particularly alteration of attention, and verbal fluency [98] and therefore may lead to treatment failure [82].

Few studies showed good efficacy of Zonisamide (ZNS) in patients with JME. One particular study by Kothare and colleagues looked at 15 patients with JME: 13 patients received ZNS as first monotherapy and 2 as add-on therapy. There were 80% of responders in the monotherapy group. 69% of patients were GTCS-free, 62% were seizure-free for MJ, and 38% were seizure-free for absences. The daily dose ranged between 200 and 500 mg [99].

Another study showed that ZNS treatment led to more than 50% reduction of seizure frequency in 83.3% of treated patients for GTCS and in 100% for MJ and absences [100].

Obeid and colleagues reported that Clonazepam was effective in controlling myoclonic jerks, but not the GTCS in JME patients [101]. In a later study, Panayiotopoulos [102] found both Clonazepam and Acetazolamide useful adjunctive drugs in JME, particularly if absences and myoclonus are associated. Mantoan and colleagues [92] found that Clonazepam can be combined with LTG in JME in order to avoid the myoclonic effects of LTG. The same authors, among others, have shown that few AEDs like Carbamazepine, Oxcarbazepine, and Phenytoin can exacerbate absences and myoclonus and even induce status epilepticus and should not be used in JME, although these drugs are able to control tonic-clonic seizures associated with JME when these are refractory to other medication [92,103-105].

Lacosamide may be effective in JME. However, larger, controlled studies confirming the usefulness of this AED are lacking [106].

### 12.2. Discontinuing AEDs in patients with JME

In a recent interesting study, Geithner and colleaguers [107] followed 31 patients with JME for as long as 25 years: Of these 31 patients, 67.7 % became seizure free. In 6 of these patients (28.6%), AEDs were discontinued with no more seizures. The most important factor that increased the chance for complete seizure freedom after stopping the AED was complete remission of the GTCS under a single anti-epileptic drug. However, the occurrence of a photo-paroxysmal response increased the risk of seizure recurrence after stopping AEDs. The authors concluded that in order to maintain seizure freedom, lifelong antiepileptic drug treatment is not necessarily required by all patients with JME.

## 13. Other treatments and approaches

### 13.1. Dietary therapy for JME

Ketogenic diet has been utilized for a number of conditions. Recently, Kossoff and colleagues [108] have looked at the effectiveness of the use of diet for treatment for AED-resistant JME. The investigators used a modified Atkins diet as an adjunct therapy to treat 8 adolescents and adults patients with JME. Six (75%) of these patients had more than 50% seizure reduction after one month, five patients (63%), had a greater than 50% seizure improvement after three months; three patients reported increasing seizures frequency during periods of noncompliance. The authors concluded that the modified Atkins diet can be a useful therapy for young patients with AED-resistant JME. However, more patients need to be studied to assess which JME patients may benefit from this therapy.

### 13.2. Vagal Nerve Stimulation (VNS)

Only one study [109] reported on the role of VNS in drug resistant JME. In this study, 12 patients with drug resistant IGE were offered VNS. Among these patients 7 patients had JME: 5 of them responded to VNS and had reduced AED-treatment at follow up. Kostov and colleagues [109] concluded that adjunctive VNS therapy is a favourable treatment option for drug-resistant IGE patients.

### 13.3. Lifestyle, psychiatric treatment

In managing JME, life style is considered an important part of the treatment and this aspect should be discussed with the patient in order to obtain good seizure control. Patients should avoid all precipitating factors such as fatigue, sleep deprivation, alcohol and unnecessary drug intake. In particular, patients should avoid any potentially dangerous activities in the awakening period, such as taking a bath without observation, for example.

There is also a high prevalence of psychiatric disorders in patients with JME, such as mood, anxiety, and personality disorders. Early recognition and treatment of these disturbances and psychosocial difficulties play an important role in the prognosis of JME [110].

## 14. Conclusion

JME is a common form of IGE with a characteristic clinical and electroencephalographic profile. Usually, a sleep EEG or an EEG on awakening confirm the clinical suspicion. Despite this distinct clinical and EEG trait, JME is often not recognized as such; this might result in serious consequences for the sufferers: in particular, if potentially aggravating AEDs are used, especially Carbamazepine, Oxcarbazepine, Phenytoin; but also, in some patients, Lamotrigine, which might exacerbate absences and myoclonus. These AEDs are therefore contraindicated, although they can improve control of tonic-clonic seizures when these are refractory to other medications. The following AEDs should not be used in JME: Gabapentin, Pregabalin, Tiagabine, and Vigabatrin; they can worsen seizures (Tiagabine and Vigabatrin might induce absence status epilepticus).

Beside the pharmacological treatment, management of JME should also include the patient's lifestyle, with avoidance of sleep deprivation, alcohol excess and the treatment of the cognitive and psychiatric problems that is often associated with JME.

With correct diagnosis and appropriate AED treatment (such as Valproate, Levetiracetam), a small but important group of patients will be able to come off medication not requiring therefore lifelong AED medications.

In refractory cases of JME, modified Atkins diet might be useful. Vagus Nerve Stimulation, Callosotomy and Deep brain stimulation are rarely contemplated.

Neuroimaging, using advanced imaging techniques, suggests subtle structural and functional changes, mainly within the frontal lobes, in patients with JME.

These changes correlate with the observed neuropsychological deficits (frontal lobe dysfunction) in patients with JME.

Genetically, JME is the most common cause of hereditary grand mal seizures and has both Mendelien (dominant or recessive trait) and complex genetic inheritance. During the last two decades a lot of discoveries have been made in this field. Finding more chromosome loci and more epilepsy-causing mutations for JME will continue to provide definitive evidence of the complex nature of this disease and of the existence of specific diseases within JME. Future

AEDs should be designed to counter major genes that cause JME.

## Author details

Boulenouar Mesraoua[1*], Dirk Deleu[1,2], Hassan Al Hail[1,2], Gayane Melikyan[1,2] and Heinz Gregor Wieser[3]

*Address all correspondence to: boulenouar.mesraoua@wanadoo.fr

1 Department of Medicine /Neurology-Neurophysiology, Hamad Medical, Corporation, Doha, Qatar

2 Department of Clinical Neurology and Neurosciences, Weill Cornell Medical, College WCMC, Doha, Qatar

3 Department of Epileptology and Electroencephalography, Neurology Clinic, University of Zurich, Switzerland

## References

[1] Herpin T. *Des acces incomplets d'epilepsie*. Paris: Bailliere 1867.

[2] Rabot L. *De la myoclonie épileptique. Medical thesis, Paris.* Georges Carre et C.Naud,editeurs.1899.

[3] Janz D, Christian W. Impulsive-petit mal. J Neurol (Z Nervenheilkd) 1957; 176: 344–386.

[4] Castells C, Mendilaharsu C. La epilepsia mioclónica bilateral y consciente. Acta Neurol Latinoamer. 1958;4:23–48.

[5] Delgado-Escueta AV, Enrile-Bacsal F. Juvenile myoclonic epilepsy of Janz. Neurology. 1984;34(3):285-294.

[6] Commission on Classification and Terminology of the International League Against Epilepsy.Proposal for revised classification of epilepsies and epileptic syndromes. Epilepsia 1989;30:389-399.

[7] Durón RM, Medina MT, Martínez-Juárez IE, Bailey JN, Perez-Gosiengfiao KT, Ramos-Ramírez R, López-Ruiz M, Alonso ME, Ortega RH, Pascual-Castroviejo I, Machado-Salas J, Mija L, Delgado-Escueta AV. *Epilepsia*. 2005;46 Suppl 9:34-47.

[8] Benamer HT, Grosset DG. A systematic review of the epidemiology of epilepsy in Arab countries. Epilepsia. 2009;50(10):2301-2304.

[9] Tsuboi T. *Primary generalized epilepsy with sporadic myoclonias of myoclonic petit mal type*. Stuttgart: Thieme; 1977.

[10] Murthy JM, Yangala R, Srinivas M. The syndromic classification of the International League Against Epilepsy: a hospital-based study from South India. *Epilepsia* 1998; 39: 48–54.

[11] Jain S, Tripathi M, Srivastava AK, Narula A. Phenotypic analysis of juvenile myoclonic epilepsy in Indian families. *Acta Neurol Scand* 2003; 107: 356–362.

[12] Nicoletti A, Reggio A, Bartoloni A, Failla G, Sofia V, Bartalesi F, et al. Prevalence of epilepsy in rural Bolivia. *Neurology* 1999; 53: 2064–2069.

[13] Hauser WA, Annegers JF, Kurland LT. Incidence of epilepsy and unprovoked seizures in Rochester, Minnesota:1935-1984.Epilepsia 1993;34:453-468.

[14] Panayiotopoulos CP, Obeid T, Tahan AR. Juvenile myoclonic epilepsy: a 5-year prospective study. Epilepsia. 1994 Mar-Apr;35(2):285-296.

[15] Camfield CS, Striano P, Camfield PR. Epidemiology of juvenile myoclonic epilepsy. Epilepsy Behav. 2013;28 Suppl 1:S15-17.

[16] Mullins GM, O'Sullivan SS, Neligan A, McCarthy A, McNamara B, Galvin RJ, Sweeney BJ. A study of idiopathic generalised epilepsy in an Irish population. Seizure. 2007;16(3):204-210.

[17] Janz D. *Die Epilepsien*. Stuttgart:Thieme,1969.

[18] Obeid T, Panayiotopoulos CP. Juvenile myoclonic epilepsy: a study in Saudi Arabia. *Epilepsia*. 1988;29(3):280-282.

[19] Mukhin KIu, Levin PG, Nikanorova MIu. Juvenile myoclonic epilepsy. Zh Nevrol Psikhiatr Im S S Korsakova. 1995;95(3):17-21.

[20] Murthy JM, Rao CM, Meena AK.Clinical observations of juvenile myoclonic epilepsy in 131 patients: a study in South India. Seizure. 1998;7(1):43-47.

[21] Gram L, Alving J, Sagild JC, Dam M. Juvenile myoclonic epilepsy in unexpected age groups. Epilepsy Res. 1988;2(2):137-140.

[22] Gilliam F, Steinhoff BJ, Bittermann HJ, Kuzniecky R, Faught E, Abou-Khalil B. Adult myoclonic epilepsy: a distinct syndrome of idiopathic generalized epilepsy. Neurology. 2000;55(7):1030-1033.

[23] Jacob S, Martin D, Rajabally YA. Juvenile myoclonic epilepsy in an elderly patient. Age Ageing. 2006;35(2):194-196.

[24] Wolf P, Goosses R. Relation of photosensitivity to epileptic syndromes.J Neurol Neurosurg Psychiatr 1986;49:1368-1391.

[25] Lu Y, Waltz S, Stenzel K, Muhle H, Stephani U. Photosensitivity in epileptic syndromes of childhood and adolescence. Epileptic Disord. 2008;10(2):136-143.

[26] Serafini A, Rubboli G, Gigli GL, Koutroumanidis M, Gelisse P. Neurophysiology of juvenile myoclonic epilepsy. Epilepsy Behav. 2013;28 Suppl 1:S30-39.

[27] Létourneau K, Cieuta-Walti C, Deacon C. Epileptiform asymmetries and treatment response in juvenile myoclonic epilepsy. Can J Neurol Sci. 2010;37(6):826-830.

[28] Jayalakshmi SS, Srinivasa Rao B, Sailaja S. Focal clinical and electroencephalographic features in patients with juvenile myoclonic epilepsy. Acta Neurol Scand. 2010;122(2):115-123.

[29] Salas Puig J, Tuñón A, Vidal JA, Mateos V, Guisasola LM, Lahoz CH. [Janz's juvenile myoclonic epilepsy: a little-known frequent syndrome. A study of 85 patients]. Med Clin (Barc). 1994;103(18):684-689.

[30] Larch J, Unterberger I, Bauer G, Reichsoellner J, Kuchukhidze G, Trinka E. Myoclonic status epilepticus in juvenile myoclonic epilepsy. Epileptic Disord. 2009;11(4): 309-314.

[31] Janz D. Juvenile Myoclonic Epilepsy with typical absences. In:Dunca JS, Panayiotopoulos CP, ed. *Typical Absences and Related Syndromes*. London: Churchill,1994.

[32] Syvertsen MR, Markhus R, Selmer KK, Nakken KO. Juvenile myoclonic epilepsy. Tidsskr Nor Laegeforen. 2012;132(14):1610-1613.

[33] Panayiotopoulos CP, Obeid T, Waheed G. Differentiation of typical absence seizures in epileptic syndromes. A video EEG study of 224 seizures in 20 patients. Brain. 1989;112 (Pt 4):1039-1056.

[34] Martínez-Juárez IE, Alonso ME, Medina MT, Durón RM, Bailey JN, López-Ruiz M, Ramos-Ramírez R, León L, Pineda G, Castroviejo IP, Silva R, Mija L, Perez-Gosiengfiao K, Machado-Salas J, Delgado-Escueta AV. Juvenile myoclonic epilepsy subsyndromes: family studies and long-term follow-up. Brain. 2006;129(Pt 5):1269-1280.

[35] Kasteleijn-Nolst Trenité DG, Schmitz B, Janz D, Delgado-Escueta AV, Thomas P, Hirsch E, Lerche H, Camfield C, Baykan B, Feucht M, Martínez-Juárez IE, Duron RM, Medina MT, Rubboli G, Jerney J, Hermann B, Yacubian E, Koutroumanidis M, Stephani U, Salas-Puig J, Reed RC, Woermann F, Wandschneider B, Bureau M, Gambardella A, Koepp MJ, Gelisse P, Gurses C, Crespel A, Nguyen-Michel VH, Ferlazzo E, Grisar T, Helbig I, Koeleman BP, Striano P, Trimble M, Buono R, Cossette P, Represa A, Dravet C, Serafini A, Berglund IS, Sisodiya SM, Yamakawa K, Genton P. Consensus on diagnosis and management of JME: From founder's observations to current trends. Epilepsy Behav. 2013;28 Suppl 1:S87-90.

[36] Beniczky S, Guaranha MS, Conradsen I, Singh MB, Rutar V, Lorber B, Braga P, Fressola AB, Inoue Y, Yacubian EM, Wolf P. Modulation of epileptiform EEG discharges in juvenile myoclonic epilepsy: an investigation of reflex epileptic traits. Epilepsia. 2012;53(5):832-839.

[37] Genton P, Salas Puig X, Tunon A, Lahoz C, Del Socorro Gonzales Sanchez M. Juvenile myoclonic epilepsy and related syndromes: clinical and neurophysiological aspects. In: Malafosse A, Genton P, Hirsch E, Marescaux C, Broglin, Bernasconi R, eds. *Idiopathic generalized Epilepsies*. London:John Libbey,1994, p.253-265.

[38] Santiago-Rodríguez E, Harmony T, Cárdenas-Morales L, Hernández A, Fernández-Bouzas A. Analysis of background EEG activity in patients with juvenile myoclonic epilepsy. Seizure. 2008;17(5):437-45.

[39] Delgado-Escueta AV. Advances in genetics of juvenile myoclonic epilepsies. Epilepsy Curr 2007;7:61–67.

[40] Terwilliger JD, Göring HH. Gene mapping in the 20th and 21st centuries: statistical methods, data analysis, and experimental design. Hum biology, 72(1). Detroit, Michigan: Wayne State University Press; 2000. p. 63–132.

[41] Escayg A, De Waard M, Lee DD, et al. Coding and noncoding variation of the human calcium-channel beta4-subunit gene CACNB4 in patients with idiopathic generalized epilepsy and episodic ataxia. Am J Hum Genet 2000;66(5):1531–1539

[42] Kapoor A, Satishchandra P, Ratnapriya R, et al. An idiopathic epilepsy syndrome linked to 3q13.3-q21 and missense mutations in the extracellular calcium sensing receptor gene. Ann Neurol 2008;64(2):158–167.

[43] Cossette P, Liu L, Brisebois K, et al. Mutation of GABRA1 in an autosomal dominant form of juvenile myoclonic epilepsy. Nat Genet 2002;31(2):184–189.

[44] Dibbens LM, Feng HJ, Richards MC, et al. GABRD encoding a protein for extra- or peri-synaptic GABAA receptors is a susceptibility locus for generalized epilepsies. Hum Mol Genet 2004;13(13):1315–1319.

[45] Suzuki T, Delgado-Escueta AV, Aguan K, et al. Mutations in EFHC1 cause juvenile myoclonic epilepsy. Nat Genet 2004;36(8):842–849.

[46] Noebels JL, Avoli M, Rogawski MA, Olsen RW, Delgado-Escueta AV. The next decade of research in the basic mechanisms of the epilepsies. In: Noebels JL, Avoli M, Rogawski MA, Olsen RW, Delgado-Escueta AV, editors. *Jasper's basic mechanisms of the epilepsies*. Oxford University Press; 2012. p. 3–11.

[47] Pal DK, Evgrafov OV, Tabares P, Zhang F, Durner M, Greenberg DA. BRD2 (RING3) is a probable major susceptibility gene for common juvenile myoclonic epilepsy.Am J Hum Genet 2003;73(2):261–270.

[48] Mas C, Taske N, Deutsch S, et al. Association of the connexin36 gene with juvenile myoclonic epilepsy. J Med Genet 2004;41(7):e93.

[49] Greenberg DA, Cayanis E, Strug L, et al. Malic enzyme 2 may underlie susceptibility to adolescent-onset idiopathic generalized epilepsy. Am J Hum Genet 2005;76(1):139–146.

[50] Roach JC, Glusman G, Smit AF, et al. Analysis of genetic inheritance in a family quartet by whole-genome sequencing. Science 2010;328(5978):636–639.

[51] Robinson PN. Whole-exome sequencing for finding de novo mutations in sporadic mental retardation. Genome Biol 2010;11(12):144.

[52] Robinson PN, Krawitz P, Mundlos S. Strategies for exome and genome sequence data analysis in disease-gene discovery projects. Clin Genet 2011;80(2):127–132.

[53] Feng HJ, Kang JQ, Song L, Dibbens L, Mulley J, Macdonald RL. Delta subunit susceptibility variants E177A and R220H associated with complex epilepsy alter channel gating and surface expression of alpha4beta2delta GABAA receptors. J Neurosci 2006;26(5):1499–1506.

[54] Ding L, Feng HJ, Macdonald RL, Botzolakis EJ, Hu N, Gallagher MJ. GABA(A) receptor alpha1 subunit mutation A322D associated with autosomal dominant juvenile myoclonic epilepsy reduces the expression and alters the composition of wild type GABA(A) receptors. J Biol Chem 2010;285(34):26390–26405.

[55] Mihalek RM, Banerjee PK, Korpi ER, et al. Attenuated sensitivity to neuroactive steroids in gamma-aminobutyrate type A receptor delta subunit knockout mice. Proc Natl Acad Sci U S A 1999;96(22):12905–12910.

[56] Pung T, Schmitz B. Circadian rhythm and personality profile in juvenile myoclonic epilepsy. Epilepsia, 2006; 47 (Suppl. 2): 111-114.

[57] Devinsky O, Gershengorn J, Brown E, Perrine K, Vazquez B, Luciano D. Frontal functions in juvenile myoclonic epilepsy. Neuropsychiatry Neuropsychol Behav Neurol. 1997, 10:243-246.

[58] Pascalicchio TF, de Araujo Filho GM, da Silva Noffs MA et al. Neuropsychological profile of patients with juvenile myoclonic epilepsy: a controlled study of 50 patients.Epilepsy Behav, 2007;10: 263-267.

[59] Piazzini A, Turner K, Vignoli A, Canger R, Canevini MP. Frontal cognitive dysfunction in juvenile myoclonic epilepsy. Epilepsia, 2008; 49:657-662.

[60] Wandschneider B, Kopp UA, Kliegel M et al. Prospective memory in patients with juvenile myoclonic epilepsy and their healthy siblings. Neurology, 2010; 75 (24): 2161-2167.

[61] Prevett MC, Duncan JS, Jones T, Fish DR, Brooks DJ. Demonstration of thalamic activation during typical absence seizures using H2(15)O and PET. Neurology, 1995; 45: 1396–1402.

[62] Swartz BE, Simpkins F, Halgren E et al. Visual working memory in primary generalized epilepsy: an 18FDG-PET study. Neurology, 1996; 47:12031212.

[63] Meschaks A, Lindstrom P, Halldin C, Farde L, Savic I. Regional reductions in serotonin 1A receptor binding in juvenile myoclonic epilepsy. Arch Neurol, 2005; 62 (6): 946-950

[64] Ciumas C, Wahlin TB, Espino C, Savic I. The dopamine system in idiopathic -615.615.

[65] Mory SB, Li LM, Guerreiro CA, Cendes F. Thalamic dysfunction in juvenile myoclonic epilepsy: a proton MRS study. Epilepsia, 2003; 44:1402–1405.

[66] Savic I, Osterman Y, Helms G. MRS shows syndrome differentiated metabolite changes in human-generalized epilepsies. NeuroImage, 2004; 21:163–172.

[67] Simister RJ, McLean MA, Barker GJ, Duncan JS. Proton MRS reveals frontal lobe metabolite abnormalities in idiopathic generalized epilepsy. Neurology, 2003; 61:897–902.

[68] Vollmar C, O'Muircheartaigh J, Barker GJ et al. Motor system hyperconnectivity in juvenile myoclonic epilepsy: a cognitive functional magnetic resonance imaging study. Brain, 2011; 134:1710–1719.

[69] Inoue, Y, Kubota H. Juvenile myoclonic epilepsy with praxis-induced seizures. *Juvenile myoclonic epilepsy, The Janz Syndrome*. Wrightson Biomedical Publishing, Petersfield, UK (2000), pp. 73–81.

[70] Woermann FG, Sisodiya SM, Free SL, Duncan JS. Quantitative MRI in patients with idiopathic generalized epilepsy. Evidence of widespread cerebral structural changes. Brain 1999; 121:1661–1667.

[71] Woermann FG, Free SL, Koepp MJ, Sisodiya SM, Duncan JS. Abnormal cerebral structure in juvenile myoclonic epilepsy demonstrated with voxel-based analysis of MRI. Brain,1999; 122:2101–2108.

[72] O'Muircheartaigh J, Vollmar C, Barker GJ et al. Focal structural changes and cognitive dysfunction in juvenile myoclonic epilepsy.Neurology, 2011;76:34–40.

[73] Chipaux M, Vercueil L, Kaminska A, Mahon S, Charp S. Persistence of cortical censory processing during absence seizures in human and an animal model: Evidence from EEG and intracellular recordings. PLoS One. 2013; 8(3): e58180.Published online 2013 March 4.doi: 10.1371/journal.pone. 0058180 10.1371/journal.pone.0058180.

[74] Killam EK. Photomyoclonic seizures in the baboon, *Papio papio*. Fed Proc 1979; 38:2429–2433.

[75] Naquet R, Catier J, Menini C. Neurophysiology of photically induced epilepsy in *Papio papio*. In: Meldrum BS, Marsden CD. Advances in neurology, vol 10:107-118. New York (NY): Raven Press; 1975. p.107-118.

[76] Szabó CÁ, Knape KD, Leland MM, Cwikla DJ, Williams-Blangero S, Williams JT. Epidemiology and characterization of seizures in a pedigreed baboon colony. Comp Med. 2012;62(6):535-538.

[77] Killam EK, Matsuzaki M, Killam KF. Studies of anticonvulsant compounds in the *Papio papio* model of epilepsy. In: Sabelli H, editor. Chemical Modulation of Brain Function. New York: Raven Press; 1973. p. 161–171

[78] Meldrum BS, Anlezark G, Balzamo E, Horton RW, Trimble M. Photically induced epilepsy in *Papio papio* as a model for drug studies. In: Meldrum BS, Marsden CD, editors. Advances in Neurology. Vol 10. New York: Raven Press; 1975. p.119-128.

[79] Gelisse P, Genton P, Thomas P, Rey M, Samuelian JC, Dravet C. Clinical factors of drug resistance in juvenile myoclonic epilepsy. J Neurol Neurosurg Psychiatry, 2001;70:240–243.

[80] Penry JK, Dean JC, Riela AR. Juvenile myoclonic epilepsy: long-term response to therapy. Epilepsia, 1989; 30 (Suppl. 4):19–23.

[81] Genton P, Gelisse P, Thomas P. Juvenile myoclonic epilepsy today: current definition and limits. B. Schmitz, T. Sander (eds.) *Juvenile myoclonic epilepsy: the Janz syndrome*, Wrightson Medical Publishing, Petersfield and Philadelphia (2000), pp. 11–32.

[82] Marson AG, Al-Kharusi AM, Alwaidh M, et al. The SANAD study of effectiveness of valproate, lamotrigine, or topiramate for generalised and unclassifiable epilepsy: an unblinded randomised controlled trial. Lancet, 2007; 369:1016–1026.

[83] Buchanan N. The use of lamotrigine in juvenile myoclonic epilepsy. Seizure, 1996; 5:149–151.

[84] Morris GL, Hammer AE, Kustra RP, MessenheimerJA. Lamotrigine for patients with juvenile myoclonic epilepsy following prior treatment with valproate: results of an open-label study. Epilepsy Behav, 2004; 5:509–512.

[85] Wlodarczyk BJ, Palacios AM, George TM, Finnell RH. Antiepileptic drugs and pregnancy outcomes. Am J Med Genet A 2012;158A(8):2071–2090.

[86] Lampl C, Katsarava Z, Diener HC, Limmroth V. Lamotrigine reduces migraine aura and migraine attacks in patients with migraine with aura. J Neurol Neurosurg Psychiatry 2005;76:1730–1732.

[87] Geddes JR, Calabrese JR, Goodwin GM. Lamotrigine for treatment of bipolar depression: independent meta-analysis and meta-regression of individual patient data from five randomised trials. Br J Psychiatry 2009;194:4–9.

[88] Crespel A, Genton P, Berramdane M, et al. Lamotrigine associated with exacerbation or de novo myoclonus in idiopathic generalized epilepsies. Neurology 2005;65:762–764.

[89] Berkovic SF, Knowlton RC, Leroy RF, Schiemann J, Falter U. Levetiracetam N01057 Study Group. Placebo-controlled study of levetiracetam in idiopathic generalized epilepsy. Neurology 2007;69:1751–60

[90] Noachtar S, Andermann E, Meyvisch P, et al. Levetiracetam for the treatment of idiopathic generalized epilepsy with myoclonic seizures. Neurology 2008;70:607–616.

[91] Rosenfeld WE, Benbadis S, Edrich P, Tassinari CA, Hirsch E. Levetiracetam as add-on therapy for idiopathic generalized epilepsy syndromes with onset during adolescence: analysis of two randomized, double-blind, placebo-controlled studies.Epilepsy Res 2009;85:72–80.

[92] Mantoan L, Walker M. Treatment options in juvenile myoclonic epilepsy. Curr Ttreat Options Neurol 2011;13(4):355–370.

[93] Babtain FA. Levetiracetam may worsen myoclonus in patients with juvenile myoclonic epilepsy: case reports. Clin Neuropharmacol. 2012;35(4):201-202.

[94] Kellett MW, Smith DF, Stockton PA, Chadwick DW. Topiramate in clinical practice:first year's post-licensing experience in a specialist epilepsy clinic. J Neurol Neurosurg Psychiatry 1999;66:759–763.

[95] Prasad A, Kuzniecky RI, Knowlton RC, et al. Evolving antiepileptic drug treatment in juvenile myoclonic epilepsy. Arch Neurol 2003;60:1100–1105.

[96] Biton V, Bourgeois BFD. Topiramate in patients with juvenile myoclonic epilepsy.Arch Neurol 2006;62:1705–1708.

[97] Levisohn PM, Holland KD. Topiramate or valproate in patients with juvenile myoclonic epilepsy: a randomized open-label comparison. Epilepsy Behav 2007;10:547–552.

[98] Sousa Pda S, Araújo Filho GM, Garzon E, Sakamoto AC, Yacubian EM. Topiramate for the treatment of juvenile myoclonic epilepsy. Arq Neuropsiquiatr 2005;63:733–737.

[99] Kothare SV, Valencia I, Khurana DS, Hardison H, Melvin JJ, Legido A. Efficacy and tolerability of zonisamide in juvenile myoclonic epilepsy. Epileptic Disord 2004;6:267–270.

[100] O'Rourke D, Flynn C, White M, Doherty C, Delanty N. Potential efficacy of zonisamide in refractory juvenile myoclonic epilepsy: retrospective evidence from an Irish compassionate-use case series. Ir Med J 2007;100:431–433.

[101] Obeid T, Panayiotopoulos CP. Clonazepam in juvenile myoclonic epilepsy..Epilepsia. 1989;30(5):603-606.

[102] Panayiotopoulos CP. Treatment of typical absence seizures and related epileptic syndromes.Paediatr Drugs. 2001;3(5):379-403.

[103] Thomas P, Valton L, Genton P. Absence and myoclonic status epilepticus precipitated by antiepileptic drugs in idiopathic generalized epilepsy. Brain. 2006;129(Pt 5): 1281-1292.

[104] Osorio I, Reed RC, Peltzer JN. Refractory idiopathic absence status epilepticus: a probable paradoxical effect of phenytoin and carbamazepine. Epilepsia, 2000; 7:887–894

[105] Larch J, Unterberger I, Bauer G, Reichsoellner J, Kuchukhidze G, Trinka E. Myoclonic status epilepticus in juvenile myoclonic epilepsy. Epileptic Disord. 2009;11(4): 309-314.

[106] Afra P, Adamolekun B. Lacosamide treatment of juvenile myoclonic epilepsy. Seizure. 2012;21(3):202-204.

[107] Geithner J, Schneider F, Wang Z, Berneiser J, Herzer R, Kessler C, Runge U. Predictors for long-term seizure outcome in juvenile myoclonic epilepsy: 25-63 years of follow-up. Epilepsia. 2012;53(8):1379-1386.

[108] Kossoff EH, Henry BJ, Cervenka MC. Efficacy of dietary therapy for juvenile myoclonic epilepsy. Epilepsy Behav. 2013;26(2):162-164.

[109] Kostov H, Larsson PG, Røste GK. Is vagus nerve stimulation a treatment option for patients with drug-resistant idiopathic generalized epilepsy? Acta Neurol Scand Suppl. 2007;187:55-58.

[110] de Araujo Filho GM, Yacubian EM. Juvenile myoclonic epilepsy: psychiatric comorbidity and impact on outcome. Epilepsy Behav. 2013;28 Suppl 1:S74-80.

Chapter 5

# Juvenile Myoclonic Epilepsy — A Maturation Syndrome Coming of Age

Russell D. Pella, Lakshmi Mukundan and
C. Akos Szabo

Additional information is available at the end of the chapter

http://dx.doi.org/10.5772/58372

## 1. Introduction

Juvenile Myoclonic Epilepsy (JME) was first described by Herpin [1] in the middle of the $19^{th}$ century, reporting seizure symptoms he observed in his adolescent son. While Rabot also described myoclonia in 1899 [2], it was not until 1957 when Janz and Christian [3] provided a detailed explanation of the complete syndrome, which was subsequently called by Antonio Delgado-Escueta [4] as "JME of Janz". Since then, JME has become a well-defined epilepsy syndrome that is recognized as one of the most common forms of genetic generalized epilepsy (GGE). JME is primarily characterized by the hallmark manifestation of myoclonic seizures (mainly upon awakening), although patients also often present with a combination of absence and generalized tonic-clonic seizures (GTCS) as well. Other prevalent features of JME include various types of reflex seizures, particularly with photosensitivity. Electrophysiologically, there are prominent generalized ictal and interictal discharges on scalp electroencephalography (EEG). In contrast to other GGE syndromes, such as Childhood Absence Epilepsy (CAE), and contrary to earlier assumptions, JME has been shown to be associated with cognitive and behavioral problems, a lifetime risk of continued seizures, and medication resistance.

In this chapter, we will review the electroclinical definition of JME along with treatments for its associated seizure types, cognitive and behavioral complications, and underlying pathophysiology. In contrast to long-standing assumptions that JME is due to frontal lobe hyperexcitability that primarily involves corticothalamic pathways, recent literature suggests that JME likely reflects an underlying developmental disorder affecting multiple brain regions.

## 2. Clinical definition

In an historical context, the syndrome has been previously referred to as "impulsive petit-mal" epilepsy or Janz syndrome [5]. The International League Against Epilepsy (ILAE) later designated the term "Juvenile Myoclonic Epilepsy" in 1975. Under the revised Classification of Epilepsies and Epileptic Syndromes [6], the ILAE defined JME as follows:

Impulsive petit mal appears around puberty and is characterized by seizures with bilateral, single or repetitive, arrhythmic, irregular myoclonic jerks, predominantly in the arms. Jerks may cause some patients to fall suddenly. No disturbance of consciousness is noticeable. Often, there are GTCS and, less often infrequent absences. The seizures usually occur shortly after awakening and are often precipitated by sleep deprivation.

While the duration of myoclonic seizures are routinely short to the effect that it is often not possible to determine if individuals lose awareness during episodes, repetitive or clustered myoclonic seizures may be associated with altered levels of consciousness. Although GTCS and/or absence seizures occur less frequently in JME, repetitive and clustered seizures can lead to secondarily generalized tonic-clonic episodes. Whereas it is nearly pathognomonic for seizures to occur upon awakening and are precipitated by sleep deprivation, an additional diurnal pattern may also be apparent as myoclonic seizures can occur late in the afternoon or in the evening as well. Electroclinically, interictal EEG patterns consist of generalized 4-6 Hz spike-or polyspike-and-wave discharges, and a relatively high prevalence of photosensitivity.

## 3. Epidemiology

JME accounts for approximately 4-11% of all epilepsies [7-12] with an incidence of 0.1 to 0.2 occurrences in 100,000 per year. In terms of gender effects, early studies reported that the incidence of JME was greater in male patients than females [13]; however, gender differences have not been consistent across studies, as some studies [9] have shown an equal distribution, and other researchers [10, 14-16] have reported a higher proportion (60%) of females. It is also a typically occurring GGE [12], with prevalence in GGE groups ranging from 5.5% [17] to representing the majority (45.5%) of patients [18]. Prevalence of JME as a proportion of GGE may also differ across the lifespan in some individuals as approximately 15% patients with childhood absence and juvenile absence syndromes develop JME as they age, especially those patients with photosensitive spike-and-wave findings on EEG [4, 19]. As there is a strong genetic component to JME, its prevalence is high among family members of JME patients as well, which has been shown to be the case for certain ethnic groups. Studies from Saudi Arabia, Turkey, and Iran have shown JME patients have a family association with rates that vary from 42.3 to 48.9% [18, 20, 21].

Not only are prevalence rates sensitive to genetic influences, the proportion of patients in particular clinic settings may vary as well. For example, the prevalence in primary epilepsy clinics is high, while it tends to be much lower in tertiary referral centers, due in large part

http://dx.doi.org/10.5772/58372

because primary centers achieve successful treatment via antiepileptic medications. Moreover, JME is often under-diagnosed and misdiagnosed [22, 23], which can present a challenge to estimating its prevalence. As an illustration, a French study [24] showed that the prevalence of JME in one geographic area was 0% between 1986 and 1994, which rose to nearly 50% between 1996 and 2000, likely due to increased recognition of the syndrome. Moreover, variations in clinical onset can present a challenge to estimating the prevalence of JME. Although JME the peak age of onset is in adolescence (12-18 years), similar to most GGE syndromes [10, 25], the age of onset can vary from 8 to 36 years. There are also *de novo* diagnoses in early adulthood, and JME can, though rarely, begin or be reactivated in advanced ages. For instance, a case report presented two patients from Turkey in whom JME began after the age of 70 [26].

## 4. Clinical diagnosis

**Myoclonic Seizures**: As noted above, myoclonic movements are one of the main symptoms of JME and consist of generalized seizures which are brief, irregular jerks of the head, trunk, and limbs that can be either symmetric or asymmetric, and may involve isolated regions of the body or the whole body. They usually predominate in the upper limbs (mostly distal muscles), although they occasionally involve muscles in the abdomen, paraspinal distributions, and lower extremity [27]. It is not uncommon for myoclonic seizures to be so subtle or brief that they are perceived as benign "inner shocks". As subtle as the may be, patients tend to more readily notice asymmetric jerks involving the dominant upper extremity because such movements noticeably impair daily functioning. The jerks, if violent, may cause the patient to drop or throw objects, or fall to the floor-which may be mistaken for nonpathological clumsiness. Myoclonic seizures can manifest as discrete, single events or they may occur in clusters. They are usually not associated with loss of consciousness but it is not uncommon for patients to lose awareness during the jerks. Moreover, clusters of myoclonic events can evolve into a GTCS and cause post-ictal confusion. They can occur during transition to sleep or during awakening from sleep, usually in the early morning hours [28]. This early morning pattern of seizures is associated with an increase in cortical excitability during that time of day, which has been noted in other patients with GGE, but to a greater extent in JME patients [29]. Developmentally, these jerks often subside in the fourth decade of life, although GTCS or absence seizures tend to persist [30].

**Generalized Tonic-Clonic Seizures**: About 80% of the patients with JME have GTCS. These seizures are also more likely to occur if precipitated by sleep deprivation or alcohol intake. Usually, these events are immediately preceded by a series of myoclonic jerks and associated tongue biting prior to generalization. Because, as noted above, myoclonic events are often insidious, patients first seek treatment following an initial GTCS and are then subsequently diagnosed with JME. Typically, upon further evaluation, such patients acknowledge also that they have experienced myoclonic jerks, a description which adds additional clinical support for a JME diagnosis. Due to this association, some authors have proposed that clinicians should strongly consider a JME diagnosis, until proven otherwise, when teens present with an initial unprovoked GTCS [31].

**Absence Seizures**: Absence seizures are characterized by a brief loss of awareness (i.e., a few seconds) without any motor manifestations. Such seizures are relatively uncommon in patients with JME. For example, Janz reported that 28% of his patients with JME had absence seizures [3, 13]. Commonly, children who initially experience absence seizures may develop myoclonus or GTCS within 1 to 9 years of their seizure onset, and then may subsequently be diagnosed with JME. The absence seizures of JME differ from those of other GGE, such as CAE or Juvenile Absence Epilepsy (JAE) as they are shorter in duration and associated with a lesser degree of altered consciousness [32].

**Myoclonic Status Epilepticus**: Myoclonic status epilepticus is rare in JME and can present in a variety of ways. Namely, patients can display prolonged myoclonic events, a combination of myoclonic seizures and GTCS, or GTCS that that can follow prolonged absence seizures. Risk factors for this clinical phenomenon include AED withdrawal, sleep deprivation, alcohol intake, and suboptimal therapy [33, 34].

**Precipitating Factors**: As sleep deprivation is the usual precipitant of seizures in JME [35], adolescents or a young adults often experience myoclonic or GTC seizures precipitated by late night studying or socialization. Because of its strong association with precipitating seizures in JME patients, sleep deprivation is typically employed as an activation procedure to provoke characteristic EEG changes that are diagnostically relevant (4-6Hz generalized polyspike-and-wave discharges). Also associated with sleep patterns, sudden and provoked awakenings, pose additional increased risk for seizures in JME. Two-thirds of patients with JME have at least one provoking factor [28].

## 5. Reflex seizures associated with JME

Reflex seizures are temporally preceded by some type of external stimuli and may occur exclusively, or in conjunction with, spontaneous seizures. Common external triggers include alcohol use, flashing lights, heat, bathing, and eating. They can also be less frequently elicited by internal stimuli such as stress, fever, hyperventilation, thinking, fatigue, menstrual cycle, and sleep. The proclivity of certain stimuli for seizure provocation is often age-dependent. For instance, fever is a more common provoking stimulus in children than in adults.

Regardless of the suspected triggers, it is import to obtain a detailed history from patients and family members to determine if there is a reflex component to seizures. This should include querying about specific triggers, seizure semiology (partial or generalized), family history of reflex seizures, and whether unprovoked seizures occur as well. As patients rarely lose awareness with myoclonic seizures, they may have adequate awareness of their subjective triggers. Knowing the patterns of responses in patients along with prevalence of triggers is, therefore, key to initially investigating a reflex component prior to enlisting formal testing. Moreover, identification of triggering factors leads to finding ways to avoid precipitants and

http://dx.doi.org/10.5772/58372

develop nonpharmaceutical therapeutic interventions, which is important for the treatment of reflex seizures, in addition to an AED regimen [36].

Using a questionnaire, da Silva Sousa [35] found 23% of JME patients surveyed reported having reflex seizures with thinking and concentration, 20% with praxis, 11% with speaking in public, 15% with visual stimuli, 7% with reading, 5% with calculating and writing, 5% with music, and 3% with drawing. In a subsequent study by the same group, patients were continuously monitored for 4 to 6 hours by video-EEG while neuropsychological and physiological triggers were presented [37]. These triggers had a provocative effect in 38%, with praxis being most common trigger. There were also inhibitory effects of tasks in over 90%. 40% of the patients had no effects on ictal or interictal epileptic discharges. A more recent multicenter, video-EEG study, controlling for spontaneous fluctuation of ictal and interictal epileptic discharges, found the provocative effect of neuropsychological and physiological triggers was decreased from 22 to 18%, while the rate of inhibition was decreased from over 90% to 29% [38]. The inhibition was thought to be a non-specific effect of arousal and mental activation, while the provocative triggers were task-specific.

In clinical practice, if reflex epilepsy is suspected from convergence of other clinical data as noted above, clinical procedures can often helpful in identifying suspected triggers. Detailed neuropsychological testing can be conducted with patients who have seizures induced by thinking and completing complex mental activities [39]. Also, functional MRI can be used to study signal changes in the networks involved in generating reflex seizures during tasks.

Because the prevalence of photosensitivity in JME ranges from 25 to 40% [40], clinicians often opt to elicit reflex responses via intermittent photic stimulation (ILS) with concomitant routine EEG with video. Photic stimulation typically consists of flashing light at various frequencies in front of the patient, with seizures being most likely to be elicited by frequencies of 12-18Hz in individuals who are photosensitive. Various visual patterns may be used to provoke seizures as well, and include alternating, oscillating, black or white, or linear responses. Beyond formal testing procedures via ILS, photoparoxysmal and convulsive responses along with ictal and interictal epileptic discharges can be triggered by reading and praxis in patients with JME [13, 40, 41]. Patients who are photosensitive may have the following responses: photic driving, non-convulsive photoparoxysmal episodes, or photoconvulsive responses. Notably, photosensitivity in JME patients is increased in the early morning hours soon after awakening, which is consistent with the diurnal pattern of seizures in these patients [42]. There are no differences in age or gender of patients with reflex seizures in general; however, in patients with photosensitivity there is a clear predominance in adolescence and in the female sex [42].

In terms of the biological mechanisms of reflex seizures in JME patients, particular cerebral regions have been implicated. Myoclonic seizures in JME are expressed in the primary motor and supplementary motor cortices [43], which are extensively connected to the primary sensory and association cortices. It is not clear whether visual stimuli generates a response from the sensory cortices that propagates to functionally connected cortical and subcortical structures, or if a there is parallel synchronization of larger networks induced by the stimuli [44, 45]. In the case of photosensitivity, visual stimulation likely synchronizes frontoparietal cortices prior to the onset of the epileptiform discharge [44-46], a phenomenon suspected on the basis of large cortical blood flow increases even before the appearance of hyperventilation-induced generalized spike-and-wave discharges in absence epilepsy [47, 48]. Given the variety of aforementioned triggers, it is expected that reflex seizures are provoked by stimulation of different sensory cortices, such as the primary somatosensory cortex, primary auditory cortex, or related association areas. For instance, somatosensory or cognitively evoked seizures are evoked in physiologically activated cortical areas that overlap with hyperexcitable cortices giving rise to ictal or interictal epileptic discharges [49].

## 6. Electrographic findings

Given the rate of reflex seizures in JME, it is clinically indicated to conduct a sleep-deprived EEG with activation procedures, such as photic stimulation and hyperventilation. Although a positive EEG highly supports a clinical diagnosis of JME, a negative EEG does not definitively rule out the diagnosis. When examining EEG results, there are typical EEG patterns that occur in JME that should be considered.

**Interictal Pattern**: The interictal background activity is usually considered normal in patients with JME [4, 7, 50, 51]. However, Genton et al. [52] found that routine EEGs were normal in 27% of cases and misleading or nonspecific in 20% of cases; although, 54% showed generalized interictal epileptic discharges (IEDs). Other studies have reported percentages of abnormalities, consisting of generalized polyspike-and-wave or spike-and-wave complexes in 75 to 85% of probands [9, 53, 54]. The typical interictal EEG finding in JME is a 3.5-6 Hz spike-and polyspike-and-wave pattern with a frontocentral predominance that lasts up to 15 to 20 seconds [50, 55]. While a pattern of more prolonged IEDs with a 2-3 Hz frequency is not specific to patients with absence seizures, there may be an increased risk of clinical absence seizures if a 3Hz spike-and-wave interictal pattern is noted. IEDs are primarily detected during drowsiness and or sleep onset, or upon awakening [56, 57] (see Figures 1-3), during ILS (photoparoxysmal response) (see Figure 4), and hyperventilation. They may be associated with myoclonic seizures or GTCS (photoconvulsive response). Family members of JME are often affected or carry the abnormal EEG traits, as 80% percent of symptomatic siblings and 6% of asymptomatic siblings have diffuse 4-6 Hz spike-and-wave complexes [58]. Photoparoxysmal EEG response is another heritable trait of JME, and can be found in 20 to 60% of near relatives of probands [59].

**Ictal Pattern**: The ictal EEG during myoclonic jerks (see Figure 5) typically reveals a sudden onset of brief (less than 0.5 sec) bursts of 10-16 Hz polyspike-and-wave discharges often followed by 1-3 Hz slow waves [60]. The number of spike-and-wave complexes ranges from 5 to 20 per discharge and correlates with the intensity, rather than the duration of each seizure [3]. Absence seizures in patients with JME are usually associated with 3 Hz spike-and-wave activity, which is sometimes preceded by 4-6 Hz polyspike-and-wave discharges that decrease in frequency to 3 Hz as the patient loses consciousness. These spike-and-wave discharges are usually shorter in duration than those observed in childhood and juvenile absence epilepsies [61].

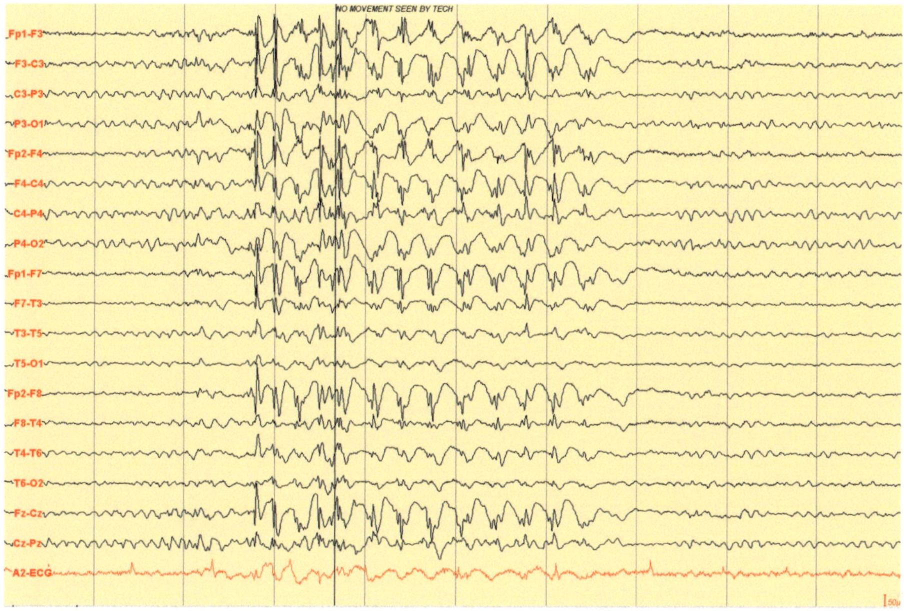


Legend: The discharge is originally 5 Hz, slowing down to 3 Hz. No clinical signs were noted.

**Figure 1.** Generalized Interictal Epileptic Discharge in Drowsiness

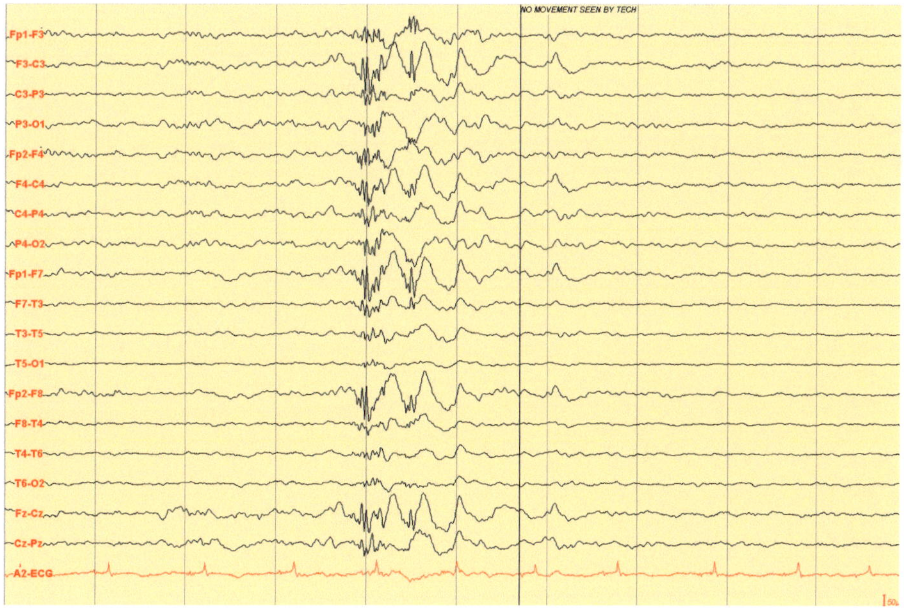


**Figure 2.** Generalized Polyspikes Activated in Stage II Sleep

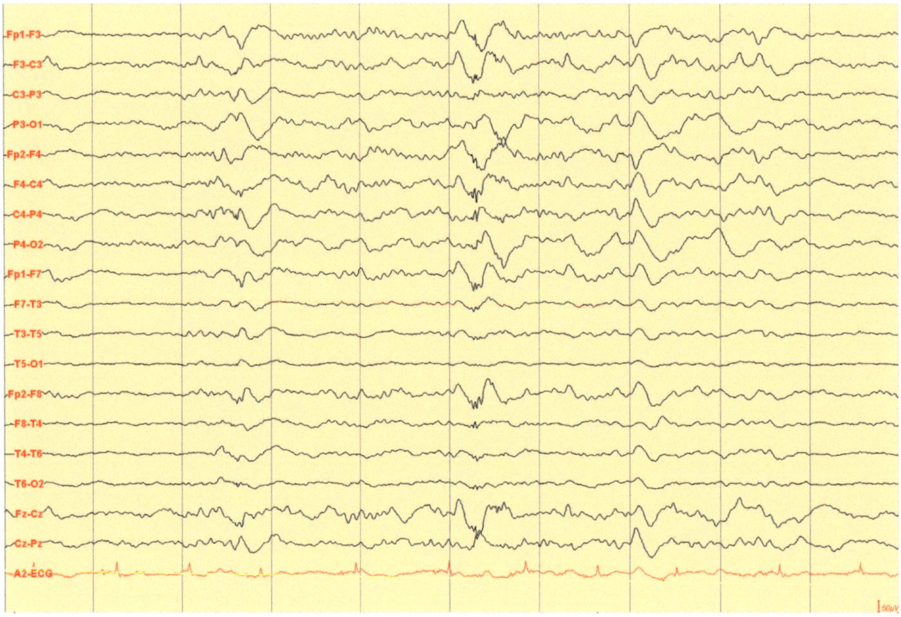


**Figure 3.** Fragmented Spikes in Slow-Wave Sleep

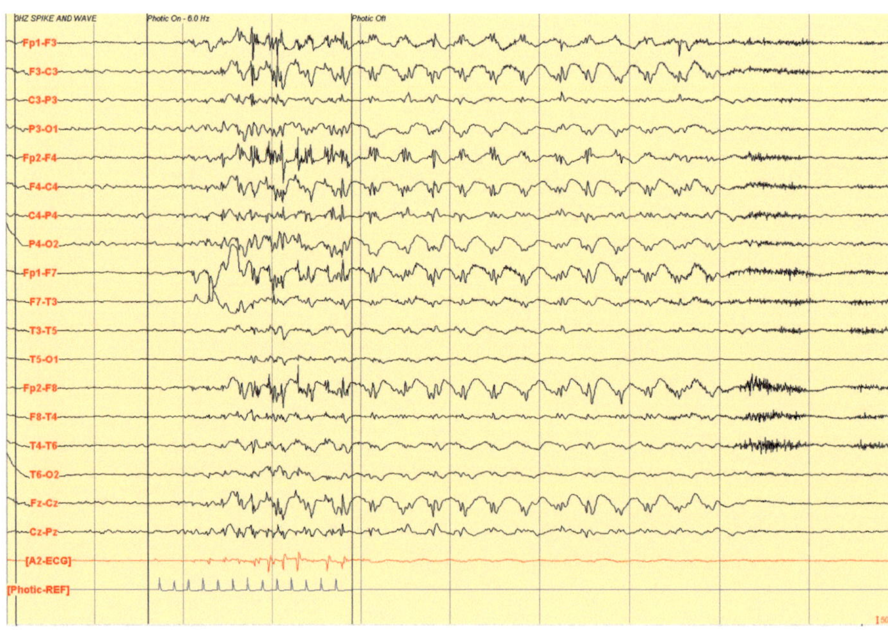


**Figure 4.** Photoparoxysmal Response

http://dx.doi.org/10.5772/58372

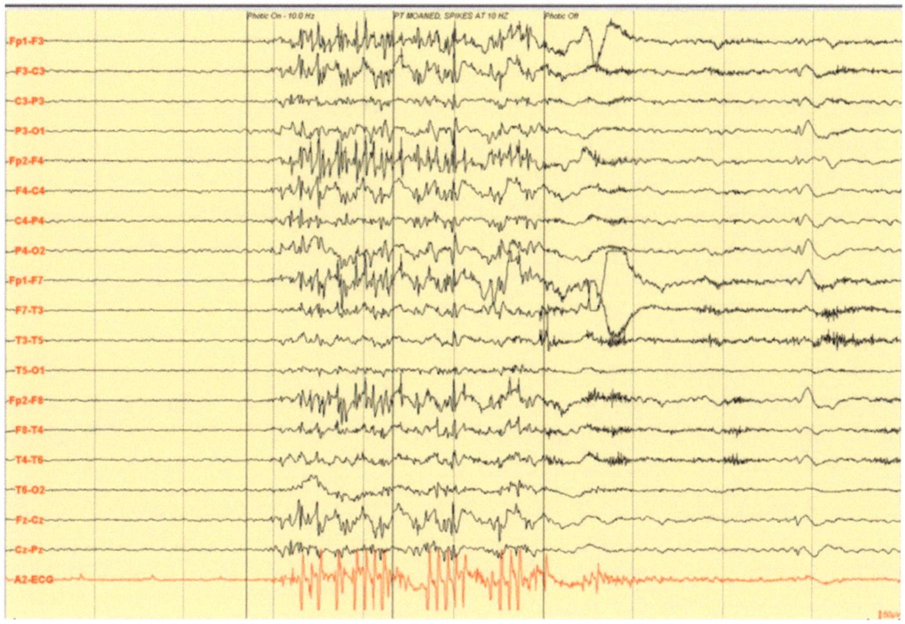


Legend: Note movement artifact on ECG channel.

**Figure 5.** Myoclonic Seizure

## 7. Focal features of EEG

Accurate identification of EEG patterns is crucial to avoid misdiagnosis of JME (e.g., misidentified as a partial seizure). As an illustration, some researchers have reported that clinical and EEG asymmetries can lead to a delay in the diagnosis of JME by an average of 2 years [62]. Thus, it is important to understand that although JME is classified as a GGE, myoclonic seizures and GTCS may be associated with a combination of focal clinical and/or EEG findings. However, focal abnormalities are rarely found on routine neurological exams or MRI studies. Further, the challenges of making a diagnoses by electroclinical data is best exemplified by evidence of versive seizures or circling seizures that are associated with generalized discharges on EEG [63, 64].

Data from JME patients in the UK found that 36.7% of EEGs had focal slow waves, spikes, and sharp waves as well as focal onset of a generalized discharge. In more than half of the JME patients, at least one EEG showed focal abnormalities [65]. 18 patients with JME were studied in Poland which demonstrated that 8 patients had focal abnormalities on EEG that previously led to misdiagnosis [66]. Further, a large study of JME patients in India [67] found asymmetrical clinical presentations in 17% of patients and focal EEG abnormalities were seen over 45% of patients. Of these patients, focal findings seen on EEG included amplitude asymmetry or a lateralized onset of generalized discharges in 45% and independent focal EEG abnormalities in 33% of patients. A video-EEG study from the Cleveland Clinic Foundation [68] reported that of the patients with only JME, 54% exhibited either focal semiologic or electroencephalographic features or a combination thereof. Focal myoclonic seizures (i.e., unilateral myoclonic jerks, version, and asymmetric limb posturing) were recorded in six patients whose ictal EEG showed a generalized seizure pattern. Two patients had lateralized upper extremity extensor posturing ("sign of four") with evolution of a GTCS. They also reported that one patient had primary GTCS presenting after successful resection of a parietal tumor, and two patients had temporal lobe epilepsy in addition to JME. Taken together, the findings of focal or multifocal clinical or EEG abnormalities suggest the possibility of a symptomatic etiology, and may reflect an underlying genetically mediated neurodevelopmental disorder.

It has also been suggested that asymmetric EEG findings in such patients increase the probability of decreased response to standard antiepileptic therapies. A group of Canadian researchers [69] presented data showing 90% of their JME patients had generalized epileptiform discharges on EEG and 57% of patients had asymmetric EEG abnormalities, including focal epileptic discharges or slowing. Of their patients, over 39% were medically refractory to at least one AED, and a poor treatment response was observed to a greater degree in patients with EEG asymmetries. Nevertheless, the clinical significance of focal clinical or EEG abnormalities has not received much attention in the literature.

## 8. Differential diagnosis

**Doose Syndrome**: Myoclonic-astatic epilepsy was first described by Dr. Hermann Doose in 1970 [70]. The ILAE classified it formally as symptomatic generalized epilepsy, and it was renamed as epilepsy with myoclonic-atonic seizures. It is much less prevalent than JME (1 to 2% of all epilepsies) and has a younger onset age (7 months to 6 years, peak of 2 to 4 years) with a greater prevalence among males. Unlike JME, there is an association with cognitive impairment, which is severe in some cases of Doose Syndrome. This syndrome is characterized by symmetric myoclonic jerks followed by sudden atonia. During a myoclonic atonic seizure, the EEG consists of irregular generalized 2-3Hz or more spike-and polyspike-and-wave discharges. The atonic seizures coincide with subsequent slow waves. The interictal EEG background similarly consists of frequent 2-3 Hz generalized spike-and polyspike-and-wave discharges, contrasted 4-6 Hz discharges in JME. Up to 50% of patients may achieve seizure freedom and continue to have normal development, but the remainder may have severe impairments in cognitive functioning, behavioral problems, and ataxia [71].

**Childhood Absence Epilepsy (CAE)**: CAE is a type of GGE that has an age of onset of 4 to 10 years with a peak between 5 to 7 years [72]. The characteristic seizure type is an absence seizure during which there is brief loss of consciousness for a few seconds with associated behavioral arrest, eyelid myoclonia, and rare hand and facial automatisms [73]. They may occur multiple times a day. While absence seizures may also occur in JME, their frequency and duration is far less than in CAE. The interictal EEG of CAE is characterized generalized spike-and polyspike-and-wave discharges of 3Hz frequency [6]. Prognosis for CAE is considered excellent, with possibility of remission upon discontinuation of antiepileptic medications. In patients with CAE, the development of GTCS or myoclonic seizures may indicate a poorer prognosis, and increases the probability that such patients will develop JME.

**Juvenile Absence Epilepsy (JAE)**: Similar to JME, JAE is a GGE that mainly presents during adolescence, usually at the age of 10 to 17 years. Although the central semiology is signified by absence seizures, there is often less impairment of consciousness than in CAE, and the seizures do not tend to occur in clusters. On the other hand, GTCS seizures are more common (reported in 80% of patients) with JAE, similar to JME. While some authors have contended that myoclonic jerks do not occur in JAE [32], the ILAE has included them in the definition of JAE [6]. As such, 15 to 25% of JAE patients have myoclonic jerks-they are infrequent, mild, and of random temporal distribution [74]. In comparison to JME, myoclonic movements usually occur in the afternoon rather than in the morning after awakening.

**Myoclonic Absence**: This rare epilepsy syndrome begins in preadolescence and is characterized by prolonged absence seizures lasting 10-60 seconds in duration with rhythmical bilateral myoclonias affecting the extremities, often synchronously with 3 Hz polyspike-and-wave discharges [75]. This syndrome is associated with cognitive dysfunction and is medically refractory in a majority of the patients.

**Generalized Tonic-Clonic Seizures Upon Awakening**: The onset of this epilepsy syndrome is also in adolescence [6]. As in JME, the GTCS tend to occur upon awakening. There is still debate on whether this syndrome should be classified with JME, due to electroclinical similarities and possible genetic associations [76]. However, patients with GTCS Upon Awakening do not exhibit myoclonic or absence seizures.

**Progressive Myoclonic Epilepsy (PME)**: Progressive myoclonic epilepsies are a group of genetic disorders which are characterized by myoclonus, seizures, and advancing cognitive and neurological decline [77]. Examples of PMEs include Lafora disease, neuronal ceroid lipofuscinosis, Unverricht Lundborg disease, myoclonic epilepsy with ragged red fibers (MERRF), Gaucher disease, and dentatorubral pallidoluysian atrophy. They can start in childhood, adolescence, or young adulthood and tend to be associated with multiple neurological and cognitive deficits. The myoclonus associated with these disorders can be induced by action or stimulation and tends to be multifocal. ILS tends to activate myoclonus at frequencies below 6 Hz. These diseases are associated with progressive neurological deterioration along with worsening seizure control. The diagnosis is more difficult to determine, primarily during the early stages of the disease when seizure severity is limited [78].

## 9. Treatment of JME

Response to treatment is drug-specific in the case of JME. Valproic acid (VPA) is considered the first line of therapy in patients with GGE, especially JME; however, it is not FDA approved for this condition [79, 80]. Rather than randomized control trials, evidence supporting the choice of VPA has primarily come from observational studies. While VPA is a broad spectrum antiepileptic medication, it is very effective in 80% of patients [81]. Nonetheless, even in seizure-free patients, discontinuation of VPA leads to a very high rate of relapses [9]. A chart review from Duke University [82] of 33 JME patients identified resistance to VPA in 30%. The VPA refractory patients had a higher prevalence of EEG asymmetries (40% vs. 10%), atypical seizure characteristics including auras and postictal confusion (30% vs. 4%), and intellectual deficiency (20% vs. 0%). Clinical characteristics combined with EEG data may help in predicting which JME patients will respond favorably to VPA.

More recent support of VPA's relative efficacy over other antiepileptic medications was demonstrated The Standard and New Anti-Epileptic Drugs trials (SANAD) [83]. SANAD was a concurrent pragmatic parallel group, open-label, randomized trial which examined seizure control, tolerability, quality of life, and economic outcomes of standard antiepileptic medications used for GGE. VPA was compared with lamotrigine (LTG) and topiramate (TPM), and was significantly better for time to treatment failure (i.e., seizure control, side effects, addition of another AED). For time to 12-month remission, VPA was significantly better than LTG for GGE. Overall, VPA was better tolerated than TPM and more efficacious than LTG. In children with GGE, a retrospective observational study [84] examined the long-term effectiveness of LTG compared to VPA monotherapy in newly diagnosed patients. After 12 months of treatment, 69% patients continued LTG compared to 89% that adhered to their VPA regimen; after 24 months rates were 57% and 83%, respectively. Valproate showed equal efficacy in all GGE syndromes, whereas LTG showed better efficacy in CAE and JAE syndromes, than in JME.

Although VPA can be used safely by a number of patients, including those with comorbid psychiatric disease or underlying psychiatric vulnerability [85], there are some limitations. In particular, there is a high risk of teratogenicity as well as delayed cognitive development in children (pregnancy class D) [86, 87]. Exposure to VPA in utero has been associated with maladaptive behavior and decreased socialization skills in childhood [88, 89] as well. In cases where VPA is used during pregnancy, either because of unplanned pregnancy or because alternative treatment options of equivalent efficacy are unavailable, appropriate counseling, precautionary measures, and monitoring should be provided. Although LTG may have a lower efficacy, and has even been reported to increase myoclonic seizures in some patients, it is a popular alternative to VPA in women in childbearing age for the treatment of JME.

LTG and TPM have FDA approval for primary GTCS in the United States. Hence, more studies, albeit with small numbers of participants, have been conducted with those agents. An open-label study [90], designed to evaluate LTG monotherapy as a possible alternative in patients with JME who previously failed VPA, found LTG to be as effective and better tolerated compared with VPA. A small (*N*=28) randomized, controlled trial [91] compared TPM and VPA in adolescents and adults during a 12-week maintenance period, which resulted in 67% seizure freedom in the TPM group and 57% in the VPA group. The researchers concluded that TPM may be an effective, well-tolerated alternative to VPA as the groups had similar rates of adverse effects. Similarly, a prospective, open-label, randomized observational study in Korea [92] compared the efficacy and tolerability between TPM and VPA in patients with JME and did not find differences in efficacy, although the side effect profile of TPM was more favorable.

Clonazepam (CLN) was one of the first medications to be approved for myoclonic seizures and GTCS. While VPA is as an effective treatment for most patients with JME, one study showed [9] additional efficacy with CLN adjunctive therapy to control myoclonic seizures. However, CLN monotherapy did not consistently prevent GTCS. The authors concluded that adding CLN may allow the dose of VPA to be reduced in patients demonstrating dose-dependent VPA side effects. The addition of CLN can also be utilized in combination with other antiepileptic medications as well.

Levetiracetam (LVT) was more recently approved for the treatment of myoclonic seizures associated with JME. Similar to CLN, it can reduce the myoclonic seizures and GTCS [93], possibly with even better long-term efficacy and fewer side effects. Recently, several studies found LVT to be effective in all types of seizures in JME as concluded in a randomized double-blinded placebo controlled trial [94, 95]. Hence, LVT is another worthy alternative to VPA, especially in women of childbearing age. Other medications that may be effective in JME include zonisamide (ZNS), felbamate (FBM), lacosamide (LCM), and clobazam (CLB); however, systematic trials of those medications are lacking. Nonetheless, a retrospective chart review demonstrated that ZNS monotherapy was effective and well-tolerated in JME patients [96].

In contrast to effective AED treatments, there are a number of antiepileptic medications that may aggravate certain seizure types associated with JME, particularly myoclonic and absence seizures. Among commonly prescribed anticonvulsants, carbamazepine appears to have the strongest aggravating potential for myoclonic seizures, whereas the aggravating effect of phenytoin (PHT) is less prominent [97]. A retrospective study of patients with GGE, showed that oxcarbazepine (OXC) also aggravated myoclonic and absence seizures, and increased interictal epileptic discharges on EEG [98]. A retrospective study of adult GGE patients who developed video-EEG documented status epilepticus were all found to be taking either CBZ, PHT, vigabatrin (VGB), or gabapentin (GBP). Potential precipitating factors included dose increase of CBZ or PHT, or the initiation of CBZ, VGB, or GBP, as well as a decrease of phenobarbital (PB) dosage. Withdrawal of the aggravating agents and adjustment to medication resulted in full seizure control in that study. Therefore, it is important to make the appropriate diagnosis and medication selection to avoid potentially worsening certain seizure types. It is also important to consider the age, gender, comorbid conditions, and drug interactions when initiating antiepileptic therapy in JME patients.

## 10. Lifestyle modifications

As with a number of other chronic medical conditions (i.e., diabetes, hypertension, obesity), lifestyle factors have an important role to play in managing epilepsy disease burden. To illustrate the problems of how psychosocial factors influence the treatment of JME, Baykan et al. [30] followed 48 patients with JME for a mean of 20 years and found that 16.7% of patients had "pseudo-resistance" due to medication non-adherence and lifestyle related issues. Sharpe and Buchanan [99] studied 36 patients with JME at a tertiary referral hospital and found that most patients with poor seizure control had provoked seizures only, which prompted the authors to emphasize the importance of lifestyle changes in managing seizures in JME.

Thus, perhaps the most obvious lifestyle factors worth noting in terms of JME pertain to patients' exposure to stimuli s associated with increased seizure frequency (i.e., fatigue, stress, sleep deprivation, alcohol intake, drug use, photic stimulation, stress, AED withdrawal/non-compliance) [9, 100-103]. From a disease management perspective, behavioral medicine strategies that are promoted by integrated, multidisciplinary teams may prove useful [104]. Elements of such programs often include steps to improve medication adherence, treat psychological problems and manage stress, implement sleep hygiene interventions, and develop methods to cope with exposure to possible seizure triggers [104-106]. Moreover, administration of psychoeducation should be considered a minimum standard of care by providing patients with information regarding identifiable triggers, risks of sleep deprivation, possible problems with exposure to stroboscopic flashes, relationship of alcohol intake and seizure control, as well as ways to increase AED adherence to prevent further seizures [107]. Although, some authors [103] have suggested that it is more efficient to use higher doses of AEDs in the initial treatment of JME, rather than implement behavior change strategies, systematic study of the effects of multidisciplinary treatment on lifestyle factors in epilepsy has remained an understudied area. However, initial studies in JME have suggested that there may be a desirable effect on seizure control, and likely improvements in quality of life. If some of these measures are put into place, it is likely that the number of patients with refractory seizures will be reduced [108, 109].

## 11. Cognitive and behavioral tasks

There have been studies suggesting that cognitive tasks can provoke or inhibit of epileptiform discharges in patients with JME [110]. These uncommon precipitating factors, such as mental and motor hand tasks, may be under-recognized in JME [35]. A Japanese study involving 480 patients with epilepsy tested the effect of cognitive tasks on EEG (i.e., neuropsychological EEG activation) consisting of reading, speaking, writing, written calculation, mental calculation, and spatial construction. They found that these cognitive tasks have an inhibitory effect on EEG discharges in the majority of epilepsy patients (64%), although they have a provocative effect in other patients (8%). The seizures caused by the provocative effect of these tasks were found to be precipitated by action programming or thinking activity (linguistic and praxic) [111]. Following this, Beniczky et al. [38] studied 60 patients with JME and found that the provocative effect of the cognitive tasks is task-specific, whereas the inhibitory effect seems to be related to cognitive activation in general. Another study involving 76 JME patients had similar results, suggesting that the inhibitory effect of these tasks support non-pharmacologic therapeutic interventions in JME [37].

## 12. Non-medical treatments

Alternative surgical therapies do not represent standard of care and are not approved for JME. However, in medically-refractory patients, clinicians often employ additional treatment options.

**Vagal Nerve Stimulator**: While vagal nerve stimulators (VNS) are approved for the treatment of medically refractory partial seizures, its use in JME has not been well documented. Nonetheless, the most recent AAN guidelines on VNS for epilepsy [112] asserted that VNS was effective in children with medically-refractory GGE. A greater than 50% reduction in seizure frequency was reported in 55% of the 470 children with partial or generalized epilepsy in the reviewed literature (13 class III studies). Based on those findings, the authors concluded that VNS could be considered adjunctive treatment. No such recommendations have been made regarding adult treatment as more trials are needed in that population.

One of the few JME specific studies of VNS [113] indicated that adjunctive therapy with VNS might be considered a favorable option for treatment of refractory cases of GGE, and may be the only alternative to refractory JME. A total of 12 cases with GGE and VNS implantation were followed for a mean of 23 months. It was found that the rate of seizure total reduction was 61%: a 62% reduction of GTCS, 58% reduction in absences, and 40% reduction of myoclonic seizures. Five of the seven patients in that study who had a JME diagnoses were responders, two of which became seizure free. One patient in that study was diagnosed with JME and failed conventional treatment that included 8 AEDs and the ketogenic diet. Subsequent to VNS implantation, she had >75% reduction in GTCS along with >50% reduction in absence and myoclonic seizures.

**Corpus Callosotomy**: There are even fewer studies that have examined the effectiveness of corpus callosotomy for medically refractory cases of JME. A case series (*N*=11) from Brazil [114] included GGE patients who underwent extensive one-stage callosal transections. At least 75% reduction in frequency of GTCS was noted in all patients; three patients had complete remission of absences and the other patients had a >90% reduction in the frequency of absence seizures. Only one patient who had myoclonic seizures prior to surgery remitted completely after surgery. Postoperative EEG recordings showed disruption of bilateral synchrony of interictal epileptic discharges in all patients, with minimal neurological deficits. While the authors reported that the patients only showed a decline of 4 points in mean FSIQ standard score, a more detailed examination of the patients' psychosocial status and cognitive functioning was not reported. A Japanese case report [115] of medically refractory JME showed that an anterior corpus callosotomy resulted in a desynchronization of generalized spike-and-wave discharges and a complete resolution of myoclonic seizures. An anterior corpus callosotomy may be effective for seizure reduction in some cases of refractory GGE due to the disruption of interhemispheric synchronization [116].

## 13. Prognosis

As with most GGE syndromes, JME responds well to an appropriate AED regimen, demonstrating a 70 to 80% response rate [117]. However, the long-term prognosis remains AED-dependent as there is an 80% recurrence risk after AED withdrawal. It is thought that CAE typically has a better prognosis and spontaneous remission that JME even when untreated, although this can be complicated by the development of myoclonus. In fact, a Canadian study [19] noted that 65% of children with CAE experienced remission after 10 to 18 years after their diagnosis, although 44% of patients who did not experience remission developed JME.

A long-term population-based study [14] that included 24 patients (majority women) who developed JME by 16 years of age were followed for up to 25 years after seizure onset. Eight patients (36%) developed convulsive status epilepticus and three patients (12.5%) had intractable seizures. In this cohort, 17% of patients had full remission of all seizure types and only myoclonic seizures persisted in 13%. Thus, about one-third of patients had remission of disabling seizures without the need for continuation of AEDs. Moreover, although JME rarely undergoes spontaneous remission, there may be an age effect that increases the probability of spontaneous remission as individuals reach the fourth decade of life. In another long-term study [30, 118] that followed 48 patients with JME (mean age of 40 years) for approximately 20 years, the authors indicated that seizure severity and myoclonic frequency seemed to change across the lifespan. For instance, myoclonia went into remission in the fifth decade of life in 54% of patients. Five patients discontinued AED treatment and six patients had lower AED dosage; 10 out of 11 of these patients did not relapse during the mean follow up of 8 years.

One of the longest outcome studies of JME [119] followed 31 patients for an average of 39 years. Over two-thirds became seizure free, a third of whom could be taken off AEDs altogether. Predictors of poor outcome were identified as presence of GTCS preceded by myoclonic seizures, longer duration of epilepsy, and the need for AED polytherapy. Predictors of seizure freedom included complete control of GTCS. Other factors may also influence remission as well. In a study of 32 JME patients in Japan, those with focal interictal epileptic discharges on EEG and or discharges activated by cognitive stimuli, had the least favorable outcome [120]. However, value of EEG in predicting outcome of JME remains controversial [121]. A study from Brazil [122] showed that patients with persistent seizures had an earlier age of onset, higher prevalence of personality disorders, and higher incidence of sensitivity to praxis-and verbally-induced ictal or interictal epileptic discharges.

Hence, it appears that a subgroup of patients can be identified as being at a higher risk of refractory seizures. JME patients with an earlier onset of epilepsy, GTCS evolving after a build-up of myoclonic seizures, focal EEG abnormalities or sensitivity to cognitive activation, and cognitive or behavioral problems may be more likely to remain refractory to antiepileptic medications. Ideally, the significance of these factors needs to be validated prospectively by a multicenter, multinational study.

http://dx.doi.org/10.5772/58372

## 14. Cognitive dysfunction

Consideration of neuropsychological findings in JME has a brief history beginning with largely anecdotal reports suggesting patients displayed a combination of behaviors consistent with a "frontal syndrome" [3] in the context of normal intellectual functioning. As the following literature illustrates, the initial assertions that JME patients display a unitary "frontal syndrome" are challenged by data from a number of novel empirical designs. For instance, consideration of non-frontal factors is considered germane in light of the involvement of multiple brain regions that extend beyond structural frontal anatomy in JME [123]. Several themes of investigation have surfaced regarding cognition including delineation of neuropsychological profiles in JME, group performance differences, disease specific and non-disease factors that may affect cognitive functioning, along with neuroanatomical and functional correlates of task performance. Because of the idiopathic nature of JME and its strong genetic linkage, family studies of cognition have surfaced as well.

Given that JME has been conceptualized as a disorder affecting frontal brain regions, some of the first empirically-driven studies focused on the relationship of JME and performance on tasks that theoretically require significant frontal lobe recruitment. In this vein, JME groups have been assessed by a number of clinical and basic research cognitive measures that task frontal systems. For instance, a small study showed that a group of adult JME patients ($n$=9) demonstrated an abnormal number of errors and omissions on a task that requires individuals to identify previously learned complex visual information [124]. Such match-to-sample measures often have immediate and delayed portions that task visual working memory systems, which an aspect of executive functioning. The magnitude of abnormality was generally not as large as performance by a group of patients with frontal lobe epilepsy (FLE; $n$=15), although it was typically lower than control participants' ($n$=14) proficiency. In a follow-up study with a similar task, researchers from that group [125] reported that adult JME patients ($n$=9) had fewer correct responses than controls ($n$=14) on the same delayed match to sample paradigm. The JME patients performed normally on the immediate match to sample tasks, which generally relates to attentional functioning, but slower psychomotor speed performance than controls. However, it is not clear if the same patient and control groups were used across both of the studies, which decreases the generalizability of the findings. Nevertheless, these two projects served as stringboards for subsequent applied and experimental designs.

While it is important that the latter studies showed commonalities between theoretically similar patient groups (i.e., groups with suspected frontal lobe abnormalities) in order to provide additional validity evidence for a "frontal" construct, it is also relevant to include data from groups that are theoretically divergent (i.e., controls, various non-frontal clinical samples). As such, neurocognitive performance of a mixed sample [126] of patients diagnosed with temporal lobe epilepsy (TLE; $n$=15) and JME ($n$=15) was contrasted on a number of tasks related to frontal lobe functions. JME patients had impairments on a wide range of cognitive measures, as defined as >1.5 standard deviations below the mean for the normative comparison group. Of the 15 JME patients used in their analyses, one patient showed no impairment on testing, while half of the remaining sample had impairments on <3 tests and the other half was impaired on >4 tests. Performance was most frequently impaired on a task that requires one to identify relationships between objects on a conceptual level, which was significantly discrepant from the TLE group. Additionally, patients were administered a task that requires individuals to quickly and sequentially alternate responses according to specific rules (i.e., Trailmaking Test Part B). On this task, the JME participants had lower scores than the TLE patients.

Additional studies have contrasted JME groups on test batteries that primarily include executive function tasks that have a relationship with frontal lobe involvement. Piazzini et al. [127] reported data from an adult sample of patients with JME ($n$=50), TLE ($n$=40), FLE ($n$=40), and controls ($n$=40). In that study, JME patients performed statistically significantly lower than TLE patients and controls, yet scored similarly to FLE on the Wisconsin Card Sorting Test and a semantic verbal fluency test. A group in Austria [128] published data from two studies [129, 130] that included a decision-making task and a number of neuropsychological variables in patients. Their results showed that JME patients performed similarly to mesial TLE patients on nearly all variables except they had lower semantic verbal fluency scores and slower psychomotor speed. The groups were otherwise similar on measures of verbal attention, verbal working memory, cognitive flexibility, planning, along with abstraction and categorization.

Other results [131] indicated that children with CAE ($n$=28), another GGE, largely performed worse than individuals with JME ($n$=11) on tasks of visual sustained attention, the Stroop Test, and Trailmaking Test. No differences were noted for verbal memory. Another group [132] sampled children and adolescents who had a similar level of normal intellectual functioning and were diagnosed with either recent-onset JME ($n$=20) or recent-onset benign childhood epilepsy with centrotemporal spikes (BCECTS; $n$=12). A sample of first-degree cousins ($n$=51) were also utilized as control participants. The researchers focused their cognitive assessment on objective measures of executive functioning that included subtests from the Delis-Kaplan Executive Function System (D-KEFS). They also examined behaviors that were subjectively rated by parents on the Behavior Rating Inventory of Executive Function (BRIEF). In their samples, the JME group performed significantly poorer than the control group on the D-KEFS Inhibition subtest. Parent report on the BRIEF indicated the JME group had more pathological ratings on the Behavioral Regulation and Metacognition scales than controls as well. However, there were no significant differences between the BCECTS and JME group on any measures. The latter study highlights that executive differences exist early in the disease course of JME, although the magnitude of executive dysfunction may not be larger than that in other genetic epilepsies.

There have also been reports [133] of cognitive functioning in mixed samples of children and adolescents with normal intellectual functioning diagnosed with GGE (i.e., JME & Absence) and genetic localization-related epilepsies (i.e., BECTS & non-BECTS). Although that study included 26 JME patients and a number of other patients with IPE, the JME group was compared only with the 72 healthy children, which indicated that the JME group performed below the mean of the control group in all cognitive domains (e.g., intelligence, academic, language, memory, executive functioning, fine motor dexterity and speed, cognitive processing speed). In particular, JME patients' lowest performance was in arithmetic, inhibition, concept formation, psychomotor speed, and fine motor dexterity and speed. While visual inspection of the data suggested that JME patients performed lower than the group with absence seizures on a measure of language fluency, the group with absence seizures appeared to perform lower in all other domains. However, statistical group comparisons were not conducted to differentiate any of the clinical groups from one another.

Just as there has been variability in finding group differences between patient samples, findings from studies contrasting JME patients from healthy controls have been mixed and contradictory. In addition to the few clinical comparison group designs noted above, a number of other studies have shown that JME patients have statistically significantly lower proficiency than controls across a host of measures including intelligence [134, 135], verbal IQ [135], working memory [135], digit span [136-139], sustained attention [134], processing speed [134, 135], Trailmaking Parts A and B [137-139], Trailmaking Part B [140], mental flexibility [134, 137, 141, 142], response inhibition [134, 143], Stroop Test [137, 139, 140], Stroop Interference [136], speeded color reading [143], verbal abstraction [141], concept formation [136, 137, 142], perseverations[140], clock and cube drawing [143], sematic fluency [136, 138, 140, 143-145], phonemic fluency [137, 138, 140, 141, 144, 145], naming [135, 141], verbal learning and memory [138, 140, 143], visual memory [140, 141], and prospective memory [136]. On the other hand, a number of studies have also failed to find group differences on tasks of IQ [138, 143, 144], verbal intelligence [141, 145], auditory working memory and attention span [141, 143-146], spatial span [144, 145], spatial working memory [147], psychomotor speed [146], motor speed [134], mental flexibility [146], Trailmaking Parts A and B [145], response inhibition[141, 144, 145], figural fluency [134, 145], semantic fluency [141, 146], phonemic fluency [146], reading [134], naming [143], language comprehension[144], line orientation [143], facial recognition [143], memory [134], verbal learning [141, 144, 145], visual memory [144, 145], and design learning [141]. Another more recent study [148] reported no difference between JME patients and controls on a range of neuropsychological measures. Beyond functioning on objective cognitive tasks, another avenue for research is to examine patients' perceptions of cognitive dysfunction. Such a design has shown the JME patients rate a higher level of self-reported executive dysfunction than controls [144].

Given that the variability in findings across studies likely relates to a number of primary and secondary factors, the influence of the effects of such factors is also likely varied. For instance, one small study [128] showed that it may be important to consider the influence of seizure frequency on JME patients' abilities. In general, the JME patients in their sample did not perform differently from healthy controls on tasks of verbal attention, verbal working memory, or phonemic verbal fluency. On the other hand, controls consistently performed higher than JME patients on tasks related to psychomotor speed, cognitive flexibility, categorical verbal fluency, planning, along with abstraction and categorization. The patients who had been seizure free ($n$=11) for a year were also compared with patients who continued to have seizures ($n$=11). The groups showed no differences on any of the cognitive testing, although JME patients scored significantly lower on three of four indices of a decision-making task than controls; seizure status in the JME group was related to performance. However, when compared with controls, the patients who were not seizure-free also showed additional significant differences on measures of cognitive flexibility, planning, and facets of a decision-making task than the seizure-free patients. Another study [146] examining performance on the same decision-making task reported no difference between controls and JME patients as a group, although fewer patients who were not seizure-free showed improvement and learning on the task over time. These findings suggest that seizure frequency may be an important modulator of cognitive abilities in JME patients.

Overall ability level, such as intelligence, has been shown to relate to performance across cognitive domains [149, 150]. This is likely to be the case with JME patients as well and may influence the significance level of research findings, particularly in small samples. As an example, heterogeneous cognitive performance in JME patients was noted in a published abstract [151], indicating that drug-resistant patients who performed in the impaired range on tasks of executive functioning had a high rate of impairments on tests in other domains. However, there were only limited published data available to review from that report. Other variables have been associated with cognitive performance in JME patients as well such as age of epilepsy onset [138], duration of epilepsy [135, 138], and educational level [135]. Additionally, JME patients on a regimen of multiple AEDs may perform worse than patients on monotherapy on tests of psychomotor speed, cognitive flexibility, and phonemic fluency [128]. On the other hand, some studies have shown no relationship with cognitive performance in JME groups related to duration of epilepsy [127], education level [138], the frequency of seizures [127, 138], treatment status [127], the type of seizures [127], age [138], sex [138], family history [138], or previous intake of an AED [138].

In addition to demography, there may be biological determinants of particular cognitive performance profiles in JME. In a magnetic resonance spectroscopy (MRS) study of the brain [152], researchers indicated JME patients with reduced frontal N-Acetylaspartic acid had lower mental flexibility compared with JME patients with normal levels. Other biological influences may relate to cognitive functioning as well. In JME patients, frontal and thalamic volumes have been associated with executive task performance [132]. Those researchers did not show similar anatomical relationships with cognitive performance in patients diagnosed with other epilepsies [132]. Similar work [141] has revealed an association of fractional anisotropy (FA), a measure of white matter integrity on diffusion tensor imaging (DTI), in anterior supplementary motor area (SMA) regions with scores on a picture naming task. FA values in the posterior cingulate region and corresponding gray matter volume (GMV) negatively predicted scores on the Trailmaking task. However, no other FA values were correlated with any other clinical variables or neuropsychological testing scores [141]. In terms of functional paradigms, fMRI patterns have differed for JME patients versus controls during tasks that require a high level of attention, concentration, and working memory, suggesting motor cortex involvement even though there were no group differences in the actual outcome of the task [147]. This suggests that there may be alterations in or pathological changes to cerebral regional recruitment during task performances as a function of disease state.

Regarding EEG studies, mixed findings have been demonstrated for the relationship of EEG patterns on cognitive performance. One group [143] examined the influence of paroxysmal EEG findings on task performance, but did not indicate a relationship. A non-significant relationship between cognition and abnormal EEG findings has also been demonstrated elsewhere [137, 138]. In contrast, JME patients with at-rest epileptiform discharges ($n$=11) [142] have been shown to demonstrate worse abstract reasoning concept formation, and mental flexibility.

In addition to demonstrating divergence and convergence amongst clinic groups and anatomy considerations, the genetic involvement of JME may also elucidate predictable patterns of cognitive performance. In a unique approach to demonstrating potential genetic vulnerabilities that underlie cognitive performance, there is evidence [131] that first degree relatives of JME patients have lower performance on tasks of sustained attention than relatives of individuals diagnosed with CAE or TLE. Another small sample of young adults with JME performed similarly to their siblings on a number of tasks related to executive, language, verbal memory, phonemic and semantic fluency, and general intellectual functioning [144]. However, the JME group scored in a range suggesting more self-reported symptoms related to behavioral, motivational, cognitive, and emotional factors than their sibling [144]. Somnez et al. [143] also reported a number of similarities in cognitive performance for patients with JME who had a relative with epilepsy and those who did not have a relative with epilepsy. Nevertheless, the authors indicated that patients with a family history of seizures were found to be "less successful in general cognitive evaluation," in particular on a spatial perceptual task, forward auditory digit span repetition, and speeded reading measure. A more recent sibling study of cognitive differences in JME patients [136] included data indicating that a group of siblings showed no significant differences on any cognitive measure when compared with the JME group. Moreover, the siblings' performance was not discrepant from a group of healthy controls except for an aspect that indicates JME participants generated more responses that were counter to the test rules on a measure of prospective memory (i.e., remembering to remember). Taken together, findings in these studies may reflect a familial genetic vulnerability for subtle cognitive abnormalities in JME family members who do not have a history of seizures.

In general, the aggregate of cognitive studies in JME patients has indicated that JME patients do not typically perform at the same level as individuals from control or normative groups. Moreover, review of the literature suggests that more studies show JME patients have poorer performance on frontally mediated tasks (i.e., processing speed, response inhibition, and verbal fluency) Than those that do not show such difference. So, too, virtually no study has shown that JME groups perform better than controls on any number of tasks. However, the totality of results has not been consistent and a number of studies have shown that patients with JME sometimes display cognitive abnormalities in other neuropsychological domains as well. Additionally, the literature is weighted toward including mainly tasks related to frontal lobe functioning at the expense of investigating other cognitive domains, which results in literature bias. There are also conflicting data across the studies comparing controls with JME patients to the degree that it is difficult to definitively conclude there is a specific JME cognitive endophenotype, such as a "frontal syndrome." It is likely that, based on the number of inconsistent results for both fontal and non-frontal task performances, any given JME patient may display a range of cognitive abnormalities-the expression of which is likely dependent upon various factors that have not been adequately described in the literature. As such, it will be important for researchers to continue to investigate etiological contributors to cognitive functioning in JME patients that account for the influence of psychosocial variables, neurobiological functions, and various other metrics of individual differences and disease characteristics.

## 15. Psychiatric complications

Psychiatric disorders are common in the general population [153] and the presence of a neurological disorder has been associated with increased prevalence [154]. Patients with epilepsy, as a group, have a high rate of psychopathology [155] and present with unique psychiatric problems that may complicate proper diagnosis and treatment. For instance, patients with epilepsy may experience psychiatric symptoms that are caused by, maintained, or exacerbated by discrete epileptiform activity. The most striking incidence of this relates to peri-ictal states that cause intense psychiatric reactions that include any number of symptoms consistent with anxiety, fear and panic, negative emotionality, and even psychotic phenomena [156]. Moreover, psychiatric factors in epilepsy have been related to challenges that stem from the functional impact of the disease that serve to restrict, modify, or impair individuals' functional status [157]. In that regard, limitations, such as restricted job duties and driving cessation, have been related to a higher incidence of mental health and psychological adjustment challenges along with lower rated quality of life [158].

While much is known about the psychosocial aspects of some common epilepsy syndromes (i.e., TLE), there has been less prominent study in individuals who have been diagnosed with other less frequently occurring forms of epilepsy. Within the JME literature, there are oft mentioned assertions that individuals with JME possess characteristics including irresponsibility, labile behavior, poor discipline, quixotic temperaments, emotional regulation difficulties, and egocentrism, although few empirical data have directly addressed these psychological traits in a systematic fashion [159]. Early retrospective study identified that there was a high portion (36.4%) of "character neurosis disorder" in JME patients [160]. Other indications [161] were that approximately 29% had some type of psychiatric disorder. Applying personality typologies that are specific to epilepsy patients has proven challenging [162], although research-driven examinations have begun to more clearly identify the types of interictal psychiatric complications that frequently occur in patients with JME. Within that empirical approach, researchers have reported that a combination of personality features, psychiatric symptoms, and contemporary psychiatric disorders occur in JME patients. To a lesser extent, relationships with extra-disease factors, such as social functioning have also been investigated in JME patients.

In terms of research approaches, differential prevalence designs have shown that JME patients have a higher rate of psychiatric diagnoses than controls. Standardized diagnostic interviews, such as the Structured Clinical Interview for DSM-IV [163], have demonstrated rates of psychopathology in JME patient clinical samples ranging from 35 to 62% [164-167]. Lifetime prevalence of Axis I (30%) and Axis II (26%) diagnoses has also been reported to be high, and a number (47%) have current or lifetime prevalence of some type of disorder. Concerning specific disorders in studies where standardized diagnostic interviews were used, results are as follows: anxiety disorders (21-23.8%) [165, 167], generalized anxiety disorder (19-23%) [165, 166, 168], depression (17-20.9%) [165-168], and somatoform disorders (5.6-7%) [165, 166, 168]. Those studies also indicated that less than 5% of patients have a substance abuse disorder, psychotic disorder, obsessive compulsive disorder, dysthymia, specific phobias, or attention deficit disorder [165-168]. Beyond Axis I disorders, it has been noted that 9 to 20% of samples have met criteria for a personality disorder according to structured diagnostic interview [166, 167]. Histrionic, paranoid personality, and borderline personality disorders were most prevalent. There has been a high rate of Axis I (11.3-19%) [164, 165] and Axis II (23%) [164] psychiatric comorbidity in those patients who have one diagnosis as well.

Across other various diagnostic schemes, such as retrospective chart review and clinical judgment, prevalence of any mental disorder varies from 26.5 to 34% [109, 169, 170]. Different researchers have also reported various combinations of mental disorders in samples such that 25.3% in one study were determined to have either an anxiety, phobia, or somatization disorder and 18.1% had a mood disorder [140]. Frequency of specific diagnoses for depression (8.6-14.5%) [109, 170] and anxiety disorders 15.5% [109] have also been reported. Similar to studies using structured diagnostic interviews, other designs have indicated a low prevalence (<5%) for psychotic disorders [169] and obsessive compulsive disorder [109, 169]. However, some data points have been outliers in that less than 5% of samples have been diagnosed with depression and anxiety disorders as well as depression [169]. In addition to diagnostic rates, patients with JME have scored significantly higher than controls on measures of symptoms related to depression, anxiety [171], internalization (24%), and externalization (16%) [172]. Prevalence of personality disorders in studies with less controlled diagnostic procedures have varied widely and have comprised 3 to 14% of samples [109, 169, 170]. Data from the aforementioned studies have indicated that groups of patients with JME have a higher incidence of mental disorders than members of control groups. Such studies have tended not to report comorbidity rates.

In addition to comparing control groups, research has been conducted examining differential prevalence of mental health problems in JME groups with that found in other groups of epilepsy patients. In a study of 157 patients with GGE [173], there were no significant differences in rates of psychiatric diagnoses between JME patients (23%) and diagnoses occurring in other GGE syndromes. In comparison to patients with partial epilepsies, varying degrees of concordance have been noted. In larger samples [168], similar numbers of psychiatric disorders have been found in JME patients (49%) and patients with refractory MTS (50%). In contrast, an early study using a structured diagnostic interview [174] found only 22% of the JME patients met criteria for a mental disorder, while 55% of patients with TLE met criteria for a mood disorder. Comparison of prevalence has suggested that JME patients have a lower rate of psychotic disorders than patients with MTS, although anxiety disorder may be more prevalent in JME groups (23%). That research group [165] also compared a larger cohort and did not find differing levels of psychiatric diagnoses (i.e., mood, anxiety or somatoform) between JME patients and those with MTS, although the presence of psychotic disorders was associated with MTS group membership. Indeed, that group also published data [175] from the same sample that indicated significantly more MTS patients (11.6%) had at least two core symptoms of psychosis compared with 4.8% of those with JME. However, the proportion of individuals with post-ictal psychosis or interictal psychosis was similar between the groups. With regard to symptom severity, a sample of 20 JME and 20 TLE patients had scores on measures of stress and depression [176] that were similar.

Further research on various broad spectrum measures of personality traits [172] has revealed a number of disparate findings. Study on JME patients' responses on metrics of personality characteristics has indicated JME patients [172] experience significantly higher levels of a "repressive defensiveness" trait than the test normative sample, although a number of other variables were not significantly elevated. Other research [177] has compared responses of JME patients and controls on the Minnesota Multiphasic Personality Inventory [178], and another group [176] did not find group differences on a Five Factor personality inventory in JME and TLE patients. Another study [171] of personality features in JME explored possible endophenotypic expressions of personality features that are not part of contemporary psychiatric diagnostic categorization. As such, compared with controls, JME patients were shown to have a high rate of novelty seeking, low rate of harm avoidance, and low rate of self-directedness (e.g., lack of goal direction and incongruent habits). Taken together, these disparate and non-significant findings have led to conclusions that there may not be a specific personality profile associated with JME [172, 176].

Finding factors that influence or modify the expression of mental health problems in JME has also proven elusive. In a retrospective chart review [169] of 155 JME patients, it was shown that psychiatric disorders were more prevalent in patients with medically resistant seizures (58.3%) as opposed to non-resistant (19%). Similarly, anxiety disorders in JME patients have been associated with lack of seizure control and a history of having several lifetime GTCSs [167]. Moreover, other psychiatric and personality disorders have been associated with seizure frequency [166]. In contrast, others have not found associations of frequency of psychiatric diagnoses with duration of epilepsy [164, 168], type of seizures [164, 168], seizure frequency [164, 168], or number and type of AEDs [168]. Researchers [164] have also not found factors associated with psychiatric comorbidity including age and medication adherence.

Regarding personality features [177], patient endorsements on measures of personality have not been related to age of epilepsy, diagnosis onset, or seizure frequency. In JME patients with a personality disorder compared with JME patients without a personality disorder, there have been no group associations with disease duration [166], age of onset, or "adequate treatment" [179-181]. However, there have been significant associations of personality facets, such as novelty seeking, with early age of epilepsy onset and higher frequency of myoclonic seizures [171]. Disease chronicity has also been shown to relate to personality features, such as restraint [172].

In contrast to describing risk factors, researchers have identified potentially protective factors against anxiety and personality disorders that include being treated with an AED for more than 2 years [166, 168]. Although the direction of causality for this association is not known, it

http://dx.doi.org/10.5772/58372

is likely that lower psychological stress promotes adherence, and the effect of being adherent leads to better seizure control, which also likely results in better psychosocial functioning and fewer psychiatric symptoms. As psychological stress has been inconsistently shown to be related with seizure control, researchers have implemented psychological interventions expecting this will affect seizure outcome. In one such study [109], 58 JME patients with uncontrolled seizures receiving a "rational AED regimen" participated in psychological intervention with the goal to eliminate seizure precipitants. Treatment modalities in that study included an anti-stress program or Cognitive Behavioral Therapy intervention. The results from that study indicated that patients showed a reduction of seizure activity across three time epochs as treatment progressed and this also coincided with ratings of psychiatric functioning. This is particularly relevant as there are indications that psychological factors might relate to seizure control and perceptions of seizure control. For instance, in a survey of JME patients, 62 of 75 respondents (83%) reported that they viewed stress as the most frequent seizure precipitant [35]. These findings provide preliminary evidence that medication adherence and psychological treatment may have important roles in influencing emotional well-being in these patients.

Along with rates of psychiatric problems, other functional outcomes have been examined. For example, long-term follow-up of JME patients [14] has indicated that 65% to 77% of patients reported being "very satisfied" with their health, work, friendships, and social life. However, 61% were prescribed a psychiatric medication, nearly one third were unemployed, and 74% reported at least one "unfavorable" social outcome (i.e., school suspensions, truancy, fighting, criminal offense, social isolation, social impulsiveness, unplanned pregnancy). The authors noted that such outcomes occur at similar rates as other childhood onset epilepsies. Likewise, other data has indicated that only 22.9% of patients may find employment [140]. JME patients [182] also have adjustment challenges in work and family life. De Araujo Filho et al. [166] noted that JME patients, compared with controls, also had a lower DSM-IV Global Assessment of Functioning score and they reported a higher number of psychosocial problems.

As neurobiological theories of psychiatric disorders continue to be advanced [183, 184], there have also been more studies of cerebral functioning in patients with epilepsy and comorbid psychiatric conditions [185]. In that vein, a group from Brazil has been at the forefront of examining characteristics of brain structure and function in JME patients with and without psychiatric diagnoses. In particular, they [180] have shown small to medium correlations for NAA/Cr in the left medial primary motor region and right thalamus, and small to medium correlations for GLX/Cr in the right medial primary motor and left lateral primary motor areas for psychiatric JME patients. That group [179] also showed that 16 JME patients with a cluster B personality disorder had reduction of GMV in right thalamus compared with JME patients with no psychiatric problems. The JME group with personality disorders also had bilateral increases in GMV in the middle frontal gyrus and right orbitofrontal cortex, and decreased white matter in the posterior corpus callosum. Additional study [181] of cluster B personality disorders in patients with JME showed bilateral morphological changes in thalamic, frontal, and limbic structures compared with JME patients with no psychiatric disorder.

Overall, the current data do not consistently support a specific combination of behavioral or psychiatric symptoms among JME patients suggestive of a personality syndrome. In general, rates of psychiatric diagnoses in JME patients have been shown to be higher than the general population. Further, preliminary data suggest that JME patients have a higher rate of anxiety symptoms and less prevalence of psychotic symptoms compared with TLE patients. However, while there are a number of studies of psychiatric functioning in other epilepsies, such as TLE, psychiatric data in JME patients remain inadequate. Moreover, a number of studies typically lack satisfactory sample size, control and differing patients groups, psychosocial contributors, neuroimaging (i.e., structural and functional) studies, or neurophysiological metrics. Additionally, there is an absence in the literature regarding genetic, sibling, or family studies that address psychiatric variables. Such data will further help advance causal models of psychopathology and influence treatment implications. Although many details regarding psychiatric problems in JME are not well defined, the combined data has immediate implications for clinical care.

## 16. Pathophysiology

Based upon evidence of broad cognitive dysfunction and psychiatric problems along with the findings that there are various seizures triggers and EEG findings, mounting evidence has suggested that JME affects the brain in a generalized manner. Nonetheless, the contention that JME is a disease of the frontal lobe and or thalamus has been asserted consistently by the literature, even in the face of contradictory neuroimaging and neuropsychological findings. Recent developments in genetics and neuroimaging have opened new avenues to understand the mechanisms underlying GGEs and expand the conceptualization of JME.

**Genetics**: The genetic inheritance of JME is complex as about 20 chromosomal loci have been linked to the disease. While several channelopathies have been associated with multiple GGE phenotypes, particular mutations have also been specifically implicated in JME. For instance, GABRA1 mutation reducing GABA-receptor function and expression [186], and EFHC1 mutations affecting mitotic spindle organization [187, 188], are both candidate mechanisms.

Autosomal dominant inheritance of a GABRA-1 mutation was reported in a large French-Canadian family with seven members exhibiting JME [186]. GABRA1 on chromosome 5q34 encodes for the $\alpha_1$-subunit of the $\gamma$-aminobutyric acid receptor subtype A (GABA$_A$-receptor), linked to a chloride-ionophore. Activation of this receptor allows chloride to enter the neuron, leading to hyperpolarization. The GABRA1 mutation leads to a loss of function of the GABA$_A$-receptor in vitro, either due to altered gating properties or decreased expression of receptors at the cell surface, resulting in decreased inhibition.

EFHC1 is located on chromosome 6p11 and was the first gene to be identified with specific linkage to JME [58]. EHFC1 mutations are more commonly associated with JME than channelopathies, accounting for 9% of JME cases [189]. They are inherited in an autosomal dominant fashion, but also account for singletons and sporadic cases. The putative role of EFHC1 is evolving, but appears to affect neuronal division and migration during corticogenesis [187,

190]. EFHC1 was originally shown to be a microtubule–associated protein (MAP), localized to the centrosome and the mitotic spindle. Loss of function of EFHC1 disrupts mitotic spindle organization and impairs radial migration of neurons to the developing cortex. Moreover, recent evidence points to impaired radial glia scaffold formation and altered morphology of both radially (mainly excitatory) and tangentially (mainly inhibitory) migrating neurons. This disruption of migration results in fewer neurons and impaired cortical migration of both excitatory and inhibitory neurons. Impaired morphology, i.e., the inability to transform from multipolar to bipolar cells, leads to the neurons remaining in the intermediate zone.

Susceptibility genes may also play an important role in the expression of epilepsy phenotypes. Although channelopathies only account for 3% of JME [187], their genetic associations with epilepsy are likely important markers for disease characteristics. GABRA2 mutations, in addition to SCN1A and SCN1B mutations, have been liked to pedigrees with generalized epilepsy with febrile seizures, whose members may present with many types of epilepsy syndromes, including JME [186].

There are also several susceptibility single-nucleotide polymorphism (SNP) loci that contribute to the risk of developing GGE and JME, which has been suggested by genome-wide association analyses. For GGE, in general, at least three candidate genes have been noted: 1) VRK (2p16.1): a loci that may affect cortical development, 2) SCNA1 (2q24.3): a loci for coding the $\alpha$-subunit of the neuronal voltage-gated sodium channel, mutations of which are the most common cause of channelopathy-related epilepsy, and 3) PNPO (17q21.32): a loci that is involved in pyridoxine oxidation to an active-cofactor involved in neurotransmitter metabolism. Susceptibility candidates for specific GGE syndromes, such as absence epilepsy or JME, were also identified. The linkage associated with JME is at 1q43, flanked by two regions with high recombination rates and covering the CHRM3 gene, which codes for the M3 muscarinic acetylcholine receptor. While nicotinic acetylcholine receptor mutations have been linked to autosomal dominant frontal lobe epilepsy, it is not clear how muscarinic receptors contribute to epileptogenesis.

In summary, the genetic effects of JME are manifold and various genotypes can result in the same epilepsy syndrome. This may be due to a combination of factors, including the lack of accurate epilepsy syndrome differences in subphenotypes or failure to identify epigenetic mechanisms.

**Pathology**: Brain pathology studies in cases of JME are rare (i.e., approximately 20 cases) as patients usually have an average life expectancy. Meencke and Janz [191] published brain pathology in eight patients with GGE, most of whom had myoclonic seizures. They found microdysgenesis in seven patients, defined as persistent uni-or bipolar cells in the subpial region, increased cell density in the stratum moleculare, portrusions of neurons against the pial membrane, containing well-differentiated neurons, columnar neuronal architecture, and increased neurons in the subcortical white matter. Atypical neurons were also found in the hippocampus and cerebellum [192]. There was no evidence of gliosis or chronic neuronal loss. Those findings were criticized for a lack of control group pathology samples [193], and the authors [191] published an additional number of brains from a mixed group of GGE. Using that latter group of specimens, the authors reported that there was a frontal-to-parietal decrease of "microdysplasia", which was absent in the occipital lobes. In 2000, another group compared five GGE patients, who died of SUDEP, to a nonepileptic control group [194]. This study did not find an increase of the "microdysplasia". Based upon the small numbers of patients with heterogeneous electroclinical features and syndromal diagnoses, it is difficult to draw any conclusions regarding pathology findings that are specific to JME.

**Neuroimaging**: In contrast to pathology studies, a number of morphometric imaging approaches since the 1990s have contributed to the neurobiological understanding of GGE and JME. Such studies have demonstrated differences between JME patients and normal controls, and suggest that there are subtle pathological changes difficult to assess by routine histology. Further, while JME manifests as generalized electroclinical activity, structural imaging has indicated circumscribed, multiregional abnormalities. Similarly, functional imaging studies have also demonstrated a predominance of symmetrical multiregional involvement (i.e., cerebral blood flow [CBF] or blood oxygen level dependent [BOLD] changes) prior to and during absence seizures or prolonged spike-and-wave discharges.

*Structural Neuroimaging:* Routine clinical structural imaging scans, whether computerized tomography (CT) or magnetic resonance imaging (MRI), are considered normal in individuals with JME. In contrast, more sensitive techniques, such as voxel-based morphometry (VBM), can show subtle variations in gray matter concentration or volume that differ in JME patients compared with healthy controls. An early quantitative MRI volumetric study in GGE participants showed increases in normalized gray matter volume (GMV) compared to healthy controls, but also possible differences between other epilepsy syndromes as well [195]. In a subsequent VBM study, statistical parametric mapping demonstrated increased mesial frontal GMV in JME patients. Individual patients also demonstrated additional changes in parieto-temporal regions along with decreases frontally in individual patients analyzed by a volume of interest method [195].

More recent VBM studies in JME patients have demonstrated multiregional volumetric and morphometric brain changes [196, 197]. Lin et al. [196] showed increased right frontal GMV and but reduced bilateral thalami, insula, and cerebellar hemisphere volume. Differences in GMV as a function of photosensitivity were also detected, as there were decreases in occipital lobe, left inferior frontal lobe, and hippocampal GMV in photosensitive patients compared with non-photosensitive JME patients. Altered cortical morphology differences in terms of surface area metrics, and not cortical thickness, was noted in the left hemisphere: insula, cingulate, occipital pole, middle temporal, and fusiform gyri [197]. Similar differences were also noted in the right hemisphere: insula, inferior temporal gyrus, and precuneus. Mean cortical curvature measurements in JME patients were also different from controls in the bilateral insula, left cingulate, and right inferior temporal gyrus.

Diffusion tensor imaging (DTI) provides a complementary approach to traditional morphometric and volumetric analyses, as it accounts for white matter connectivity in cortical and subcortical regions. The most important parameters assessed by DTI are fractional anisotropy (FA) and mean diffusivity (MD), which provide information about the microstructural integrity of the white matter pathways. In particular, FA is influenced by myelin integrity and fiber density, whereas MD is correlated with microscopic membrane disruption and extracellular fluid accumulation.

http://dx.doi.org/10.5772/58372

It has been shown that JME patients have symmetric FA reductions in the bilateral superior and anterior corona radiata, genu and body of the corpus callosum, along and with middle and superior frontal tracts [139]. White matter tract MD increases coincided with the FA reduction, although the left superior parietal lobe was also affected. There were no increases in FA or decreases in MD found in JME patients. Frequency of GTCS was correlated with the severity of the microstructural changes.

Several studies have shown similar reductions of FA [141] and increases in MD [198], mainly concentrated in thalamocortical pathways and the corpus callosum; pathways that are responsible for propagation and synchronization of spike-and-wave discharges. These DTI changes have been correlated with GMV reduction in the supplementary motor area (SMA) and posterior cingulate cortices. The SMA FA was correlated with deficits for word naming tasks and expression, while the posterior cingulate changes predicted cognitive inhibition scores [141] and deficits in frontal lobe executive motor functioning. A subsequent study by the same group demonstrated a correlation between DTI and fMRI-based measures of structural and functional connectivity between prefrontal cognitive cortex and motor cortices [198]. Connectivity was decreased between prefrontal and frontopolar regions and increased between occipital cortex and SMA. The authors suggested that the frontal connectivity changes may be related to cognitive deficits seen in JME, and occipitomotor connectivity may be related to photosensitivity, although association with these clinical phenotypes were not investigated.

The neuroanatomic basis of GMV and DTI changes is not clear. It is suspected that the decreased connectivity may be due to altered cortical organization, but as many of these changes are correlated with disease activity, degenerative changes cannot be ruled out. So far, the studies have shown disparate findings, thus, leading to an inconclusive picture of cerebral differences in JME. However, the aggregate of study results suggests diffuse abnormalities with morphological changes that include regions beyond the frontal lobes. Moreover, pathological studies are needed to confirm changes in cortical organization and connectivity. To improve on this knowledge base, an animal model of JME could be utilized to investigate a number of clinicopathological data points including morphometric changes and histopathologic findings.

*Functional – Interictal PET, MRS, EEG-fMRI*: Interictal PET studies using flourodeoxyglucose (FDG) evaluate metabolism over a 40-minute uptake time. While FDG-PET is not ideal for evaluating transient ictal state or interictal discharges it does evaluate a more chronic functional status. In an early study, Swartz et al. [125] indicated that JME patients showed decreased metabolism in the prefrontal and premotor cortices, which was associated with behavioral and cognitive dysfunction. Increased metabolism in the bilateral thalami was correlated with increased ictal or interictal epileptic discharges in another study of JME patients [199]. These studies demonstrated predominantly frontal lobe dysfunction, and an obvious activation of the thalami by ictal and interictal epileptic discharges. The absence of cortical activations suggests more fluctuation in metabolism due to epileptic discharges, decreased neuronal activity in the interictal state, or a general decrease of metabolism due to the disruptive effect of epileptic discharges on cognitive activity.

Magnetic resonance spectroscopy (MRS) can be used to measure neuronal function and concentrations of neurotransmitters in the cerebral cortex and subcortical structures. One study showed reduced N-acetyl aspartate (NAA) in the medial frontal lobes (and not in the occipital lobes) of individuals with JME [200]. The same group [152] showed NAA reduction was specific to JME as compared to individuals diagnosed with GTCS upon Awakening, but was similar to patients with CAE and JAE, [201]. However, investigation of cortical regions was limited and did not involve the lateral frontal, sensorimotor, or parietal areas.

Functional connectivity is a novel method evaluating regional covariance of BOLD signal over time. Most of the functional connectivity data is derived from generalized spike-and-wave discharges in individuals with absence epilepsy. Such studies have examined BOLD changes before, during, and after spike-and-wave discharges. While the thalamocortical networks projecting mainly to the medial frontal lobe are briefly activated during these discharges, more sustained activations occur in the parietal lobes prior to the discharges. As the discharges evolve, increases in BOLD dissipate, resulting in decreased frontoparietal BOLD signal. The brief discharges associated with JME are less likely to cause similar degrees of BOLD changes. Therefore, evaluating more stable measures such as functional connectivity, can provide more information regarding the networks.

One group has contributed greatly to evaluating connectivity both structurally and functionally in individuals with JME [141, 198]. As mentioned above, they have demonstrated that functional connectivity is closely correlated with structural connectivity in the medial frontal cortices in individuals with JME [198]. They showed that during verbal fluency tasks, there is diffuse symmetrical activation of the SMA, bilateral inferior frontal gyri, left premotor area, left thalamus and bilateral putamen, as well as bilateral ventral visual areas. The thalamic region, showing altered connectivity on DTI, was connected to cortices largely overlapping the areas of functional activation. Overall, the individuals with JME showed increased task-dependent connectivity with respect to the frontal cortices compared to controls. Based upon these findings, the authors concluded that the thalamus serves an important function in increased frontal connectivity and coherence.

However, another study of functional connectivity in GGE, which included a large number of patients with JME, showed that there was no alteration of thalamic or mesial frontal connectivity when no discharges occurred [202]. Although the investigative techniques have differed between studies and selection biases may be present, the latter study suggests that generalized spike-and-wave discharges may be generated by healthy networks in response to abnormal connectivity and cortical synchronization in disparate brain regions. Indeed, another study showed alterations in default mode network (DMN) connectivity (in the absence of ictal or interictal epileptic discharges) between posterior and anterior nodes in patients with GGE [203]. Given the current limitations, additional methods, such as intracranial electrophysiological recordings and subsequent pathological examination would help elucidate the mechanisms underlying the CBF changes.

**Transcranial Magnetic Stimulation**: Transcranial magnetic stimulation (TMS) leads to a brief depolarization of cortical neurons, and can target brain regions such as the motor and language cortices. Motor cortex stimulation, such as the primary hand motor area, is conducted frequently as the stimulus direction and intensity can be optimized to specifically activate hand muscles. TMS can be applied using single or paired pulses. Single pulse TMS is used to measure the resting motor threshold and the cortical silent period (CSP). Furthermore, the paired pulse stimulation paradigm is used measure short-latency intracortical inhibition and facilitation.

Early studies demonstrated an increase in CSP in individuals with GGE, suggesting increased intracortical inhibition [204]. Sleep deprivation in individuals with GGE resulted in a greater change in cortical excitability compared to patients with partial epilepsies or controls, based upon paired-pulse studies [205]. ILS appears to decrease CSP in healthy controls, but not in photosensitive or nonphotosensitive individuals with GGE [206]. These findings are somewhat counterintuitive, as it would be expected that hyperexcitability would be observed in the motor cortices of GGE patients.

Antiepileptic medications can reduce cortical excitability, but there also seems to be variable findings across studies of various GGE syndromes and stimulation protocols. A meta-analysis, showed normal rMT threshold in all GGE patients, except for individuals with JME, who had decreased motor thresholds compared to healthy controls [207], suggesting a degree of cortical excitability in the patients. In a subsequent study, the same group showed that rMT was decreased for photosensitive individuals with GGE compared to those without, although this may be due to syndrome-specific bias, as photosensitivity is most commonly associated with JME [208].

**Animal Models**: Due to its high rate of photosensitivity, the bamboo species, *Papio hamadryas papio*, has long been investigated as a model for epileptic photosensitivity [209]. Researchers established the electrophysiological mechanisms of the model by intracranial depth electrode recordings during photoepileptic responses that occurred with or without lesioned pathways. The ictal and interictal epileptic discharges have been shown to conform to a generalized cerebral distribution, and photic stimulation in the baboons elicits both myoclonic seizures and GTCS. Unlike humans, who respond to ILS at different frequencies (i.e., 12-18Hz), this species is maximally photosensitive at frequencies of 20-25 Hz. However, like humans, they are the most sensitivity to seizure activity in mornings, shortly after awakening. In order to study the seizures in a controlled way, researchers typically elicit activation by ILS, often after pretreatment with allylglycine, a GABA synthesis inhibitor, or by administering proconvulsants (e.g., pentylenetetrazol) [210].

Observations of the baboon's photoparoxysmal responses have indicated that there are generalized spike-and-wave complexes that occur maximally in the frontocentral regions, particularly in the mesial frontal surfaces [43]. During ILS, the occipital driving responses remain localized to occipito-parietal areas with occipital IEDs being rare and temporally unsustained. Fischer-Williams's et al. [43] early study showed that subcortical structures, thalamus, basal ganglia, and brainstem were only secondarily affected by frontocentral IEDs, usually in association with high-amplitude or repetitive discharges. The amygdala, hippocampus, and uncus were not involved in the photoepileptic responses.

In addition, paroxysmal visually evoked potential (PVEP) studies, using single flash stimuli administered after a 10 second train of 25 Hz ILS, have enabled researchers to track discrete electrophysiological events through cortical-subcortical networks. In one study, earliest activation of frontocentral IEDs occurred 20 to 30 milliseconds after the flash, with a subsequent eyelid or facial myoclonus occurring 10 to 12 milliseconds later [211]. Motor symptoms tend to occur only when the amplitude of the cortical discharges exceeded 200 microvolts [212]. Following a PVEP, activation of thalamic nuclei (e.g., ventralis lateralis, centrum medianum, & lateralis posterior) occurs at cortical discharge amplitudes exceeding 400 microvolts. Silva-Barrat et al. [213] also demonstrated that photosensitive and asymptomatic control baboons show difference PVEP responses in peristriate and parietal regions, but not in the striate or cortical responses. It was also shown that frontocentral ictal or interictal epileptic discharges were not activated by ILS following bilateral occipital lobe ablation, while destruction of the superior colliculus or pulvinar unilaterally caused only transient suppression of photoepileptic responses [214]. In terms of other lesion study, corpus callosum sectioning in combination with unilateral stimulation of the occipital lobe resulted in frontocentral discharges and seizures that remained ipsilateral to the activated occipital lobe [215].

Although the epilepsy in these baboons appears similar to JME in many ways, early researchers failed to establish a model for JME from characteristics seen with the baboons. However, the phenotypic expression of several hundred baboons has recently been described clinically and with scalp EEG [216]. The onset of the syndrome typically occurs in adolescence, there is a preponderance of myoclonic and GTCS, and there is a similar diurnal pattern (i.e., predominately in the morning) to JME [217]. The myoclonic seizures tend to affect the face and arms. The GTCS tend to occur in a sporadic fashion, and can be triggered by handling or ketamine. The interictal epileptic discharges are generalized in distribution with 4-6 Hz in frequency, although 2-3 Hz discharges have been noted in younger baboons [218]. Younger baboons can also present with absence seizures, and even spike-and-wave stupor [216]. Although the baboon model of epilepsy and JME share many clinical characteristics, including a genetic etiology, the mode of inheritance and underlying genotypes are not known.

Szabó et al. [219, 220] have also evaluated structural and functional neuroimaging in a large baboon colony. Their findings indicated that while routine structural MRI studies are normal, morphometric analyses revealed decreased central and intraparietal sulci along with cingulate sulci in baboons that showed IEDs [221]. The smaller sulcal areas may reflect an underlying developmental anomaly, resulting in decreased U-fibers, rather than a process due to aging or degenerative disease.

Functional studies have aimed to describe epileptic neuronal networks. IED rate and associated cerebral blood flow changes on $H_2{}^{15}O$-PET have demonstrated co-activations of the premotor, perirolandic, insular-parietal, and occipital cortices, areas that are also observed in human GGE networks [222]. Resting state fMRI has shown altered connectivity of the motor, but not the visual cortices, and DMN connectivity was altered in the epileptic baboons as well [223]. This group has also used intracranial video-EEG in combination with depth, grid, and strip electrodes in order to improve spatial resolution [224]. Results from this approach indicated that frequent multiregional IEDs occur, which appears to trigger generalized discharges. The researchers also recorded myoclonic and GTCS, most of which were triggered multifocally. The focal regions that were most active, according to invasive monitoring, included the parieto-occipital region, parietal lobe, premotor area, and orbitofrontal cortex. This was unexpected as the scalp EEG demonstrated only generalized discharges, and the multifocal discharges appeared to occur both parietally and frontally, reflecting a more diffuse pathology.

Previous examinations of pathology have not shown cortical abnormalities in these animals, suggesting developmental changes or seizure-induced injury [225]. Comparison of neuronal counts from the molecular layer, cortical layer 6 and subcortically, in three sulci, namely the cingulate, intraparietal, and lunate sulcus between asymptomatic photosensitive and healthy control baboons did not demonstrate cortical neuronal reductions, but did detect increased numbers of neurons in the subcortical white matter of the anterior cingulate sulcus [226]. Neuronal flow cytometry, on the other hand, demonstrated cortical neuronal reductions, particularly in the frontal regions related to motor functions and somatosensory cortex [227]. The motor cortices most involved were the face and hand regions, areas most commonly involved in the manifestation of myoclonic seizures. The decreased cell counts are likely to result in decreased functional connectivity, especially in the motor cortices. The electrophysiological effects of the neuron reductions are unclear, but may be related to the extent that inhibitory and excitatory neurons are affected. More detailed pathology for specific types of neurons needs to be pursued, and a qualitative evaluation of cortical organization also needs to be considered.

## 17. Conclusions

JME is a commonly occurring and electroclinically well-defined neurological disease. As with most GGEs, it is relatively straightforward to diagnose and treat; however, there is still uncertainty regarding features which may influence a poor response to antiepileptic medications and complicate overall prognosis. Compared to other types of GGE, JME appears to have a structural etiology. Thus, although it shares commonalities with other GGEs, JME breaks that mold, as the condition has been associated with multiregional and/or asymmetric electroclinical findings. While the nature and development of that structural and functional etiology does not appear to be frontal lobe specific, research has yet to define a replicable model for the disease. Moreover, as JME sub-syndromes [228] may have overlapping features with other epilepsies, establishment of cognitive signs and symptoms that are specific to JME has continued to elude researchers.

As a result of the apparent time-course of syndrome onset along with differing prevalence and prognosis across the lifespan, in conjunction with the subtle cerebral abnormalities as noted above, some cases of JME likely reflect a set of underlying neurodevelopmental processes. Nevertheless, there remains a lack of knowledge regarding histopathology, electroclinicial characteristics and propagation patterns, genetic contributors to phenotypic expression, and disease biomarkers. Given its similarities to human GGE syndromes (i.e., similar seizure types and electrophysiological characteristics) and methodical constraints in human research, recent contributions from the animal literature suggest that continued investigation of these factors in a baboon population will offer a unique avenue to further refine human models of JME, particularly in the setting of photosensitivity. Furthermore, pharmacological development would likely be assisted by employing trials of agents in animals that show behaviors consistent with human phenotypes.

In terms of treatment, various combinations of agents have been shown effective, although some cases of JME result in intractability. Overall, regardless of intractability, JME has the potential for negatively impacting quality of life. For instance, recent studies indicated discernible interictal effects on cognition and behavior, even in patients that are relatively well-controlled. Thus, contrary to previous assertions and clinical lore, JME does not appear to be a "benign" disease.

## Author details

Russell D. Pella, Lakshmi Mukundan and C. Akos Szabo*

*Address all correspondence to: szabo@uthscsa.edu

University of Texas Health Science Center at San Antonio, USA

## References

[1] Herpin TH. Des asces incomplets de l'epilepsie. Paris, France: J Balliere et Fils; 1867.

[2] Rabot LDL. De la myoclonie épileptique. Paris, France: Thèse; 1899.

[3] Janz D, Christian W. Impulsiv-petit mal. Deutsche Zeitschrift fu°r Nervenheilkunde 1957;176:346–86.

[4] Delgado-Escueta AV, Enrile-Bacsal F. Juvenile myoclonic epilepsy of Janz. Neurology. 1984;34(3):285-94.

[5] Lund M, Reintoft H, Simonsen N. [A controlled social and psychological investigation of patients with juvenile myoclonic epilepsy]. Ugeskrift For Laeger. 1975;137(41): 2415-8.

[6] Commission on Classification and Terminology of the International League Against Epilepsy. Proposal for revised classification of epilepsies and epileptic syndromes. Epilepsia. 1989;30(4):389-99.

[7] Genton P, Gelisse P. Juvenile myoclonic epilepsy. Archives of Neurology. 2001;58(9): 1487-90.

[8] Hauser WA, Annegers JF, Rocca WA. Descriptive epidemiology of epilepsy: contributions of population-based studies from Rochester, Minnesota. Mayo Clinic Proceedings Mayo Clinic. 1996;71(6):576-86.

[9] Panayiotopoulos CP, Obeid T, Tahan AR. Juvenile myoclonic epilepsy: a 5-year prospective study. Epilepsia. 1994;35(2):285-96.

[10] Figueredo R, Trevisol-Bittencourt PC, Ferro JB. [Clinical-epidemiological study of patients with juvenile myoclonic epilepsy in Santa Catarina State, Brazil]. Arquivos De Neuro-Psiquiatria. 1999;57(2B):401-4.

[11] Kun LN, Ling LW, Wah YW, Lian TT. Epidemiologic study of epilepsy in young Singaporean men. Epilepsia. 1999;40(10):1384-7.

[12] Murthy JM, Yangala R, Srinivas M. The syndromic classification of the International League Against Epilepsy: a hospital-based study from South India. Epilepsia. 1998;39(1):48-54.

[13] Janz D, Durner M. Juvenile Myoclonic Epilepsy. In: Engel J, Pedley, T.A., editor. Epilepsy: a comprehensive textbook. Philadelphia, PA: Lippincott-Raven Press; 1998. p. 2389–400.

[14] Camfield CS, Camfield PR. Juvenile myoclonic epilepsy 25 years after seizure onset: a population-based study. Neurology. 2009;73(13):1041-5.

[15] Camfield CS, Striano P, Camfield PR. Epidemiology of juvenile myoclonic epilepsy. Epilepsy & Behavior. 2013;28(Suppl 1):S15-S7.

[16] Mullins GM, O'Sullivan SS, Neligan A, McCarthy A, McNamara B, Galvin RJ, et al. A study of idiopathic generalised epilepsy in an Irish population. Seizure. 2007;16(3): 204-10.

[17] Manonmani V, Tan CT. A study of newly diagnosed epilepsy in Malaysia. Singapore Medical Journal. 1999;40(1):32-5.

[18] Asadi-Pooya AA, Emami M, Sperling MR. A clinical study of syndromes of idiopathic (genetic) generalized epilepsy. Journal of the Neurological Sciences. 2013;324(1-2): 113-7.

[19] Wirrell EC, Camfield CS, Camfield PR, Gordon KE, Dooley JM. Long-term prognosis of typical childhood absence epilepsy: remission or progression to juvenile myoclonic epilepsy. Neurology. 1996;47(4):912-8.

[20] Obeid T, Panayiotopoulos CP. Juvenile myoclonic epilepsy: a study in Saudi Arabia. Epilepsia. 1988;29(3):280-2.

[21] Sozmen V, Baybas S, Dirican A, Koksal A, Ozturk M. Frequency of epilepsies in family members of patients with different epileptic syndromes. European Neurology. 2011;65(1):4-9.

[22] Grünewald RA, Chroni E, Panayiotopoulos CP. Delayed diagnosis of juvenile myoclonic epilepsy. Journal of Neurology, Neurosurgery & Psychiatry. 1992;55(6):497-9.

[23] Gunatilake SB, Seneviratne SL. Juvenile myoclonic epilepsy: a study in Sri Lanka. Seizure. 2000;9(3):221-3.

[24] Genton P, Gelisse P, Thomas P. Juvenile myoclonic epilepsy today. In: Schmitz B, Sander, T, editor. Juvenile Myoclonic Epilepsy: The Janz Syndrome. Berlin, Germany: Hippokrates-Verlag; 2000. p. 11-32.

[25] Serratosa JM, Delgado-Escueta AV. Juvenile Myoclonic Epilepsy. In: Wyllie E, editor. The treatment of epilepsy: principles and practices. Philadelphia, PA: Lea & Febiger; 1993. p. 552-70.

[26] Tóth V, Rásonyi G, Fogarasi A, Kovács N, Auer T, Janszky J. Juvenile myoclonic epilepsy starting in the eighth decade. Epileptic Disorders: International Epilepsy Journal With Videotape. 2007;9(3):341-5.

[27] Hirano Y, Oguni H, Funatsuka M, Imai K, Osawa M. Differentiation of myoclonic seizures in epileptic syndromes: A video-polygraphic study of 26 patients. Epilepsia. 2009;50(6):1525-35.

[28] Kasteleijn-Nolst Trenité DGA, de Weerd A, Beniczky S. Chronodependency and provocative factors in juvenile myoclonic epilepsy. Epilepsy & Behavior. 2013;28(Suppl 1):S25-S9.

[29] Badawy RA, Macdonell RA, Jackson GD, Berkovic SF. Why do seizures in generalized epilepsy often occur in the morning? Neurology. 2009;73(3):218-22.

[30] Baykan B, Altindag EA, Bebek N, Ozturk AY, Aslantas B, Gurses C, et al. Myoclonic seizures subside in the fourth decade in juvenile myoclonic epilepsy. Neurology. 2008;70(22 Part 2):2123-9.

[31] Genton P, Thomas P, Kasteleijn-Nolst Trenité DGA, Medina MT, Salas-Puig J. Clinical aspects of juvenile myoclonic epilepsy. Epilepsy & Behavior. 2013;28(Suppl 1):S8-S14.

[32] Sadleir LG, Scheffer IE, Smith S, Carstensen B, Carlin J, Connolly MB, et al. Factors influencing clinical features of absence seizures. Epilepsia. 2008;49(12):2100-7.

[33] Larch J, Unterberger I, Bauer G, Reichsoellner J, Kuchukhidze G, Trinka E. Myoclonic status epilepticus in juvenile myoclonic epilepsy. Epileptic Disorders: International Epilepsy Journal With Videotape. 2009;11(4):309-14.

[34] Crespel A, Velizarova R, Genton P, Gélisse P. [Juvenile myoclonic epilepsy with recurrent myoclonic status: efficacy of valproate]. Thérapie. 2009;64(5):321-3.

[35] da Silva Sousa P, Lin K, Garzon E, Sakamoto AC, Yacubian EMT. Self-perception of factors that precipitate or inhibit seizures in juvenile myoclonic epilepsy. Seizure. 2005;14(5):340-6.

[36] Wolf P. From precipitation to inhibition of seizures: rationale of a therapeutic paradigm. Epilepsia. 2005;46 Suppl 1:15-6.

[37] Guaranha MS, da Silva Sousa P, de Araújo-Filho GM, Lin K, Guilhoto LM, Caboclo LO, et al. Provocative and inhibitory effects of a video-EEG neuropsychologic protocol in juvenile myoclonic epilepsy. Epilepsia. 2009;50(11):2446-55.

[38] Beniczky S, Guaranha MSB, Conradsen I, Singh MB, Rutar V, Lorber B, et al. Modulation of epileptiform EEG discharges in juvenile myoclonic epilepsy: an investigation of reflex epileptic traits. Epilepsia. 2012;53(5):832-9.

[39] Matsuoka H, Takahashi T, Sasaki M, Matsumoto K, Yoshida S, Numachi Y, et al. Neuropsychological EEG activation in patients with epilepsy. Brain: A Journal of Neurology. 2000;123(2):318-30.

[40] Kasteleijn-Nolst Trenité DGA. Provoked and reflex seizures: surprising or common? Epilepsia. 2012;53 Suppl 4:105-13.

[41] Mayer T, Wolf P. Perioral reflex myoclonias in reading epilepsy and juvenile myoclonic epilepsy. In: Wolf P IY, Zifkin B editor. Reflex Epilepsies: Progress in Understanding. New Barnet, Herts, UK: John Libbey; 2004. p. 62–3.

[42] Kasteleijn-Nolst Trenité DG. Photosensitivity in epilepsy. Electrophysiological and clinical correlates. Acta Neurologica Scandinavica Supplementum. 1989;125:3-149.

[43] Fischer-Williams M, Poncet M, Riche D, Naquet R. Light-induced epilepsy in the baboon, Papio papio: Cortical and depth recordings. Electroencephalography & Clinical Neurophysiology. 1968;25(6):557-69.

[44] Parra J, Kalitzin SN, Iriarte J, Blanes W, Velis DN, Lopes da Silva FH. Gamma-band phase clustering and photosensitivity: is there an underlying mechanism common to photosensitive epilepsy and visual perception? Brain: A Journal of Neurology. 2003;126(Pt 5):1164-72.

[45] Varotto GELSGF. Enhanced frontocentral EEG connectivity in photosensitive generalized epilepsies: A partial directed coherence study. Epilepsia. 2012;53(2):359-67.

[46] Moeller F, Siebner HR, Ahlgrimm N, Wolff S, Muhle H, Granert O, et al. fMRI activation during spike and wave discharges evoked by photic stimulation. NeuroImage. 2009;48(4):682-95.

[47] Bai X, Vestal M, Berman R, Negishi M, Spann M, Vega C, et al. Dynamic time course of typical childhood absence seizures: EEG, behavior, and functional magnetic resonance imaging. The Journal of Neuroscience: The Official Journal of The Society for Neuroscience. 2010;30(17):5884-93.

[48] Benuzzi F, Mirandola L, Pugnaghi M, Farinelli V, Tassinari CA, Capovilla G, et al. Increased cortical BOLD signal anticipates generalized spike and wave discharges in

adolescents and adults with idiopathic generalized epilepsies. Epilepsia. 2012;53(4): 622-30.

[49] Ferlazzo E, Zifkin BG, Andermann E, Andermann F. Cortical triggers in generalized reflex seizures and epilepsies. Brain: A Journal of Neurology. 2005;128(Part 4):700-10.

[50] Asconapé J, Penry JK. Some clinical and EEG aspects of benign juvenile myoclonic epilepsy. Epilepsia. 1984;25(1):108-14.

[51] So GM, Thiele EA, Sanger T, Schmid R, Riviello JJ, Jr. Electroencephalogram and clinical focalities in juvenile myoclonic epilepsy. Journal of Child Neurology. 1998;13(11): 541-5.

[52] Genton P, Gonzalez Sanchez MS, Saltarelli A, Bureau M, Dravet C, Roger J. [Misleading aspects of the standard electroencephalogram in juvenile myoclonic epilepsy: a retrospective study of 56 consecutive cases]. Neurophysiologie clinique=Clinical neurophysiology. 1995;25(5):283-90. Epub 1995.

[53] Baise-Zung C, Guilhoto LMFF, Grossmann RM. Juvenile myoclonic epilepsy: non-classic electroencephalographical presentation in adult patients. European Journal of Neurology. 2006;13(2):171-5.

[54] Syvertsen MR, Markhus R, Selmer KK, Nakken KO. [Juvenile myoclonic epilepsy]. Tidsskr Nor Laegeforen. 2012;132(14):1610-3.

[55] Genton P, Puig A, Tunon C, Lahoz CH, Sanchez M. Juvenile myoclonic epilepsy and related syndromes: clinical and neurophysiological aspects. In: Malafosse A, Genton, P, Hirsch, E, Marescaux, C, Broglin, D, Bernasconi, R, editor. Idiopathic generalized epilepsies: clinical, experimental and genetic aspects. London, UK: John Libbey and Company Ltd; 1994. p. 253–65.

[56] Fittipaldi F, Currà A, Fusco L, Ruggieri S, Manfredi M. EEG discharges on awakening: a marker of idiopathic generalized epilepsy. Neurology. 2001;56(1):123-6.

[57] Janz D. Epilepsy with impulsive petit mal (juvenile myoclonic epilepsy). Acta Neurologica Scandinavica. 1985;72(5):449-59.

[58] Delgado-Escueta AV, Greenberg DA, Treiman L, Liu A, Sparkes RS, Barbetti A, et al. Mapping the gene for juvenile myoclonic epilepsy. Epilepsia. 1989;30 Suppl 4:S8.

[59] Koeleman BPC, de Kovel CGF, Kasteleijn-Nolst Trenité DGA. Photoparoxysmal EEG response and genetic dissection of juvenile myoclonic epilepsy. Epilepsy & Behavior. 2013;28(Suppl 1):S69-S71.

[60] Serafini A, Rubboli G, Gigli GL, Koutroumanidis M, Gelisse P. Neurophysiology of juvenile myoclonic epilepsy. Epilepsy & Behavior. 2013;28(Suppl 1):S30-S9.

[61] Panayiotopoulos CP. A clinical guide to epileptic syndromes and their treatment 2nd ed. United Kingdom: Springer; 2007.

[62] Lancman ME, Asconapé JJ, Penry JK. Clinical and EEG asymmetries in juvenile myoclonic epilepsy. Epilepsia. 1994;35(2):302-6.

[63] Lancman ME, Asconapé JJ, Golimstok A. Circling seizures in a case of juvenile myoclonic epilepsy. Epilepsia. 1994;35(2):317-8.

[64] Ferrie CD. Idiopathic Generalized Epilepsies Imitating Focal Epilepsies. Epilepsia. 2005;46:91-5.

[65] Aliberti V, Grünewald RA, Panayiotopoulos CP, Chroni E. Focal electroencephalographic abnormalities in juvenile myoclonic epilepsy. Epilepsia. 1994;35(2):297-301.

[66] Niedzielska K, Kuran W, Romaniak A. [Clinical-electrophysiological pattern of juvenile myoclonic epilepsy]. Neurologia I Neurochirurgia Polska. 1995;29(5):675-85.

[67] Jayalakshmi SS, Srinivasa Rao B, Sailaja S. Focal clinical and electroencephalographic features in patients with juvenile myoclonic epilepsy. Acta Neurologica Scandinavica. 2010;122(2):115-23.

[68] Usui N, Kotagal P, Matsumoto R, Kellinghaus C, Lüders HO. Focal Semiologic and Electroencephalographic Features in Patients with Juvenile Myoclonic Epilepsy. Epilepsia. 2005;46(10):1668-76.

[69] Létourneau K, Cieuta-Walti C, Deacon C. Epileptiform asymetries and treatment response in juvenile myoclonic epilepsy. The Canadian Journal of Neurological Sciences Le Journal Canadien Des Sciences Neurologiques. 2010;37(6):826-30.

[70] Kelley SA, Kossoff EH. Doose syndrome (myoclonic-astatic epilepsy): 40 years of progress. Developmental Medicine & Child Neurology. 2010;52(11):988-93.

[71] Caraballo RH, Chamorro N, Darra F, Fortini S, Arroyo H. Epilepsy with myoclonic atonic seizures: an electroclinical study of 69 patients. Pediatric Neurology. 2013;48(5):355-62.

[72] Fisher RS, Boas WvE, Blume W, Elger C, Genton P, Lee P, et al. Epileptic Seizures and Epilepsy: Definitions Proposed by the International League Against Epilepsy (ILAE) and the International Bureau for Epilepsy (IBE). Epilepsia. 2005;46(4):470-2.

[73] Adie WJ. Pyknolepsy: a form of epilepsy in children, with a good prognosis. Sect Neur. 1924;17:19-25.

[74] Panayiotopoulos CP. Typical absence seizures and related epileptic syndromes: Assessment of current state and directions for future research. Epilepsia. 2008;49(12): 2131-9.

[75] Tassinari CA, Bureau M. Epilepsy with myoclonic absences. In: Roger J DC, Bureau M, Dreifuss FE, Wolf P, editor. Epileptic syndromes in infancy, childhood and adolescence. London, England: John Libbery Eurotext; 1985. p. 121-9.

[76] Durner M, Sander T, Greenberg DA, Johnson K, Beck-Mannagetta G, Janz D. Localization of idiopathic generalized epilepsy on chromosome 6p in families of juvenile myoclonic epilepsy patients. Neurology. 1991;41(10):1651-5.

[77] Berkovic SF, So NK, Andermann F. Progressive myoclonus epilepsies: clinical and neurophysiological diagnosis. Journal of Clinical Neurophysiology. 1991;8(3):261-71.

[78] Shahwan A, Farrell M, Delanty N. Progressive myoclonic epilepsies: a review of genetic and therapeutic aspects. Lancet Neurology. 2005;4(4):239-48.

[79] French JA, Kanner AM, Bautista J, Abou-Khalil B, Browne T, Harden CL, et al. Efficacy and tolerability of the new antiepileptic drugs I: treatment of new onset epilepsy: report of the Therapeutics and Technology Assessment Subcommittee and Quality Standards Subcommittee of the American Academy of Neurology and the American Epilepsy Society. Neurology. 2004;62(8):1252-60.

[80] Auvin S. Treatment of Juvenile Myoclonic Epilepsy. CNS Neuroscience & Therapeutics. 2008;14(3):227-33.

[81] Calleja S, Salas-Puig J, Ribacoba R, Lahoz CH. Evolution of juvenile myoclonic epilepsy treated from the outset with sodium valproate. Seizure. 2001;10(6):424-7.

[82] Fernando-Dongas MC, Radtke RA, VanLandingham KE, Husain AM. Characteristics of valproic acid resistant juvenile myoclonic epilepsy. Seizure. 2000;9(6):385-8.

[83] Marson AG, Al-Kharusi AM, Alwaidh M, Appleton R, Baker GA, Chadwick DW, et al. The SANAD study of effectiveness of valproate, lamotrigine, or topiramate for generalised and unclassifiable epilepsy: An unblinded randomised controlled trial. The Lancet. 2007;369(9566):1016-26.

[84] Mazurkiewicz-Bełdzińska M, Szmuda M, Matheisel A. Long-term efficacy of valproate versus lamotrigine in treatment of idiopathic generalized epilepsies in children and adolescents. Seizure. 2010;19(3):195-7.

[85] Ben-Menachem E, Schmitz B, Tomson T, Vajda F. Role of valproate across the ages. Treatment of epilepsy in adults. Acta Neurologica Scandinavica. 2006;114:14-27.

[86] Adab N, Kini U, Vinten J, Ayres J, Baker G, Clayton-Smith J, et al. The longer term outcome of children born to mothers with epilepsy. Journal of Neurology, Neurosurgery & Psychiatry. 2004;75(11):1575-83.

[87] Morrow J, Russell A, Guthrie E, Parsons L, Robertson I, Waddell R, et al. Malformation risks of antiepileptic drugs in pregnancy: a prospective study from the UK Epilepsy and Pregnancy Register. Journal of Neurology, Neurosurgery & Psychiatry. 2006;77(2):193-8.

[88] Vinten J, Bromley RL, Taylor J, Adab N, Kini U, Baker GA. The behavioral consequences of exposure to antiepileptic drugs in utero. Epilepsy & Behavior. 2009;14(1):197-201.

[89] Duncan S. Teratogenesis of sodium valproate. Current Opinion in Neurology. 2007;20(2):175-80.

[90] Morris GL, Hammer AE, Kustra RP, Messenheimer JA. Lamotrigine for patients with juvenile myoclonic epilepsy following prior treatment with valproate: results of an open-label study. Epilepsy & Behavior. 2004;5(4):509-12.

[91] Levisohn PM, Holland KD. Topiramate or valproate in patients with juvenile myoclonic epilepsy: A randomized open-label comparison. Epilepsy & Behavior. 2007;10(4):547-52.

[92] Park KM, Kim SH, Nho SK, Shin KJ, Park J, Ha SY, et al. A randomized open-label observational study to compare the efficacy and tolerability between topiramate and valproate in juvenile myoclonic epilepsy. Journal of Clinical Neuroscience: Official Journal of the Neurosurgical Society of Australasia. 2013;20(8):1079-82.

[93] Specchio LM, Gambardella A, Giallonardo AT, Michelucci R, Specchio N, Boero G, et al. Open label, long-term, pragmatic study on levetiracetam in the treatment of juvenile myoclonic epilepsy. Epilepsy Research. 2006;71(1):32-9.

[94] Noachtar S, Andermann E, Meyvisch P, Andermann F, Gough WB, Schiemann-Delgado J. Levetiracetam for the treatment of idiopathic generalized epilepsy with myoclonic seizures. Neurology. 2008;70(8):607-16.

[95] Rosenfeld WE, Benbadis S, Edrich P, Tassinari CA, Hirsch E. Levetiracetam as add-on therapy for idiopathic generalized epilepsy syndromes with onset during adolescence: analysis of two randomized, double-blind, placebo-controlled studies. Epilepsy Research. 2009;85(1):72-80.

[96] Kothare SV, Valencia I, Khurana DS, Hardison H, Melvin JJ, Legido A. Efficacy and tolerability of zonisamide in juvenile myoclonic epilepsy. Epileptic Disorders: International Epilepsy Journal With Videotape. 2004;6(4):267-70.

[97] Genton P, Gelisse P, Thomas P, Dravet C. Do carbamazepine and phenytoin aggravate juvenile myoclonic epilepsy? Neurology. 2000;55(8):1106-9.

[98] Gelisse P, Genton P, Kuate C, Pesenti A, Baldy-Moulinier M, Crespel A. Worsening of Seizures by Oxcarbazepine in Juvenile Idiopathic Generalized Epilepsies. Epilepsia. 2004;45(10):1282-6.

[99] Sharpe C, Buchanan N. Juvenile myoclonic epilepsy: diagnosis, management and outcome. The Medical Journal of Australia. 1995;162(3):133-4.

[100] Mantoan L, Walker M. Treatment options in juvenile myoclonic epilepsy. Current Treatment Options In Neurology. 2011;13(4):355-70.

[101] Penry JK, Dean JC, Riela AR. Juvenile myoclonic epilepsy: long-term response to therapy. Epilepsia. 1989;30 Suppl 4:S19.

[102] Shian WJ, Chi CS. Juvenile myoclonic epilepsy. Zhonghua Yi Xue Za Zhi=Chinese Medical Journal; Free China Ed. 1994;53(6):375-8.

[103] Sokic D, Ristic AJ, Vojvodic N, Jankovic S, Sindjelic AR. Frequency, causes and phenomenology of late seizure recurrence in patients with juvenile myoclonic epilepsy after a long period of remission. Seizure. 2007;16(6):533-7.

[104] Dahl JC. A Behaviour Medicine Approach to Epilepsy-Time for a Paradigm Shift? Scandinavian Journal of Behaviour Therapy. 1999;28(3):97.

[105] Guilfoyle SM, Follansbee-Junger K, Modi AC. Development and preliminary implementation of a psychosocial service into standard medical care for pediatric epilepsy. Clinical Practice in Pediatric Psychology. 2013;1(3):276-88.

[106] Lundgren T, Dahl J, Hayes S. Evaluation of mediators of change in the treatment of epilepsy with acceptance and commitment therapy. Journal of Behavioral Medicine. 2008;31(3):225-35.

[107] Kasteleijn-Nolst Trenité DGA, Schmitz B, Janz D, Delgado-Escueta AV, Thomas P, Hirsch E, et al. Consensus on diagnosis and management of JME: From founder's observations to current trends. Epilepsy & Behavior. 2013;28(Suppl 1):S87-S90.

[108] Crespel A, Gelisse P, Reed RC, Ferlazzo E, Jerney J, Schmitz B, et al. Management of juvenile myoclonic epilepsy. Epilepsy & Behavior. 2013;28(Suppl 1):S81-S6.

[109] Martinović Z. Adjunctive behavioural treatment in adolescents and young adults with juvenile myoclonic epilepsy. Seizure. 2001;10(1):42-7.

[110] da Silva Sousa P, Lin K, Garzon E, Ceiki Sakamoto A, Yacubian EMT. Language-and praxis-induced jerks in patients with juvenile myoclonic epilepsy. Epileptic Disorders: International Epilepsy Journal with Videotape. 2005;7(2):115-21.

[111] Matsuoka H, Nakamura M, Ohno T, Shimabukuro J, Suzuki T, Numachi Y, et al. The Role of Cognitive–Motor Function in Precipitation and Inhibition of Epileptic Seizures. Epilepsia. 2005;46:17-20.

[112] Morris GL, 3rd, Gloss D, Buchhalter J, Mack KJ, Nickels K, Harden C. Evidence-based guideline update: Vagus nerve stimulation for the treatment of epilepsy: Report of the Guideline Development Subcommittee of the American Academy of Neurology. Neurology. 2013;81(16):1453-9.

[113] Kostov H, Larsson PG, Røste GK. Is vagus nerve stimulation a treatment option for patients with drug-resistant idiopathic generalized epilepsy? Acta Neurologica Scandinavica. 2007;115:55-8.

[114] Cukiert A, Burattini JA, Mariani PP, Cukiert CM, Argentoni-Baldochi M, Baise-Zung C, et al. Outcome after extended callosal section in patients with primary idiopathic generalized epilepsy. Epilepsia. 2009;50(6):1377-80.

[115] Iimura Y, Sugano H, Nakajima M, Arai H. Stepwise synchronization through the corpus callosum is one cause of myoclonic jerks. World Neurosurgery. 2012;77(2): 399.E5-8.

[116] Jenssen S, Sperling MR, Tracy JI, Nei M, Joyce L, David G, et al. Corpus callosotomy in refractory idiopathic generalized epilepsy. Seizure. 2006;15(8):621-9.

[117] Faught E. Treatment of refractory primary generalized epilepsy. Reviews In Neurological Diseases. 2004;1 Suppl 1:S34-S43.

[118] Baykan B, Martínez-Juárez IE, Altindag EA, Camfield CS, Camfield PR. Lifetime prognosis of juvenile myoclonic epilepsy. Epilepsy & Behavior. 2013;28(Suppl 1):S18-S24.

[119] Geithner JFZJRCU. Predictors for long-term seizure outcome in juvenile myoclonic epilepsy: 25-63 years of follow-up. Epilepsia. 2012;53(8):1379-86.

[120] Matsuoka H. The seizure prognosis of juvenile myoclonic epilepsy. The Japanese Journal of Psychiatry and Neurology. 1992;46(2):293-6.

[121] Seneviratne U, Cook M, D'Souza W. The prognosis of idiopathic generalized epilepsy. Epilepsia. 2012;53(12):2079-90.

[122] Guaranha MSB, Filho GMdA, Lin K, Guilhoto LMFF, Caboclo LOSF, Yacubian EMT. Prognosis of juvenile myoclonic epilepsy is related to endophenotypes. Seizure. 2011;20(1):42-8.

[123] Ristić AJ, Ostojić J, Kozić D, Vojvodić NM, Popović LM, Janković S, et al. Hippocampal metabolic dysfunction in juvenile myoclonic epilepsy: 3D multivoxel spectroscopy study. Journal of the Neurological Sciences. 2011;305(1-2):139-42.

[124] Swartz BE, Halagren E, Simpkins F, Syndulko K. Primary memory in patients with frontal and primary generalized epilepsy. Journal of Epilepsy. 1994;7(3):232-41.

[125] Swartz BE, Simpkins F, Halgren E, Mandelkern M, Brown C, Krisdakumtorn T, et al. Visual working memory in primary generalized epilepsy: an 18FDG-PET study. Neurology. 1996;47(5):1203-12.

[126] Devinsky O, Gershengorn J, Brown E, Perrine K, Vazquez B, Luciano D. Frontal functions in juvenile myoclonic epilepsy. Neuropsychiatry, Neuropsychology, & Behavioral Neurology. 1997;10(4):243-6.

[127] Piazzini A, Turner K, Vignoli A, Canger R, Canevini MP. Frontal cognitive dysfunction in juvenile myoclonic epilepsy. Epilepsia. 2008;49(4):657-62.

[128] Zamarian L, Höfler J, Kuchukhidze G, Delazer M, Bonatti E, Kemmler G, et al. Decision making in juvenile myoclonic epilepsy. Journal of Neurology. 2013;260(3): 839-46.

[129] Zamarian L, Trinka E, Bonatti E, Kuchukhidze G, Bodner T, Benke T, et al. Executive functions in chronic mesial temporal lobe epilepsy. Epilepsy Research and Treatment. 2011;Vol. 2011(ArtileID 596174):11 pages.

[130] Delazer M, Zamarian L, Bonatti E, Kuchukhidze G, Koppelstaetter F, Bodner T, et al. Decision making under ambiguity and under risk in mesial temporal lobe epilepsy. Neuropsychologia. 2010;48(194-200).

[131] Levav M, Mirsky AF, Herault J, Xiong L, Amir N, Andermann E. Familial association of neuropsychological traits in patients with generalized and partial seizure disorders. Journal of Clinical and Experimental Neuropsychology. 2002;24(3):311-26.

[132] Pulsipher DT, Seidenberg M, Guidotti L, Tuchscherer VN, Morton J, Sheth RD, et al. Thalamofrontal circuitry and executive dysfunction in recent-onset juvenile myoclonic epilepsy. Epilepsia. 2009;50(5):1210-9.

[133] Jackson DC, Dabbs K, Walker NM, Jones JE, Hsu DA, Stafstrom CE, et al. The neuropsychological and academic substrate of new/recent-onset epilepsies. The Journal of Pediatrics. 2013;162(5):1047-53.e1.

[134] Guidotti L, Seidenberg M, Pulsipher DT, Morton J, Dow C, Sheth RD, et al. Cognition and MRI volumetrics in children with recent onset juvenile myoclonic epilepsy. Epilepsia. 2006;47:110.

[135] Pascalicchio TF, de Araujo Filho GM, da Silva Noffs MH, Lin K, Caboclo LOSF, Vidal-Dourado M, et al. Neuropsychological profile of patients with juvenile myoclonic epilepsy: a controlled study of 50 patients. Epilepsy & Behavior. 2007;10(2):263-7.

[136] Wandschneider B, Kopp UA, Kliegel M, Stephani U, Kurlemann G, Janz D, et al. Prospective memory in patients with juvenile myoclonic epilepsy and their healthy siblings. Neurology. 2010;75(24):2161-7.

[137] Moschetta SP, Valente KD. Juvenile myoclonic epilepsy: the impact of clinical variables and psychiatric disorders on executive profile assessed with a comprehensive neuropsychological battery. Epilepsy & Behavior. 2012;25(4):682-6.

[138] Kim S-Y, Hwang Y-H, Lee H-W, Suh C-K, Kwon S-H, Park S-P. Cognitive impairment in juvenile myoclonic epilepsy. Journal of Clinical Neurology (Seoul, Korea). 2007;3(2):86-92.

[139] Kim JH, Suh S-i, Park S-Y, Seo W-K, Koh I, Koh S-B, et al. Microstructural white matter abnormality and frontal cognitive dysfunctions in juvenile myoclonic epilepsy. Epilepsia. 2012;53(8):1371-8.

[140] Jovic NJ, Vucinic DM, Ignjatovic PM, Kosac A. Cognitive and psychiatric issues in juvenile myoclonic epilepsy. Epilepsy & Behavior. 2013;28(2):324.

[141] O'Muircheartaigh J, Vollmar C, Barker GJ, Kumari V, Symms MR, Thompson P, et al. Focal structural changes and cognitive dysfunction in juvenile myoclonic epilepsy. Neurology. 2011;76(1):34-40.

[142] Lavandier N, de Toffol B, Gillet P, Hommet C, Corcia P, Autret A. Syndrome dysexe ′cutif, e′pilepie myoclonique juve′nile et anomalies paroxystiques intercritiques. Rev Neurol. 2002;158(Suppl. 1):164.

[143] Sonmez F, Atakli D, Sari H, Atay T, Arpaci B. Cognitive function in juvenile myoclonic epilepsy. Epilepsy & Behavior. 2004;5(3):329-36.

[144] Iqbal N, Caswell HL, Hare DJ, Pilkington O, Mercer S, Duncan S. Neuropsychological profiles of patients with juvenile myoclonic epilepsy and their siblings: a preliminary controlled experimental video-EEG case series. Epilepsy & Behavior. 2009;14(3): 516-21.

[145] Roebling R, Scheerer N, Uttner I, Gruber O, Kraft E, Lerche H. Evaluation of cognition, structural, and functional MRI in juvenile myoclonic epilepsy. Epilepsia. 2009;50(11):2456-65.

[146] Wandschneider B, Centeno M, Vollmar C, Stretton J, O'Muircheartaigh J, Thompson PJ, et al. Risk-taking behavior in juvenile myoclonic epilepsy. Epilepsia. 2013.

[147] Vollmar C, O'Muircheartaigh J, Barker GJ, Symms MR, Thompson P, Kumari V, et al. Motor system hyperconnectivity in juvenile myoclonic epilepsy: a cognitive functional magnetic resonance imaging study. Brain: A Journal of Neurology. 2011;134(Pt 6): 1710-9.

[148] Lin K, de Araujo Filho GM, Pascalicchio TF, Silva I, Tudesco ISS, Guaranha MSB, et al. Hippocampal atrophy and memory dysfunction in patients with juvenile myoclonic epilepsy. Epilepsy & Behavior. 2013;29(1):247-51.

[149] Brooks BL. A Study of low scores in Canadian children and adolescents on the Wechsler Intelligence Scale for Children, Fourth Edition (WISC-IV). Child Neuropsychology. 2011;17(3):281-9.

[150] Steinberg BA, Bieliauskas LA, Smith GE, Langellotti C, Ivnik RJ. Mayo's Older Americans Normative Studies: Age-and IQ-Adjusted Norms for the Boston Naming Test, the MAE Token Test, and the Judgment of Line Orientation Test. The Clinical Neuropsychologist. 2005;19(3-4):280-328.

[151] Walsh J, Marson AG, Smith PEM, Rees MI, Baker GA, Thomas RH. Neuroticism and executive function in drug-refractory juvenile myoclonic epilepsy. Epilepsy & Behavior. 2013;28(Suppl 1).

[152] Savic I, Österman Y, Helms G. MRS shows syndrome differentiated metabolite changes in human-generalized epilepsies. NeuroImage. 2004;21(1):163.

[153] Kessler RC, Berglund P, Demler O, Jin R, Merikangas KR, Walters EE. Lifetime prevalence and age-of-anset distributions of DSM-IV disorders in the National Comorbidity Survey Replication. Archives of General Psychiatry. 2005;62(6):593-602.

[154] Marrie RA, Fisk JD, Bo Nancy Y, Stella L, Elliott L, Caetano P, et al. Mental comorbidity and multiple sclerosis: validating administrative data to support population-based surveillance. BMC Neurology. 2013;13(1):1-8.

[155] Kessler RC, Lane MC, Shahly V, Stang PE. Accounting for comorbidity in assessing the burden of epilepsy among US adults: results from the National Comorbidity Survey Replication (NCS-R). Molecular Psychiatry. 2012;17(7):748-58.

[156] Mula M. Epilepsy-induced behavioral changes during the ictal phase. Epilepsy & Behavior. 2013.

[157] Gandy M, Sharpe L, Perry KN, Miller L, Thayer Z, Boserio J, et al. The psychosocial correlates of depressive disorders and suicide risk in people with epilepsy. Journal of Psychosomatic Research. 2013;74(3):227-32.

[158] Gilliam F, Hecimovic H, Sheline Y. Psychiatric comorbidity, health, and function in epilepsy. Epilepsy & Behavior. 2003;4 Suppl 4:S26-S30.

[159] de Araújo Filho GM, Yacubian EMT. Juvenile myoclonic epilepsy: psychiatric comorbidity and impact on outcome. Epilepsy & Behavior. 2013;28(Suppl 1):S74-S80.

[160] Lund M, Reintoft H, Simonsen N. [A controlled sociological and psychological study on patients with juvenile myoclonus epilepsy]. Der Nervenarzt. 1976;47(12):708-12.

[161] Tsuboi T. Primary generalized epilepsy with sporadic myoclonias of myoclonic petit mal type. Stuttgart: Thieme Medical Publishers; 1977.

[162] Devinsky O, Najjar S. Evidence against the existence of a temporal lobe epilepsy personality syndrome. Neurology. 1999;53(5, Suppl 2):S13-S25.

[163] First MB, Spitzer RL, Gibbon M, Williams JBW. Structured Clinical Interview for DSM-IV-TR Axis I Disorders, Research Version, Patient Edition. (SCID-I/P). New York: Biometrics Research, New York State Psychiatric Institute; 2002.

[164] Trinka E, Kienpointner G, Unterberger I, Luef G, Bauer G, Doering LB, et al. Psychiatric comorbidity in juvenile myoclonic epilepsy. Epilepsia. 2006;47(12):2086-91.

[165] de Araujo Filho G, Mazetto L, da Silva JM, Caboclo LO, Yacubian EM. Psychiatric comorbidity in patients with two prototypes of focal versus generalized epilepsy syndromes. Seizure. 2011;20(5):383-6.

[166] de Araújo Filho GM, Pascalicchio TF, Sousa PdS, Lin K, Ferreira Guilhoto LMF, Yacubian EMT. Psychiatric disorders in juvenile myoclonic epilepsy: a controlled study of 100 patients. Epilepsy & Behavior. 2007;10(3):437-41.

[167] de Araujo Filho GM, Pascalicchio TF, Lin K, Sousa PS, Yacubian EMT. Neuropsychiatric profiles of patients with juvenile myoclonic epilepsy treated with valproate or topiramate. Epilepsy & Behavior. 2006;8(3):606-9.

[168] de Araújo Filho GM, Rosa VP, Lin K, Caboclo LOSF, Sakamoto AC, Yacubian EMT. Psychiatric comorbidity in epilepsy: a study comparing patients with mesial temporal sclerosis and juvenile myoclonic epilepsy. Epilepsy & Behavior. 2008;13(1): 196-201.

[169] Gélisse P, Genton P, Thomas P, Rey M, Samuelian JC, Dravet C. Clinical factors of drug resistance in juvenile myoclonic epilepsy. Journal of Neurology, Neurosurgery & Psychiatry. 2001;70(2):240-3.

[170] Salas-Puig J, Camara da Silva A, Davet C, J R. L'épilepsie myoclonique juvenile dans la population du centre. Saint-Paul Epilepsies 1990;2:108–13.

[171] Moschetta S, Fiore LA, Fuentes D, Gois J, Valente KD. Personality traits in patients with juvenile myoclonic epilepsy. Epilepsy & Behavior. 2011;21(4):473-7.

[172] Plattner B, Pahs G, Kindler J, Williams RP, Hall RE, Mayer H, et al. Juvenile myoclonic epilepsy: a benign disorder? Personality traits and psychiatric symptoms. Epilepsy & Behavior. 2007;10(4):560-4.

[173] Akanuma N, Hara E, Adachi N, Hara K, Koutroumanidis M. Psychiatric comorbidity in adult patients with idiopathic generalized epilepsy. Epilepsy & Behavior. 2008;13(1):248-51.

[174] Perini GI, Tosin C, Carraro C, Bernasconi G, Canevini MP, Canger R, et al. Interictal mood and personality disorders in temporal lobe epilepsy and juvenile myoclonic epilepsy. Journal of Neurology, Neurosurgery & Psychiatry. 1996;61(6):601-5.

[175] de Araújo Filho GM, da Silva JM, Mazetto L, Marchetti RL, Yacubian EMT. Psychoses of epilepsy: A study comparing the clinical features of patients with focal versus generalized epilepsies. Epilepsy & Behavior. 2011;20(4):655-8.

[176] Pung T, Schmitz B. Circadian rhythm and personality profile in juvenile myoclonic epilepsy. Epilepsia. 2006;47:111-4.

[177] Karachristianou S, Katsarou Z, Bostantjopoulou S, Economou A, Garyfallos G, Delinikopoulou E. Personality profile of patients with juvenile myoclonic epilepsy. Epilepsy & Behavior. 2008;13(4):654-7.

[178] Butcher JN, Dahlstrom WG, Graham JR, Tellegen A, Kaemmer B. Manual for the restandardized Minnesota Multiphasic Personality Inventory: MMPI—2. An administrative and interpretive guide. Minneapolis: University of Minnesota Press; 1989.

[179] de Araújo Filho GM, Jackowski AP, Lin K, Guaranha MSB, Guilhoto LMFF, da Silva HH, et al. Personality traits related to juvenile myoclonic epilepsy: MRI reveals pre-

frontal abnormalities through a voxel-based morphometry study. Epilepsy & Behavior. 2009;15(2):202-7.

[180] de Araújo Filho GM, Lin K, Lin J, Peruchi MM, Caboclo LOSF, Guaranha MSB, et al. Are personality traits of juvenile myoclonic epilepsy related to frontal lobe dysfunctions? A proton MRS study. Epilepsia. 2009;50(5):1201-9.

[181] de Araujo Filho GM, de Araujo TB, Sato JR, Silva Id, Lin K, Júnior HC, et al. Personality traits in juvenile myoclonic epilepsy: evidence of cortical abnormalities from a surface morphometry study. Epilepsy & Behavior. 2013;27(2):385-92.

[182] Moschetta S, Valente KD. Impulsivity and seizure frequency, but not cognitive deficits, impact social adjustment in patients with juvenile myoclonic epilepsy. Epilepsia. 2013;54(5):866-70.

[183] Staufenbiel SM, Penninx BWJH, Spijker AT, Elzinga BM, van Rossum EFC. Hair cortisol, stress exposure, and mental health in humans: A systematic review. Psychoneuroendocrinology. 2013;38(8):1220-35.

[184] Vargas C, López-Jaramillo C, Vieta E. A systematic literature review of resting state network--functional MRI in bipolar disorder. Journal of Affective Disorders. 2013;150(3):727-35.

[185] Busch RM, Frazier T, Chapin JS, Hamrahian AH, Diehl B, Alexopoulos A, et al. Role of cortisol in mood and memory in patients with intractable temporal lobe epilepsy. Neurology. 2012;78(14):1064-8.

[186] Cossette P, Liu L, Brisebois K, Dong H, Lortie A, Vanasse M, et al. Mutation of GABRA1 in an autosomal dominant form of juvenile myoclonic epilepsy. Nature Genetics. 2002;31(2):184.

[187] de Nijs L, Wolkoff N, Coumans B, Delgado-Escueta AV, Grisar T, Lakaye B. Mutations of EFHC1, linked to juvenile myoclonic epilepsy, disrupt radial and tangential migrations during brain development. Human Molecular Genetics. 2012;21(23): 5106-17.

[188] de Nijs L, Wolkoff N, Grisar T, Lakaye B. Juvenile myoclonic epilepsy as a possible neurodevelopmental disease: Role of EFHC1 or Myoclonin1. Epilepsy & Behavior. 2013;28(s1):S58-S60.

[189] Delgado-Escueta AV. Advances in Genetics of Juvenile Myoclonic Epilepsies. Epilepsy Currents. 2007;7(3):61-7.

[190] de Nijs L, Léon C, Nguyen L, LoTurco JJ, Delgado-Escueta AV, Grisar T, et al. EFHC1 interacts with microtubules to regulate cell division and cortical development. Nature Neuroscience. 2009;12(10):1266-74.

[191] Meencke HJ, Janz D. The significance of microdysgenesia in primary generalized epilepsy: an answer to the considerations of Lyon and Gastaut. Epilepsia. 1985;26(4): 368-71.

[192] Veith G, Wicke R. Cerebrale Differenzierungsstorungen bei Epilepsie. Westdtsch Ver. 1968:515-34.

[193] Lyon G, Gastaut H. Considerations on the significance attributed to unusual cerebral histological findings recently described in eight patients with primary generalized epilepsy. Epilepsia. 1985;26:365-7.

[194] Opeskin K, Kalnins RM, Halliday G, Cartwright H, Berkovic SF. Idiopathic generalized epilepsy: lack of significant microdysgenesis. Neurology. 2000;55(8):1101-6.

[195] Woermann FG, Free SL, Koepp MJ, Sisodiya SM, Duncan JS. Abnormal cerebral structure in juvenile myoclonic epilepsy demonstrated with voxel-based analysis of MRI. Brain: A Journal of Neurology. 1999;122 (Pt 11):2101-8.

[196] Lin K, Jackowski AP, Carrete H, Jr., de Araújo Filho GM, Silva HH, Guaranha MSB, et al. Voxel-based morphometry evaluation of patients with photosensitive juvenile myoclonic epilepsy. Epilepsy Research. 2009;86(2-3):138-45.

[197] Ronan L, Alhusaini S, Scanlon C, Doherty CP, Delanty N, Fitzsimons M. Widespread cortical morphologic changes in juvenile myoclonic epilepsy: evidence from structural MRI. Epilepsia. 2012;53(4):651-8.

[198] Vollmar C, O'Muircheartaigh J, Symms MR, Barker GJ, Thompson P, Kumari V, et al. Altered microstructural connectivity in juvenile myoclonic epilepsy: The missing link. Neurology. 2012;78(20):1555-9.

[199] Kim JH, Im KC, Kim JS, Lee S-A, Kang JK. Correlation of interictal spike-wave with thalamic glucose metabolism in juvenile myoclonic epilepsy. Neuroreport. 2005;16(11):1151-5.

[200] Savic I, Lekvall A, Greitz D, Helms G. MR spectroscopy shows reduced frontal lobe concentrations of N-acetyl aspartate in patients with juvenile myoclonic epilepsy. Epilepsia. 2000;41(3):290-6.

[201] Simister RJ, McLean MA, Barker GJ, Duncan JS. Proton MRS reveals frontal lobe metabolite abnormalities in idiopathic generalized epilepsy. Neurology. 2003;61(7): 897-902.

[202] Moeller F, Maneshi M, Pittau F, Gholipour T, Bellec P, Dubeau F, et al. Functional connectivity in patients with idiopathic generalized epilepsy. Epilepsia. 2011;52(3): 515-22.

[203] McGill ML, Devinsky O, Kelly C, Milham M, Castellanos FX, Quinn BT, et al. Default mode network abnormalities in idiopathic generalized epilepsy. Epilepsy & Behavior. 2012;23(3):353-9.

[204] Macdonell RA. Prolonged cortical silent period after transcranial magnetic stimulation in generalized epilepsy. Neurology. 2001;57:706-8.

[205] Badawy RA, Curatolo JM, Newton M, Berkovic SF, Macdonell RA. Changes in cortical excitability differentiate generalized and focal epilepsy. Annals of Neurology. 2007;61(4):324-31.

[206] Groppa S, Siebner HR, Kurth C, Stephani U, Siniatchkin M. Abnormal response of motor cortex to photic stimulation in idiopathic generalized epilepsy. Epilepsia. 2008;49(12):2022-9.

[207] Brigo F, Storti M, Benedetti MD, Rossini F, Nardone R, Tezzon F, et al. Resting motor threshold in idiopathic generalized epilepsies: a systematic review with meta-analysis. Epilepsy Research. 2012;101(1-2):3-13.

[208] Brigo F, Igwe SC. Corticospinal excitability in juvenile myoclonic epilepsy: The story so far. Seizure. 2013.

[209] Killam EK. Photomyoclonic seizures in the baboon, Papio papio. Federation Proceedings. 1979;38(10):2429-33.

[210] Meldrum BS, Anlezark G, Balzamo E, Horton RW, Trimble M. Photically induced epilepsy in Papio papio as a model for drug studies. Advances In Neurology. 1975;10:119-32.

[211] Menini C, Stutzmann JM, Laurent H, Naquet R. Paroxysmal visual evoked potentials (PVEPs) in Papio papio. I. Morphological and topographical characteristics. Comparison with paroxysmal discharges. Electroencephalography & Clinical Neurophysiology. 1980;50(5-6):356-64.

[212] Silva-Barrat C, Ménini C, Bryère P, Naquet R. Multiunitary activity analysis of cortical and subcortical structures in paroxysmal discharges and grand mal seizures in photosensitive baboons. Electroencephalography & Clinical Neurophysiology. 1986;64(5):455-68.

[213] Silva-Barrat C, Ménini C. The influence of intermittent light stimulation on potentials evoked by single flashes in photosensitive and non-photosensitive Papio papio. Electroencephalography & Clinical Neurophysiology. 1984;57(5):448-61.

[214] Menini C, Silva-Barrat C. The photosensitive epilepsy of the baboon. A model of generalized reflex epilepsy. Advances In Neurology. 1998;75:29-47.

[215] Fukuda H, Valin A, Bryère P, Riche D, Wada JA, Naquet R. Role of the forebrain commissure and hemispheric independence in photosensitive response of epileptic baboon, Papio papio. Electroencephalography & Clinical Neurophysiology. 1988;69(4):363-70.

[216] Szabó CÁ, Knape KD, Leland MM, Williams JT. Electroclinical phenotypes in a pedigreed baboon colony. Epilepsy Research. 2013;105(1-2):77-85.

[217] Szabó CÁ, Knape KD, Leland MM, Cwikla DJ, Williams-Blangero S, Williams JT. Epidemiology and characterization of seizures in a pedigreed baboon colony. Comparative Medicine. 2012;62(6):535-8.

[218] Szabó CA, Leland MM, Knape K, Elliott JJ, Haines V, Williams JT. Clinical and EEG phenotypes of epilepsy in the baboon (Papio hamadryas spp.). Epilepsy Research. 2005;65(1-2):71-80.

[219] Szabó CÁ, Salinas FS, Narayana S. Functional PET Evaluation of the Photosensitive Baboon. The Open Neuroimaging Journal. 2011;5:206-15.

[220] Szabó CÁ, Narayana S, Kochunov PV, Franklin C, Knape K, Davis MD, et al. PET imaging in the photosensitive baboon: Case-controlled study. Epilepsia. 2007;48(2): 245-53.

[221] Szabó CA, Kochunov P, Knape KD, McCoy KJM, Leland MM, Lancaster JL, et al. Cortical sulcal areas in baboons (Papio hamadryas spp.) with generalized interictal epileptic discharges on scalp EEG. Epilepsy Research. 2011;93(2/3):91-5.

[222] Szabó CÁ, Narayana S, Franklin C, Knape KD, Davis MD, Fox PT, et al. "Resting" CBF in the epileptic baboon: Correlation with ketamine dose and interictal epileptic discharges. Epilepsy Research. 2008;82(1):59-65.

[223] Salinas FS, Fox PT, Szabó CA. Using resting-state fMRI to demonstrate functional connectivity differences between epileptic and control non-human primates. Presentation at: NC3R Primate Welfare Meeting; 2013; London, UK.

[224] Szabó CÁ, Salinas FS, Leland MM, Caron J-L, Hanes MA, Knape KD, et al. Baboon model of generalized epilepsy: Continuous intracranial video-EEG monitoring with subdural electrodes. Epilepsy Research. 2012;101(1/2):46-55.

[225] Ticku MK, Lee JC, Murk S, Mhatre MC, Story JL, Kagan-Hallet K, et al. Inhibitory and excitatory amino acid receptors, c-fos expression, and calcium-binding proteins in the brain of baboons (Papio hamadryas) that exhibit 'spontaneous' grand mal epilepsy. Epilepsy Research Supplement. 1992;9:141-9.

[226] Phelix CF, Szabó CÁ. Quantification of heterotopic neurons in baboons with generalized interictal epileptic discharges. Poster session presented at: American Epilepsy Society 66th Annual Meeting; 2012; San Diego, CA.

[227] Young NA, Szabó CÁ, Phelix CF, Flaherty DK, Balaram P, Foust-Yeoman KB, et al. Epileptic baboons have lower numbers of neurons in specific areas of cortex. Proceedings of the National Academy of Sciences of the United States of America. 2013.

[228] Martínez-Juárez IE, Alonso ME, Medina MT, Durón RM, Bailey JN, López-Ruiz M, et al. Juvenile myoclonic epilepsy subsyndromes: family studies and long-term follow-up. Brain: A Journal of Neurology. 2006;129(Pt 5):1269-80.

# Chapter 6

# Reflex Epilepsy

Raidah Saleem AlBaradie

Additional information is available at the end of the chapter

http://dx.doi.org/10.5772/58462

## 1. Introduction

A reflex seizure is a condition in which seizures can be provoked habitually by an external stimulus or, less commonly, internal mental processes, or by activity of the patient. It is most commonly precipitated by visual stimuli. Other somatosensory occurrences, including thinking, reading, listening to music, and eating may also induce reflex seizures [1]. Reflex epilepsies are quite uncommon, occurring in only 5% of all epilepsies. Most of these epilepsies are genetic in origin [2].

The definition of reflex epilepsy was recognized initially by on International League Against Epilepsy classification in 1989 [3]. The classification in 2001 formed reflex seizure and epilepsy definitions. Three types of reflex seizure met clinically embrace pure reflex epilepsy, reflex seizures that happen in generalized or focal epilepsy syndromes that are also connected with spontaneous seizures, and isolated reflex seizures arising in conditions that do not essentially need a diagnosis of epilepsy.

**Pathophysiology**

The occurrence of seizures in people with epilepsy is rarely predictable. Elements that aggravate seizures may differ from person to person and may include sleep deprivation, systemic illness, or ingestion of particular food products [2]. These factors typically do not activate seizures in a consistent pattern and they may lower seizure threshold in patients with unprovoked seizures. In comparison, reflex seizures denote a time-dependent response to a particular stimulus.

An example of a trigger which more reliably causes interictal epileptic discharges (IEDs) or clinical seizures in photosensitive individuals is intermittent light stimulation (ILS). 12-to 18-Hz frequencies in photic stimulation are more likely to produce seizures than others, and the degree of photosensitivity may depend on the time of day, where it is increased early in the morning [4].

Reflex seizures are also provoked by the stimulation of other primary sensory cortices, such as the primary auditory, or somatosensory cortices, and by activation of premotor, pericingulate (SMA), and parietal lobe association cortices. It is proposed that the stimulus may create an abnormal response directly in the sensory or association cortices, with a synchronized discharge spreading functionally connected cortical or subcortical structures or a physiological reaction is in charge for the initiation of synchronization of larger networks or functionally connected epileptogenic cortex. In addition, forming of real connectivity during the photoparoxysmal response indicated the frontocentral cortices were already synchronized prior to the appearance of the ictal or interictal discharge [5].

Some studies showed diminished inhibition of the motor cortex during ILS which was seen by transcranial magnetic stimulation in photosensitive individuals. This decrease seems to denote synchronization of the frontocentral cortices or the absence of an inhibitory modulation by an intervening cortical region. In addition, it was noted in patients who were studied for resective surgery that a temporal lobe epileptogenic zone was activated by ILS without clear generation of the seizure from the occipital lobe. This support that there are mechanisms other than spread being crucial to induce "reflex seizures [6]. This was also supported by animal models studies, where it demonstrated an abnormal cortical development in the baboon in the histology and the morphology with reflex seizures [7].

**Etiology**

The cause of reflex epilepsy can be acquired or inherited. Reflex seizures rarely caused by acquired cerebral lesions. The most common etiologies which affect the brain frontal, temporal or parietal association cortices are strokes, encephalitis, or cortical dysplasia. In comparison to trauma and meningitis, which incline to affect more the anterior frontal and temporal lobes. The most common signs of acquired reflex epilepsy are startle seizures, which is presented clinically by sudden myoclonic or tonic contraction of the truncal and extremity musculature. This type of seizures can be triggered by somatosensory, auditory, or proprioceptive (movement) stimuli. It is correlated by a brief electroencephalographic discharge, which might be followed by a partial seizure. The perirolandic cortices and mesial frontoparietal networks are commonly implicated in the generation of reflex seizures [8].

Whereas the inherited reflex epilepsy is supported by animal models as stated earlier such as audio genic seizures characterize genetic reflex epilepsies in predisposed strains of mice, rats, and birds [9]. Recent studies demonstrated a link of photosensitivity to bands 7q32 and 16p13 by one group [10] and was linked to 6p21 and 13q31 by another [11]. Also, children with chromosomal abnormalities have been shown to have augmented affinity to photosensitivity. Hot water epilepsy (HWE) was linked to band 4q24-q28 in one family and to band 10q21-q22 in 6 families [12]. Patients with autosomal dominant temporal lobe epilepsy have seizures triggered by speech and auditory stimuli which is associated with mutations in the LGI1 gene in chromosome 10q22-q24 [13]. Several genetic factors predispose to reflex epilepsies, some related to channelopathies, other affecting brain development such as neurodegenerative disorders. They are stated to as progressive myoclonic epilepsies which includ Unverricht-Lundborg disease, Lafora disease, neuronal ceroid lipofuscinosis, and mitochondrial ence-

http://dx.doi.org/10.5772/58462

phalomyopathies (mitochondrial disorders complicated by cognitive decline and progressive weakness) [14].

**Epidemiology**

Females have more common photosensitivity but there is no sex predilection in reflex epilepsies [1].

## 2. Precipitating factors

### 2.1. Visual stimulus

Seizures can be provoked by visual stimulus such as flickering light, removal of visual fixation or light intensity, complex visual patterns, viewing particular objects, or other visual stimuli [15]. The most common type of visually induced seizure is photosensitive seizures. Photosensitivity is an abnormal visual sensitivity of the brain in response to flickering light sources. It is expressed in the electroencephalography (EEG) as a generalized spikes/polyspikes and wave discharge (photoparoxysmal response) produced by intermittent photic stimulation, or clinical seizures in vulnerable individuals [16]. The prevalence of photosensitivity in patients with epilepsy ranges from 2% to 20%. There are three groups of Photosensitivite subjects divided based on their response to (ILS) and it is more commonly associated with idiopathic generalized epilepsy, which constitutes 20-40% of all epilepsy:

1. Individuals who develop seizures only when they are exposed to light stimulus.
2. Individuals who experience seizures with or without light stimulus.
3. Asymptomatic individuals with a photosensitive reaction on EEG.

Photosensitivity is nearly twice as common in females as in males. 25% of patients lose their photosensitivity in their 20s and 30s. Genetic tendency play an important role in photosensitivity. Regional occipital cortical hyperexcitability is noted in functional magnetic resonance imaging (MRI) and magnetoencephalography in photosensitive patients [17]. It is proposed that there is hyperexcitability of the visual cortex in photosensitive patients as noted from human and animal data. When a sufficient large area from the visual cortex is stimulated, it will lead to an epileptiform discharge and a seizure might be provoked by mechanism requires the physiologic activation of a critical area of cortical tissue especially the parvocellar more than mangocellar pathway.The most common light source that plays a role in photosensitive seizures is television more than computer monitors and video games. The reflection of sunlight directly or intermittently on a road lined with trees, lamplights, and colorful and bright blinking lights are other stimuli in photosensitive subjects [17].

The treatment of photosensitive epilepsy can be achieved with or without combined antiepileptic drugs by avoiding the stimulus, stimulus modification such as avoidance of clear sources of blinking lights and video games, avoiding extended game play, increase distance from the television set, and using a remote control are all important and useful strategies. At times, covering one eye and rotating if the screen flashes or if myoclonic jerks occur is useful. Use of 100 Hz television sets found to diminish sensitivity in many patients: the screen is naturally less provocative than a 50 Hz screen however screen content may still be stimulating. When needed, the drug of choice is valproate in monotherapy. Antiepileptic drugs that can be used in this condition such as clobazam, lamotrigine, topiramate, and levetiracetam might be helpful [18].

Photosensitive epilepsies usually carry very good prognosis, about 25% of patients with these conditions will lose their photosensitivity in their third decade. Most such patients will relapse if they discontinue the antiepileptic drugs early [19]. It worth to state that photosensitivity can be seen in idiopathic generalized epilepsies such as juvenile myoclonic epilepsy, and in crytogenic generalized epilepsies such as severe myoclonic epilepsy of infancy (Dravet syndrome), or with degenerative gray matter encephalopathies such as Lafora's disease, Unverricht-Lundborg disease, Kufs' disease, the neuronal ceroid lipofuscinoses, and in others progressive myoclonus epilepsies [19].

### 2.2. Somatosensory stimulus

It may include light touch, tapping, or immersion in hot water. The seizures can be provoked by touch may occur in infancy or childhood is called startle epilepsy or reflex myoclonic epilepsy [20]. The type of seizures inclines to be generalized and less commonly, partial-onset seizures which are activated by touch due to the activation of a sensorimotor cortex.

An important example of somatosensory stimulus induced seizure is hot water epilepsy (HWE) where seizures can be triggered by bathing with hot water pouring over the head, face, or neck. It was first described in 1945. It is the second most common type of reflex epilepsy after photosensitive epilepsy and it is considered to be rare. The seizure is induced when the individual is exposed to water warmer than 37 °C. HWE constitute 3.6-6.9% of all epilepsy cases. HWE occur mostly in infants and children and a male/female ratio of 2-3/1. But also it can be seen after 40 years of age. Familial HWE cases with more than one affected member have been reported in 7-15% of Indian patients, and 1-27% of these patients reported a history of febrile convulsions [21].

One of the important causes of HWE is the genetic etiology. Genome-wide linkage analysis of Indian families delivered proof of linkage for the disorder at 10q21.3-q22.3 and recognized a 15 Mb disease-associated haplotype in four out of six families analyzed [21]. It was noted in a study proven hyperthermic kindling in rats after several episodes of hot water stimulations, there was progressive epileptic activity displayed during lowering of rectal temperature thresholds from 41.5 to 40.0 ºC, drop in latency for developing seizures from 185 to 118 sec and increase in duration of hippocampal seizure discharge from 15 to 140 sec with gradual increase in difficulty of EEG after discharges and neuronal sprouting observed in supragranular molecular layer and in stratum lacunosum [22].

The EEG is usually normal interictally and in 15-20% might reveal diffuse abnormalities such as lateralized or localized spike discharges in the anterior temporal regions. There are technical limitations and difficulties in obtaining an ictal EEG records but a temporal lobe onset was

http://dx.doi.org/10.5772/58462

seen in few seizures. A normal computed tomography and MRI of the brain are seen in patients with HWE but in few reported cases a focal malformation of the left parietal cortex was detected with brain MRI. The underlying mechanism of HWE remains uncertain. It was suggested that repeated pouring of water (a kindling effect) and the temperature of the water (a facilitative or triggering factor) during bathing in genetically or anatomically vulnerable persons play a role in the pathophysiology and this starting stimulus is considered to be complex. It may include a mixture of factors, such as contact of scalp with hot water, the temperature of the water, and the specific cortical area of stimulation [23].

The management of HWE is to decrease the temperature of the water and altering the method of bathing. These may be enough to control the seizures. Adding an antiepileptic drug can be considered if the mentioned precautions fail. The used antiepileptic drugs include carbamazepine, phenytoin, phenobarbital, sodium valproate, oxcarbazepine, lamotrigine, clobazam, or levetiracetam. A self-abort the attack by distracting maneuvers, like listening to music, chanting the name of God, or remembering their dear ones might be used [24]. HWE usually carry good prognosis.

### 2.3. Auditory stimulation

Audiogenic seizures have been noted in many animal types and happen usually in employed mouse and rodent models of genetically determined epilepsy. Auditory stimuli are less common precipitants of reflex seizures where sounds may produce seizures in cases of startle epilepsy [2].

Musicogenic epilepsy (ME) is rare and it is considered to be reflex-evoked or sensory-evoked epilepsy. ME is a seizures induced by hearing certain sounds such as a specific type or piece of music. The seizure can be induced also while the subject is exposed to the musical trigger during sleep or merely thinking about it. In other patients, an affective component of the stimulus is obvious, in addition a nonmusical sounds, such as whirring machinery, may be actual causes in others. The prevalence is one case per 10,000,000 populations [26]. It can be underdiagnosed because the latency between stimulus and seizure onset has been found up to several minutes. It occurs more in males than females with the mean age of onset of the seizures is 14 years. This type of seizures was reported in infancy. The most common type of this seizure is a complex partial with secondary generalization with ictal EEG onset in the mesial temporal region. Cerebral single-photon emission computed tomography (SPECT) in a patient with musicogenic epilepsy demonstrated a right temporal focus. The Treatment can be achieved by the use of anti-epileptic drugs or surgical intervention for medically refractory ME with temporal lobe scars and glial temporal lesions [27].

### 2.4. Movement

Movement-induced reflex seizures such as nonketotic hyperglycemi are reflex seizures most likely to be seen by general neurologists, internists, or other medical specialists in the hospital setting which usually resolve with normalization of the metabolic disturbance. Postanoxic myoclonus (Lance-Adams) may also represent a movement-induced seizure in the medical patient population [28].

### 2.5. Complex mental processes

Some of the most unusual and intriguing disorders in neurology are the reflex epilepsies in which seizures are provoked by complex actions or mental processes. Examples of these triggers include reading, eating, micturition, tooth brushing, walking, answering the telephone, and thinking.

Reading epilepsy is an interesting syndrome which was first described in 1956 by Bickford et al. It is characterized by a feeling of movements in the jaw or throat while reading or myoclonic movements of the jaw which may lead to a generalized tonic-clonic seizure when reading remains. Jaw jerks are the most important mark of reading epilepsy but it can also manifest by an abrupt loss of consciousness, blank staring spells, paroxysmal alexia or dyslexia, and prolonged stuttering. In addition a language-related tasks other than reading, such as awkward talking, writing, difficult calculations, playing chess or card games, singing, and recitation can also induce seizures [29]

The underlying cause such as neuroanatomical and biochemical basis of reading epilepsy is not clear. The epileptogenic component of the reading process is different between patients such as eye movements, comprehension, emotional content, speech production, and proprioceptive feedback. A release of endogenous opioids during reading-induced partial seizures in areas of brain involved in normal reading has been found. This directed to the theory that there are networks of cortical areas parallel subserving both cognitive functions and epileptic activity [30]. This is evident by the data from a combined EEG/electromyography-functional MRI study which showed a network of cortical and subcortical areas that are in close proximity with functional areas relevant for language and motor functions has been shown to have significant blood oxygen level dependent changes time-locked with seizure activity [30].

It is important to differentiate between 'primary' or 'specific' reading epilepsy, with seizures only in relation to reading, from a 'secondary', non-specific variety, with seizures when reading. It can be divided into the subgroups idiopathic (primary) and less-frequently seen cryptogenic/symptomatic epilepsy. In primary reading epilepsy, only seizures produced by reading develop without spontaneous seizures.

It occurs more in males with the age of onset is in adolescence and young adulthood. A strong family history of seizures has been stated in 40-50% of patients. EEG is normal in 80% of patients interictally. A spike and wave discharges are seen in 11%, and temporal paroxysmal discharges in 5%. Ictally, 77% have epileptiform discharges consisting of short bursts of sharp waves, spikes, or spike and wave complexes that are bilateral and symmetrical in 32%, bilateral but asymmetrical in 38%, and unilateral or focal in 30%. The seizures are well controlled with valproate, clonazepam, or by modifying the stimulus. The prognosis is usually good [31]

Eating epilepsy rare and it is may cause a seizures that can be triggered by parts of anticipating food, eating itself, or the post-prandial period. It usually occurs in the second decade of life, with a male preponderance. The patients also may have spontaneous seizures. Eating epilepsy is considered symptomatic epilepsy related to localization. It is localized to temporolimbic, extralimbic, perirolandic, or suprasylvian. In the temporolimbic type, complex partial seizure develops towards the end of the meal. The extralimbic, perirolandic, and suprasylvian types are similar to simple reflex epilepsy and it has a very short latency. Clinically, they may have simple partial seizure, hemiparesis, and mental retardation. The ictal and interictal EEG findings support clinic seizures. The underlying lesion can be due to static encephalopathy, or progressive lesions, like a deep localized glioma. The seizure can be controlled by antiepileptic drug therapy such as clobazam and epilepsy surgery might be considered for refractory patients [32].

## 3. Conclusion

Various stimuli are important in aggravating reflex seizures. The physiological mechanism in reflex epilepsies is still not well defined. However, the cortical hyperexcitability is the most frequently vital factor. The hyperexcitability of different cortical areas may be due to a genetic tendency or to an acquired lesion. The diagnosis of is reflex epilepsy essential for a well understanding how brain works and management for patients. The study of these seizures will benefit the possible insights of somatomotor processing, language mechanisms and the physiology of ideation. Getting proper history and clinical data from the patient will help in better understanding and managing patients with reflex epilepsy. During the electrophysiological recording of the patient who is believed to have reflex seizures, by giving supposed stimuli, seizure type may be noted and the treatment should be controlled properly. Genetic counseling should be given in patients with strong family history of epilepsy.

## Author details

Raidah Saleem AlBaradie

Consultant Pediatric Neurologist/Epileptologist, King Fahd Specialist Hospital-Dammam, Saudi Arabia

## References

[1] Striano S, Coppola A, Del Gaudio L, Striano P. Reflex seizures and reflex epilepsies: old models for understanding mechanisms of epileptogenesis. Epilepsy Res. Jun 2012; 100(1-2):1-11.

[2] Kasteleijn-Nolst Trenité DG. Provoked and reflex seizures: surprising or common? Epilepsia. Sep 2012;53 Suppl 4:105-13.

[3] Commission on Classification and Terminology of the International League Against Epilepsy. Proposal for revised classification of epilepsies and epileptic syndromes. Epilepsia 1989;30(4):389-99.

[4] Kasteleijn-Nolst Trenite, DGA. Ictal SPECT in a case of pure musicogenic epilepsy. Photosensitivity in epilepsy: Electrophysiological and clinical correlates. Acta Neurol Scand Suppl. 1989;125:3-149.

[5] Varotto G, Visani E, Canafoglia L, Franceschetti S, Avanzini G, Panzica F. Enhanced frontocentral EEG connectivity in photosensitive generalized epilepsies: a partial directed coherence study. Epilepsia. Feb 2012;53(2):359-67. [Medline].

[6] Groppa S, Siebner HR, Kurth C, Stephani U, Siniatchkin M. Abnormal responses of motor cortex stimulation in idiopathic generalized epilepsy. Epilepsia. 2008;49:2022-2029.

[7] Young NA, Szabo CA, Phelix CF,Flaherty DK, Balaram P, Foust KB, et al. Regional cortical cell reduction in epileptic baboons. Submitted to PNAS.

[8] Garcia-Morales I, Maestu F, Perez-Jimenez MA, Elices E, Ortiz T, Alvarez-Linera J. A clinical and magnetoencephalography study of MRI-negative startle epilepsy. Epilepsy Behav. Sep 2009;16(1):166-71. [Medline].

[9] Batini C, Teillet MA, Naquet R. An avian model of genetic reflex epilepsy. Arch Ital Biol. May 2004;142(3):297-312. [Medline].

[10] Pinto D, Westland B, de Haan GJ, et al. Genome-wide linkage scan of epilepsy-related photoparoxysmal electroencephalographic response: evidence for linkage on chromosomes 7q32 and 16p13. Hum Mol Genet. Jan 1 2005;14(1):171-8. [Medline].

[11] Pinto D, Kasteleijn-Nolst Trenite DG, Cordell HJ, et al. Explorative two-locus linkage analysis suggests a multiplicative interaction between the 7q32 and 16p13 myoclonic seizures-related photosensitivity loci. Genet Epidemiol. Jan 2007;31(1):42-50. [Medline].

[12] Ratnapriya R, Satishchandra P, Dilip S, Gadre G, Anand A. Familial autosomal dominant reflex epilepsy triggered by hot water maps to 4q24-q28. Hum Genet. Nov 2009;126(5):677-83. [Medline].

[13] Brodtkorb E, Michler RP, Gu W, Steinlein OK. Speech-induced aphasic seizures in epilepsy caused by LGI1 mutation. Epilepsia. Jun 2005;46(6):963-6. [Medline].

[14] de Siqueira LF. Progressive myoclonic epilepsies: review of clinical, molecular and therapeutic aspects. J Neurol. Oct 2010;257(10):1612-9. [Medline].

[15] Chuang YC, Chang WN, Lin TK, Lu CH, Chen SD, Huang CR. Game-related seizures presenting with two types of clinical features. Seizure. Mar 2006;15(2):98-105. [Medline].

[16] Covanis A. Photosensitivity in idiopathic generalized epilepsies. Epilepsia 2005;46 Suppl 9:67-72.

[17] Wilkins AJ, Bonanni P, Porciatti V, Guerrini R. Physiology of human photosensitivity. Epilepsia 2004;45 Suppl 1:7-13.

[18] Zifkin BG, Kasteleijn-Nolst Trenité D. Reflex epilepsy and reflex seizures of the visual system: a clinical review. Epileptic Disord 2000;2(3):129-36.

[19] Lu Y, Waltz S, Stenzel K, Muhle H, Stephani U. Photosensitivity in epileptic syndromes of childhood and adolescence. Epileptic Disord 2008;10(2):136-43.

[20] Ricci S, Cusmai R, Fusco L, Vigevano F. Reflex myoclonic epilepsy in infancy: a new age-dependent idiopathic epileptic syndrome related to startle reaction. Epilepsia. Apr 1995;36(4):342-8. [Medline].

[21] Satishchandra P. Hot-water epilepsy. Epilepsia 2003;44 Suppl 1:29-32.

[22] Grosso S, Farnetani MA, Francione S, Galluzzi P, Vatti G, Cordelli DM, et al.. Hot water epilepsy and focal malformation of the parietal cortex development. Brain Dev 2004;26(7): 490-3.

[23] Ilbay G, Sahin D, Ates N. Changes in blood-brain barrier permeability during hot water-induced seizures in rats. Neurol Sci 2003;24(4):232-5.

[24] Savitha MR, Krishnamurthy B, Ashok DA, Ramachandra NB. Self abortion of attacks in patients with Hot Water Epilepsy. Indian Pediatr 2007;44(4):295-8.

[25] Mehta AD, Ettinger AB, Perrine K, Dhawan V, Patil A, Jain SK, et al. Seizure propagation in a patient with musicogenic epilepsy. Epilepsy Behav 2009;14(2):421-4.

[26] Critchley M. Musicogenic epilepsy. Brain 1937;60:13-27.

[27] [47] Tayah TF, Abou-Khalil B, Gilliam FG, Knowlton RC, Wushensky CA, Gallagher MJ. Musicogenic seizures can arise from multiple temporal lobe foci: intracranial EEG analyses of three patients. Epilepsia 2006;47(8):1402-6.

[28] Lance JW, Adams RD. The syndrome of intention or action myoclonus as a sequel to hypoxic encephalopathy. Brain. Mar 1963;86:111-36. [Medline].

[29] Bickford RG, Whean JL, Klass DW, Corbin KB. Reading epilepsy: clinical and electroencephalographic studies of a new syndrome. Trans Am Neurol Assoc. 1956;(81st Meeting):100-2. [Medline].

[30] Koepp MJ, Richardson MP, Brooks DJ, Duncan JS. Focal cortical release of endogenous opioids during reading-induced seizures. Lancet 1998;352(9132):952-5.

[31] Valenti MP, Tinuper P, Cerullo A, Carcangiu R, Marini C. Reading epilepsy in a patient with previous idiopathic focal epilepsy with centrotemporal spikes. Epileptic Disord. Sep 1999;1(3):167-71. [Medline].

[32] Senanayake N. 'Eating epilepsy'-a reappraisal. Epilepsy Res 1990;5(1):74-9.

Chapter 7

# Vagus Nerve Stimulation Therapy for Epilepsy

Reese S. Terry Jr

Additional information is available at the end of the chapter

http://dx.doi.org/10.5772/58332

## 1. Introduction

The year 2013 marked the 25th year of clinical vagus nerve stimulation (VNS). This chapter will review the preclinical history, the clinical study history, and the long-term outcomes of VNS.

Epilepsy has been treated by many strange remedies over the centuries. In 1883, Corning [1] (Corning, 1883) proposed that vagus nerve stimulation could decrease heart rate and cerebral blood flow, thereby controlling seizures. This early attempt at vagus stimulation for seizure control quickly fell out of favor.

Bailey and Bremer [2] (Bailey & Bremer, 1938) demonstrated the direct effect of vagus stimulation on the central nervous system. They found that repetitive electrical stimulation of the central end of the vagus nerve of the cat results in increased amplitude and frequency of the spontaneous potentials of the orbital surface of the frontal lobes of the cerebral cortex. Inhibition of motor activity by activation of visceral vagal afferents was first reported by Schweitzer and Wright [3] (Schweitzer A, 1937) and later confirmed by Paintal [4] (Paintal, 1973). Dell and Olsen [5] (Dell P, 1951) reported that vagus stimulation affected slow wave activity in awake cats.

These papers and others led Zabara [6] (Zabara, 1992) to investigate vagus stimulation as a potential method to treat epilepsy. Zabara stimulated the cervical vagus nerve in a strychnine dog model of status epilepsy (N=20). He reported that vagus stimulation would interrupt the strychnine-induced seizure, and the amount of seizure interruption was proportional to the length of stimulation, approximately 4 times as long as the stimulation period. Vagal stimulation terminated seizures within 0.5-5 s. Transection of the vagus distal to the stimulating electrode did not alter the antiseizure effects of vagus stimulation. The optimal stimulus parameters were estimated to be approximately 20 volts (with electrode resistance 1-5 ohms), 20-30 Hz stimulation frequency and approximately 0.2 msec pulse duration.

In 1990 Lockard [7] (Lockard JS, 1990) used an alumnia gel chronic epilepsy model in monkeys. She induced chronic epilepsy by placing alumina gel on the cortex. Previous studies showed seizure rates remained stable for at least 6 months with this model. The instrumentation triggered a 40-second burst whenever it automatically detected a seizure. The stimulation device also provided a stimulation burst once every 3 hours. The stimulation frequency was changed in 2-week intervals from 83 HZ to 143 HZ and then to a random frequency ranging from 50-250 HZ. The stimulation amplitude was increased to tolerance or 5 mA. In 1 animal, investigators reduced seizure rate to near 0, then stopped stimulation and allowed seizures to return to baseline and restarted stimulation, thereby reducing seizure rate to near 0 once more. Seizure frequency was reduced to zero in a second animal, and stability of frequency was affected in the 2 remaining animals. The 3 stimulation patterns seemed to affect seizure rate equally, but the data were not statistically significant. Although Zabara [6] was able to interrupt seizures in the strychnine dog studies, no seizures were interrupted in the Lockard alumnia gel monkey study. The delay between detection and activation was tens of seconds and may have decreased the seizure interruption effect by VNS. Heart rate and blood pressure were unaffected, and stomach ulcers were not noted. The Woodburys [8] (Woodbury, 1991) theorized that stimulation during the seizure disrupts the seizure pathways in the brain, possibly "unlearning" the seizure mechanisms. They hypothesized that stimulation absent a seizure would not have an effect on reducing future seizures, but such a hypothesis has not been tested.

## 2. Clinical studies

### 2.1. Pilot studies

The VNS pilot study (E01) included 10 patients and was based on the Zabara [6], Lockard [7] and Woodbury [8] animal studies. Because technology for seizure detection did not exist at that time, the device was programmed to stimulate periodically, and patients received a magnet to self-activate the stimulation when they experienced an aura. As might be expected, most patients did not have auras and so they did not self-activate stimulation during the seizure. Investigators hoped this periodic pattern would occasionally occur during a seizure and eventually reduce the seizure frequency. Initial stimulation parameters were typically 250 μsec pulse width, 60 sec ON, 60 minutes OFF and 50 HZ, although a stimulation frequency of 143 HZ was used on the first patient. This pilot study was a single-blind study consisting of four seizure measurement periods: pre-implant baseline, stimulation period, sham stimulation period and stimulation period. The mean percent seizure reduction was 24.3% ($p<0.049$) after 4 months of treatment. [9] (Uthman BM W. B., 1993)

The 4-patient E02 pilot study was similar to E01, except pulse width was 500 μsec, ON time 30 sec, OFF time 10 minutes, and frequency 30 HZ. The mean percent reduction after the fourth treatment period was 39.9% ($p <0.074$).

http://dx.doi.org/10.5772/58332

The mean percent seizure reduction after 14 to 35 months was 46.6% for the combined E01 and E02 patients.[9] (Uthman BM, 1993) Two patients, one who previously had 10 to 100 seizures per day before stimulation, had been seizure-free for over 1 year [10]. (Penry JK, 1990)

The stimulation parameters used in the E01 study were quite different from those in current use, and the first study patient had a remarkable response to the high frequency and long OFF-time patterns.

### 2.2. E03 and E05 pivotal studies

In the 2 randomized, blinded, active-control trials (E03 and E05), patients were randomly assigned to either of 2 treatment groups: HIGH (believed to be therapeutic) or LOW (adjusted to patient perception but programmed to 1 HZ delivered once every 90 to 180 minutes; parameters believed to be less therapeutic). Patients enrolled in the study were seen every 4 weeks during the baseline period (weeks-12 to 0). Patients meeting eligibility were implanted with the pulse generator and lead are shown in Table 1. Output was adjusted to tolerance at each visit for both groups Table 1 provides descriptions of all study patients.

| Description of Patients | | | | | | |
|---|---|---|---|---|---|---|
| | | Longitudinal | | Parallel | | |
| **Study** | **E01** | **E02** | **E04** | **E03** | **E05** | **Total** |
| No. of patients implanted | 11 | 5 | 124 | 115 | 199 | 454 |
| No. of patients stimulated | 10 | 5 | 123 | 115 | 198 | 451 |
| Age (range) | 32 (20–58) | 33 (18–42) | 24 (3–63) | 33 (13–57) | 33 (13–60) | 32 (3–63) |
| No. of females (%) | 4 (36%) | 2 (40%) | 57 (46%) | 43 (37%) | 104 (52%) | 210 (46%) |
| Years with epilepsy (range) | 22 (13-32) | 20 (5-36) | 17 (0.8-48) | 21 (4-47) | 23 (2-52) | 21 (0.8-52) |
| No. of AEDs (avg) | 1.0 | 1.0 | 2.2 | 2.1 | 2.1 | 2.1 |
| Median no. of seizures per day at baseline | 0.6 | 0.42 | 0.65 | 0.70 high/ 0.85 low | 0.58 high/ 0.51 low | - |

All patients implanted in all VNS clinical studies, N=454 [13]

**Table 1.** Description of Study Patients [13]

Two weeks after implantation, patients were randomized to the HIGH (frequency/duty cycle) or LOW (frequency/duty cycle) stimulation group, and the Pulse Generator was activated. Patients in the HIGH groups received a higher frequency, greater pulse width, and higher duty cycle of stimulation. The randomized treatment period that followed activation of the Pulse Generator lasted 14 weeks (the last 12 weeks of which were used in the efficacy analysis—the first 2 weeks were a treatment ramp-up period).

For the HIGH Group, a 30 Hz stimulation frequency was chosen based on the Woodbury rat studies [11] (Woodbury DM, 1990) and safety concerns by Agnew [12] (Agnew WF, 1990) that continuous stimulation at frequencies above 50 Hz might induce nerve damage, although the studies had shown that a 4-hour ON and 4-hour OFF at 50 HZ did not cause any damage. The E05 study used 20 Hz instead of 30 Hz. The 30 seconds ON and 5 minutes OFF was chosen as a compromise to extend battery life and achieve a 10% probability of stimulation during a seizure.

Results: The primary efficacy endpoint (percent reduction in seizure rate) was measured over 12 weeks as shown in Table 2. [13] (Cyberonics, 2013) Adverse events were assessed at each patient visit.

| **Principal Efficacy Results** | | | | | | |
|---|---|---|---|---|---|---|
| | **Longitudinal** | | | **Parallel** | | |
| **Study** | **E01** | **E02** | **E04** | **E03** | **E05** | **Total** |
| **No. of patients in efficacy analysis** | 10 | 5 | 116 | 114 | 196 | 441 |
| **Median reduction in seizures/day** | 32%* | 48% | 22%* | 23% high*/ 6% low | 23% high†/ 21% low† | - |
| **Mean reduction in seizures/day** | 24%‡ | 40% | 7%‡ | 24% high‡/ 6% low | 28% high†/ 15% low† | - |
| **Difference in mean (high/low)** | - | - | - | 17%§ (3%/31%) | 13%‖ (2%/23%) | - |
| **% with >50% response** | 30% | 50% | 29% | 30% high/ 14% low | 23% high/ 16% low | - |
| **Principal Safety Results Through Long-term Follow Up** | | | | | | |
| **Exposure (pt-yr)** | 45 | 20 | 245 | 456 | 135 | 901 |
| **SAEs¶ (high/low)** | 9%/ - | 0%/ - | 6%/ - | 5%/0% | 7%/9% | - |
| **Discontinued (LOE/AE)#** | 0/1 | 0/0 | 2/3 | 0/2 | 1/3 | 3/9 |
| **No. of explants**** | 2 | 2 | 15 | 9 | 5 | 33 |
| **Deaths: SUDEP/total††** | 0/0 | 0/0 | 3/4 | 0/3 | 1/2 | 4/9 |

All patients in efficacy analyses in all VNS clinical studies, N=441 [13]

*Within group broad analyses:*

* $P \leq 0.05$, by Wilcoxon sign rank.

† $P < 0.0001$, by anova.

‡ $P \leq 0.05$, by Student's *t-test.*

*Between group broad analyses:*

§ $P \leq 0.02$, by Wilcoxon rank sum; $P \leq 0.02$, by Student's *t-test.*

|| $P < 0.04$, by aligned ranks test; $P < 0.02$, by Student's *t-test*; $P < 3.03$, by anova.

*Safety information:*

¶ SAEs = serious adverse events.

# Discontinuing for lack of efficacy (LOE)/adverse events (AE) at one year, excluding deaths.

** Number of explants through August 1996, excluding deaths.

†† All deaths occurred by the long-term follow-up closing date of August 1996.

**Table 2.** Principal Efficacy and Safety Results [13]

# 3. Regulatory approvals

More than 70 countries have approved vagus nerve stimulation (VNS) for refractory epilepsy. Table 3 lists several countries and regions. Caregivers should be familiar with the regulatory approval status in their own countries when reading this chapter. Reimbursement levels also vary by country.

The VNS therapy was also approved for depression, but is not being commercially marketed for this indication because of lack of reimbursement.

| EPILEPSY | EU | USA | CANADA | CHINA | JAPAN |
|---|---|---|---|---|---|
| Year Approved | 1994 | 1997 | 1998 | 2008 | 2010 |
| Indicated age | All | 12 and older | All | All | All |
| Indicated Seizure Type | All | Partial | All | All | All |
| DEPRESSION | | | | | |
| Year Approved | 2001 | 2005 | 2001 | N/A | N/A |

**Table 3.** VNS Regulatory Approval Dates

## 3.1. Approved stimulation parameters

Parameter ranges currently available are provided in Table 4. The originally approved parameters included frequencies up to 143 HZ and output currents up to 12 mA, but these were later revised to those in Table 4 [13] (Cyberonics, 2013).

| Stimulation Parameters | Available Parameter Settings |
|---|---|
| Output current | 0-3.5 mA in 0.25-mA steps* ±0.25 ≤1 mA, ±10% > 1 mA |
| Signal frequency | 1, 2, 5, 10, 15, 20, 25, 30 Hz, ±6% |
| Pulse width | 130, 250, 500, 750, 1000 sec ±10% |
| Signal ON time | 7, 14, 21, 30, 60 sec ±15% or + 7 sec, whichever is greater (±15% or ±7 sec in Magnet Mode) |
| Signal OFF time | 0.2, 0.3, 0.5, 0.8, 1.1, 1.8, 3 min, and 5 to 180 min(5 to 60 in 5-min steps; 60 to 180 in 30-min steps), +4.4 / -8.4 sec |
| Magnet activation | Provided by magnet application (output current, pulse width, and signal ON time may be independently programmed for this purpose) |

*The acute phase results include seizure frequencies of the E03 Study LOW stimulation group, which included one-half the E03 patients, N=57.

Patients were permitted to change their AEDs during these long-term follow-up studies, and these changes may have contributed to the change in seizure frequency.

**Table 4.** Stimulation Parameter Ranges [13]

## 4. American academy of neurology--vagus nerve stimulation guideline

A Report of the Guideline Development Subcommittee of the American Academy of Neurology (AAN RGDS), Evidence-based guideline update, Vagus nerve stimulation for the treatment of epilepsy, was approved by the AAN Board of Directors in June 2013 and published ahead of print in August 2013 [14] (Morris 2013). This proposed guideline is referenced in several places of this chapter and is based on Class III studies. Class III studies are typically controlled trials in which outcomes are independently assessed by objective outcome measures, whereas Class I and II studies are typically prospective randomized controlled blinded trials.

Some questions from AAN Evidence Based Guideline for Vagus Nerve Stimulation are discussed here.

### 4.1. Does efficacy improve over time?

The AAN RGDS [14] analyzed 2 Class III studies [15-16], (Morris GL III, 1999) (Kuba R, 2009), that outlined VNS efficacy after stimulation for more than 6 months up to 12 years. Both reports discussed mostly adult refractory seizures patients. With time, the number of responders (subjects who had ≥50% seizure decrease) increased slightly. No controls were made for change in medications, so independent assessment of the VNS effect was impossible. A study [15] of 440 adult clinical trial patients with partial epilepsy reported the >50% seizure frequency reduction grew from 36.8% of patients at 1 year, to 43.2% at 2 years, and 42.7% at 3 years. From baseline, median seizure reductions after 1 year were 35%, after 2 years were 44.3, and after 3 years were 44%. The Kuba study evaluated 90 patients with multiple seizure types and ages 13-64 years. The responder rate (≥50% reduction in seizures) increased from 41% after 1 year, to 53.2% of 87 patients after 2 years, and to 48.9% of 85 patients after 5 years. These reductions applied to partial and generalized seizures, with reduction rates of 70% among patients with generalized tonic-clonic seizures.

The AAN RGDS concluded, based on data from these 2 Class III studies, that VNS is possibly associated with an increase in the number of patients achieving a ≥50% reduction in seizures.

### 4.2. Efficacy at 10 and 12 years

In addition to the AAN RDGS publication, two papers reviewed long term efficacy. Elliott [17] (Elliott RE, 2011) reported on a group of 65 consecutive patients with 10.4 years average follow-up. At last follow-up, average seizure frequency decrease was 76.3%. Seizure reduction increased from 35.7% after 6 months, to 65.7% after 6 years and 75.5% after 10 years of VNS, a significant reduction from baseline at each measured interval ($P<0.0001$). Toward the end of the follow-up, the data indicated a trend of increased AED burden. Seizure frequency was significantly reduced from baseline at each of the recorded intervals ($P<0.001$). Data showed a trend toward increased AED burden in the latter years of the follow-up period.

In another study of 48 patients with intractable partial epilepsy, Uthman [18] (Uthman BM R.A., 2004) discussed a seizure reduction after 1 year of 26% with increases to 30% at 5 years and 52% at 12 years.

### 4.3. Experience with young children and with other types of seizures

#### *4.3.1. Children*

Among children with epilepsy, is using adjunctive VNS to reduce seizure frequency better than not using adjunctive VNS?

After reviewing 16 studies, the AAN RGDS [14] concluded VNS is possibly effective to achieve ≥50% seizure frequency reduction (responder rate) among children. A study of 481 children found a responder rate of 55% (95% confidence interval [CI] 51%–59%), but the data were considered heterogeneous. Of the 16 studies, 2 were not included owing to a lack of responder rate or else had too many (>20%) adults. Calculations placed the pooled seizure freedom rate at 7% (95% CI 5%–10%).

#### *4.3.2. Lennox-Gastaut syndrome*

In patients with LGS, is using adjunctive VNS to reduce seizures better than not using it?

The AAN RGDS [14] included 4 Class III studies about LGS patients that focused on younger patients. They stated that VNS is possibly effective among patients with LGS. A pooled analysis showed a 55% responder rate (113 patients; 95% CI 46%–64%).

#### *4.3.3. Beyond AAN RGDS*

In addition to the AAN RGDS publication, reviews of VNS for generalized seizures, secondarily generalized seizures, and Rett Syndrome have been published.

#### *4.3.4. Generalized seizures*

Cyberonics conducted an E04 study of 114 patients to provide additional data on vagus nerve stimulation for the treatment of medically resistant seizures, including generalized seizures and children older than 2 years.

Data were available for 25 patients with generalized seizures. Median decrease in seizure frequency during stimulation was 46.6% ($p<0.01$), 44% of patients had at least a 50% reduction in seizure frequency compared with baseline. Half of these patients were aged <18 years. However, no correlation was found between response and age. The treatment response in this group was largely driven by effects on generalized tonic clonic seizures [19]. (Salinsky, 1997)

Helmers [20] (Helmers SL D. M., 2011) reported grand mal status events decreased post-VNS compared with pre-VNS (adjusted IRR=0.79, $P<0.001$) in an analysis of 1655 patients in a multistate Medicaid database (January 1997-June 2009). The generalized seizure rate was reduced by about 67% after 3 years.

*4.3.5. Secondary generalized seizures with Lennox Gastaut or Lennox Gastaut-like seizures*

Cukiert [21] (Cukiert A, 2013) reported on 47 seizure types among 24 children. After VNS, seizure frequency reduction ≥50% was noted in 35 seizure types and 17 seizure types disappeared after VNS. Atypical absence, myoclonic, and generalized tonic-clonic seizures were significantly reduced by VNS; tonic and atonic seizures did not improve. Transient seizure frequency worsening was noted in 10 of the 24 children, at a mean output current of 3.1 mA.

*4.3.6. Rett syndrome*

Wilfong [22] (Wilfong AA, 2006) reported on a series of 7 patients with Rett syndrome. At 12 months, 6 of the 7 female patients had more than 50% reduction in seizures.

### 4.4. Does VNS improve mood in epilepsy patients?

Does VNS have a beneficial mood side effect in the treatment of epilepsy? Because VNS was approved for treatment-resistant depression by the FDA in 2005 and had already obtained a CE Mark for this indication in 2001, it could have beneficial effects on the mood of epilepsy patients.

AAN RGDS [14] cites 2 Class III studies showing significant improvements in standard patient-reported mood assessment scales in adult patients with epilepsy. Test results before implantation were compared with those afterward. One study [23] (Elger G, 2000)] evaluated 11 subjects 1, 3, and 6 months after implantation. Before VNS, 7 of the 11 patients met criteria for "sub depressive mood" by the Montgomery-Åsberg Depression Rating Scale, and group mean was within the subdepressive mood range; after implantation, group mean was in the nondepressed range. Likewise, 8 of the 11 met criteria for "mild negative symptoms" by the Scale for the Assessment of Negative Symptoms before VNS. Scale and subscale scores improved at the study's 3-month follow-up ($p < 0.05$). Mood improvements were sustained at the 6-month follow-up (9 of 11 subjects).

The second study [15] (Morris GL III, 1999) evaluated 20 subjects 3 months after VNS implantation. Results for change in subject-rated scales by t tests showed improvements in the clinician-administered Cornell Dysthymia Rating Scale ($p=0.001$) and the patient self-report Beck Depression Inventory (BDI) ($p=0.045$); results on the clinician-administered Hamilton Depression Index (investigator rated) also significantly improved. The group's mean BDI score pre–VNS treatment was 12.0 ("mild mood disturbance"); this decreased to 9.4 ("non depressed") after VNS. Further, BDI scores decreased relative to those for an epilepsy control group (no therapy) studied over the same period (by repeated measures analysis of variance, $p=0.07$). This benefit was not correlated with reduced seizure frequency or with stimulation frequency or intensity.

The AAN RGDS [14] concluded VNS is possibly effective for mood improvement among adults with epilepsy.

http://dx.doi.org/10.5772/58332

## 5. Patient acceptance

Initiating VNS requires a surgical procedure to implant the generator and leads to begin the therapy. When the battery is depleted, surgery is required to replace the generator and replace the batteries. The leads are not usually disturbed during this procedure, but reattached to the new generator. Patient-perceived benefit has been hallmarked by the willingness of recipients to undergo surgical generator replacement across the history of the treatment. Morris [15] (Morris GL III, 1999) reported a 72% 3-year continuation rate among early users, and the November 11, 2013 Cyberonics Investor Conference Call placed it at 75% for the fully depleted Model 100 generators.

## 6. Battery life

The estimated battery life decreases as pulse width, frequency, output current, duty cycle, or lead impedance increases. For example, the Cyberonics Model 103 Manual [13] estimates battery life at more than 10 years for a stimulation frequency of 20 HZ, 1.5 mA output, 3 kohm impedance and 10% duty cycle for pulse widths ranging from 130 sec to 500 sec. Under the same conditions and a 50% duty cycle, the time to near end of battery life is 7.1 years, 6.2 years, and 3.7 years for pulse widths of 130 μsec, 250 μsec, and 500 μsec.

## 7. Cost effectiveness

The net cost of VNS determines medical cost reimbursement. Battery life drives cost savings after initial costs have been recouped.

Ben-Menachem [24-25] (Ben-Menachem E H. K., 2002) (Ben-Menachem E F. J., 2005) reports that implantation cost, when calculated over 8 years (battery life), is actually less than the cost of using a new AED over a similar time period. Real savings in hospital costs owing to seizures can also be expected. Average annual cost savings were approximately $3,000 per patient. This cost savings applied to all patients, whether or not they responded to VNS. These direct savings maintained over the battery life can equal or exceed the purchase price of the device.

In the Boon [26] (Boon P D. M., 2002) VNS group, mean seizure frequency decreased from 21 to 7 per month. Epilepsy-related direct medical costs (ERDMCs) in the VNS subgroup decreased from $4,826 to $2496 per year.

Bernstein [27] (Bernstein AL, 2007) compared average quarterly utilization rates of 138 patients for 12 months before device implantation with quarterly rates during 48 months of follow-up. Results during the 4th quarter of Year 4 revealed impressive decreases in utilization of all 4 measured aspects: a 91% decrease in outpatient visits, a 99% decrease in emergency department visits, a 67% decrease in hospital lengths of stay, and a 70% decrease in number of hospital admissions. Notably, no visits to the emergency department occurred during the final quarter of the study.

Helmers [20] (Helmers SL D. M., 2011) reported on 1655 patients in a pre-post analysis using multistate Medicaid data (January 1997-June 2009). After 1.5 years, she found VNS was associated with decreased resource utilization and epilepsy-related clinical events. The reduction in costs related to these hospital resources were equal to the initial costs related to the cost of purchasing and implanting the VNS System. Three years after the implant procedures the health care system had a net savings of $50,000 related to reduced hospitalizations and ER visits. Outpatient visits exceeded 50% of the total health care costs during the study. Table 5 gives the approximate net costs from Figure 2 in the Helmers paper. Using the trend line, net savings are projected for the remaining years of battery life.

| Quarter | Cumulative Investment Return | |
|---|---|---|
| 1 | -$19,000 | cost |
| 2 | -$17,000 | cost |
| 3 | -$12,000 | cost |
| 4 | -$7,000 | cost |
| 5 | -$1,000 | cost |
| 6 | +$4,000 | savings |
| 7 | +$12,000 | savings |
| 8 | +$19,000 | savings |
| 9 | +$27,000 | savings |
| 10 | +$34,000 | savings |
| 11 | +$42,000 | savings |
| 12 | +$50,000 | savings |
| Projected Savings | | |
| Year | | |
| 5 | +$100,000 | savings |
| 7 | +$150,000 | savings |
| 9 | +$200,000 | savings |

**Table 5.** Cumulative adjusted Net Healthcare Costs from Baseline

These papers indicate marked savings to health care systems throughout the world, even though seizures are controlled in only 10-15% of patients. Primary cost saving drivers are reduced ER visits, reduced hospital admissions, and reduced hospital days.

## 8. Stimulation parameter selection

Selection of the output parameters consists of 2 steps. The first is full activation so that signals are sent to the brain. The second is selection of ON time, OFF time, and stimulation frequency to obtain optimum efficacy.

The vagus nerve consists of larger myelinated A and B fibers and smaller unmyelinated C fibers. The C fibers make up about 80% of the 100,000 fibers in the vagus nerve. Of the nerves going into the brain, about 80% are afferent. The ratios of A, B, and C are assumed to be the same for afferent and efferent fibers.

The approximate currents required to stimulate the A, B, and C fibers in the Woodbury paper are given in Table 6. [11]. The diameter of the rat vagus nerve is about 0.4 mm vs 2.0 mm diameter for the human vagus nerve.

| Stimulation duration | A fiber | B fiber | C fiber |
|---|---|---|---|
| 125 μseconds | 10 μA | 50 μA | 325 μA |
| 250 μseconds | 5 μA | 25 μA | 200 μA |
| 500 μseconds | 5 μA | 10 μA | 125 μA |

**Table 6.** Stimulation Current vs Pulse Duration in Rat Vagus Nerve Model [11].

Smith [28] (Smith CD, 2001) discussed right and left cervical vagus nerve measurements of a 4-canine study using the Cyberonics 300 Series Bipolar Vagus Nerve Electrodes (Figure 1). Smith reported the average A fiber chronaxie values were 75.4 ± 24.5 μs and were 82.3 ± 23.3 μs for B fibers. The A fiber rheobase values were 0.63 ± 0.18 mA and 0.66 ± 0.22 mA for B fibers. The A fiber propagation velocities were 59.0 ± 9.6 m/s and 43.4 ± 8.0 m/s for B fibers. An effective electrode area of 5 $mm^2$ yields current densities of ±13.0 $mA/cm^2$ and 12.9 $mA/cm^2$ for A and B fibers, respectively.

Understanding how stimulation parameters affect seizure control started with the Woodbury [11] (Woodbury DM, 1990) rat models of current, pulse width, stimulation frequency, and time delay necessary to inhibit a seizure. Rat vagus nerves are much smaller than human vagus nerves, rendering it difficult to translate the current used in the rat studies to human vagus nerve stimulation. The stimulation threshold, a function of current density, is determined by ratio of the cross sectional areas of the nerves, which is the ratio of the square of the diameters. (0.4 mm vs 2 mm). Therefore, equivalent stimulation current in humans should be approximately 25 times that of rat vagus stimulation. Activation of B fibers in the rats at 250 μsec was 25 μA, which would correspond to a 0.625 mA activation in humans using the 25 multiplication factor. This is very close to the range of that clinically being used. In 2012, Helmers [29] (Helmers SL, 2012) developed a computational model of the activation of the human vagus nerve. She tied the computational model to clinical experience in Table 7, which shows the approximate vagus nerve stimulation threshold strength–duration curve. Nonetheless, 25 years of clinical vagus nerve stimulation experience has yielded little direct measurement of vagus action potential.

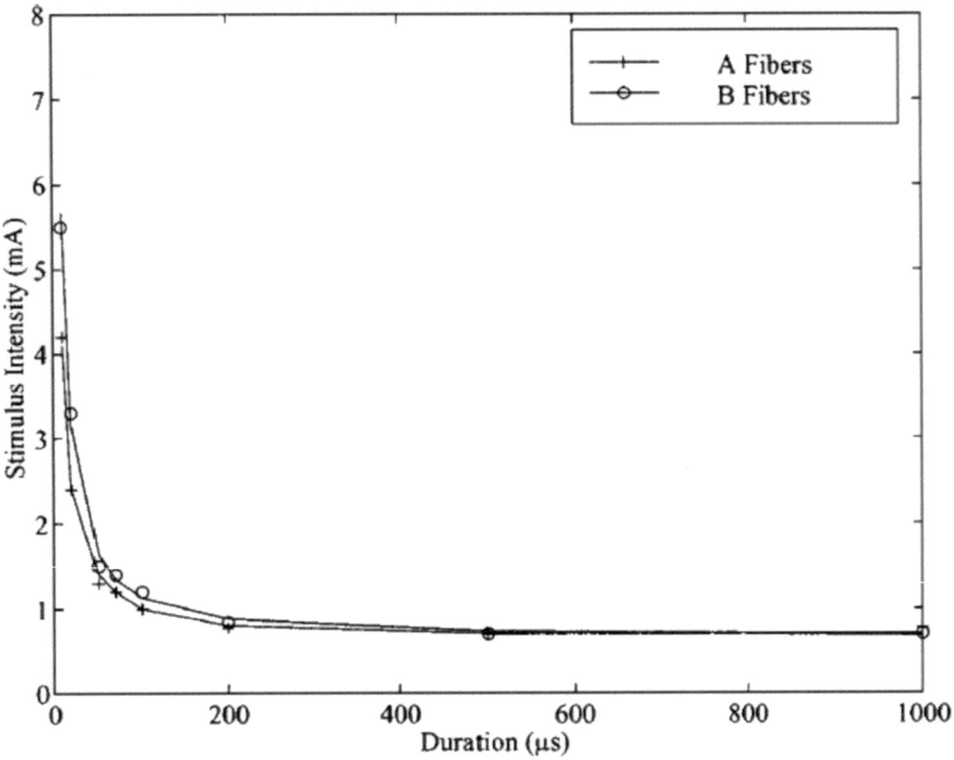


**Figure 1.** Chronixie and Propagation Velocity of Canine Vagus [28]

| Pulse Width | 6.3% range | 57.5% range | 36.2% range |
|---|---|---|---|
| 130 usec | 0.25-.1.0 mA | 1.0-2.25 mA | 2.25-3.5 mA |
| 250 usec | 0.25-0.75 mA | 0.75-1.75 mA | 1.75 -3.5 mA |
| 500 usec | .0.25-0.6 mA | 0.6-1.5 mA | 1.5-3.5 mA |

**Table 7.** Vagus Nerve Stimulation Threshold Strength–Duration Table [29]

The low number in the range is the 'minimum target' and represents the minimum output current pulse width needed to activate the nerve completely in the absence of tissue ingrowth. An increase in resistance because of tissue ingrowth may occur during the first several weeks after surgery. This increase may require additional stimulation per pulse for adequate recruitment of the vagal afferents. Very few patients (6.3%) responded with output current and pulse width falling below the 'minimum target' line. More than half of the responders (57.5%) had output current pulse width combinations above the 'minimum target' and below the high number or 'maximum target'; 36.2% of responders had output current pulse width combinations above the maximum target.

### 8.1. The stimulation S curve

Figure 2 shows that activation of the individual fibers in the vagus nerve follows the typical S curve as the current increases which assumes the activations threshold follows a normal distribution. For the A and B fibers, the 10% activation current (A) is believed to be about 0.25 mA and the 90% activation current (B) is believed to be about 1.5 mA. However, the curves have not been developed for individual patients and the curves are likely to vary among patients and may change over time for the individual patient.

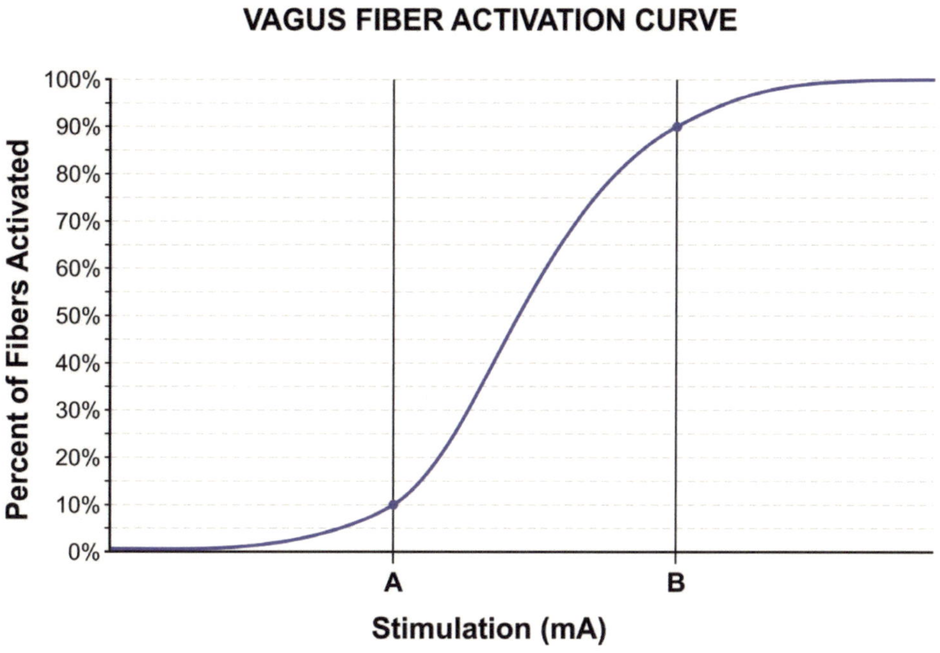


**Figure 2.** Fiber Activation Curve

### 8.2. Stimulation frequency

The typical stimulation frequency is 20 Hz, based on experience in the clinical studies and the Woodbury rat studies [11] (Woodbury DM, 1990). Woodbury reported that efficacy diminished rapidly with frequency reductions of 10 Hz and lower.

The E03 and E05 studies used 1 Hz for the placebo LOW group. Stimulation greater than 30 Hz may be equally or more effective, but battery life is shortened and patient discomfort may be increased.

### 8.3. Rapid stimulation or high duty cycle stimulation

The AAN RGDS [14] reviewed evidence on rapid stimulation cycles. Among patients receiving VNS, does rapid stimulation (usually 7 seconds "on" and 30 seconds "off") improve seizure frequency more often than standard stimulation settings (30 seconds "on" and 300 seconds "off")? In all the reviewed studies, beginning parameters were output current 0.25 mA, signal frequency 30 Hz, pulse width 250–500 ms, stimulation "on" time 30 seconds, and stimulation "off" time 300 seconds, and output current usually increased to 2–3 mA per tolerance. Schermann and collegues [30] (Scherrmann J, 2001) discussed rapid as opposed to standard stimulation settings, assessing results of 73 adult epilepsy patients with adjusted settings of standard (30 seconds "on" and 300 seconds "off", n=41) or rapid (7 seconds "on" and 30 seconds "off", n=32). After 2 years of follow-up, the standard stimulation group had greater overall seizure frequency reduction than the rapid stimulation group. In this article, the authors said a group receiving VNS at either standard or rapid stimulation did not demonstrate different response rates. For several patients, a changing to rapid stimulation several years after implantation improved outcome.

Concerning children, 2 Class III articles reported that rapid stimulation did not provide consistently better results than standard stimulation [31] (Alexopoulos AV, 2006) [32] (Shahwan A, 2009). According to the AAM RGDS, both studies lacked power to detect a difference between rapid stimulation when used either initially or after unsuccessful standard stimulation.

In contrast to the High Duty Cycle experience, very low duty cycles of 1 minute per hour have not been extensively studied, although success was observed in the E01 trial and such use would significantly extend battery life.

In summary, the physician has a range of parameter adjustments to optimize treatment for the individual patient.

## 9. Automatic seizure detection

Activation of VNS upon seizure detection was a long-time goal, but available technology limited such to self-activation by the patient or caregiver. A magnet, often worn on the patient's wrist, can be passed over the implanted generator to initiate stimulation. Morris [33] (Morris, 2003) conducted a retrospective analysis of magnet use during the E03 and E04 clinical trials of VNS. Magnet activation that aborted, decreased, terminated, or diminished a seizure was classified as an improvement; for purposes of evaluation, the patient was considered to have received a benefit. When patients in the E03 trial used magnets to activate stimulation, those in the active (HIGH) stimulation group reported more seizure improvement than those in the placebo (sham magnetic stimulation) group (P=0.0479, Fisher's test). In the E04 trial, 22% of the patients reported they were able to abort seizures and 31% reported the seizures were diminished.

Woodbury [11] (Woodbury DM, 1990) found that the length of the seizure was related to the length of delay between start of a PTZ-induced seizure in rats and start of stimulation. This rat study implied that stimulation should be initiated within the first 14 seconds of a PTZ-induced seizure. It seems intuitive that stimulation early in the seizure cycle would be most effective and may explain why Lockard [7] was not successful in terminating seizures in the primate study.

Several papers have suggested the possibility of using heart rate changes to detect seizures, including Leutmezer [34] (Leutmezer F, 2003) and [35] (Schernthaner C, 1999). Luetmezer reported that 86.9% of all seizures were accompanied by ictal-onset tachycardia and 1.4% of seizures were accompanied by bradycardia. Patients with mesial temporal lobe epilepsy (TLE) were significantly more likely to have ictal HR increase than those with non-lesional TLE or extratemporal epilepsy. Ictal-onset tachycardia was also more likely with right hemispheric seizures. Interestingly, the increased ictal HR preceded the EEG onset by a statistically significant difference (p=0.047); average of 13.7 s among TLE patients and 8.2 s among extratemporal epilepsy patients.

Cyberonics developed the Aspire SR to provide seizure detection based on heart rate changes. Boon et al [36] (Boon P, 2013) had conducted the E36 prospective multicenter clinical study in Europe to evaluate seizure detection ability of the Aspire SR. Boon [36] had presented data on 31 patients at the 2013 American Epilepsy Meeting. The study met its primary endpoint of detecting more than 80% of seizures accompanied by ictal tachycardia and false rates were low. The detections occurred close to, and in some cases, before seizure onset. The AspireSR Generator can help patients for whom are not able to apply rapid and consistent hand-held magnet activation. For example, patients whom experience seizures while sleeping would benefit from this feature. The CE Mark for the AspireSR was issued in February 2014.

Neuropace has evaluated EEG detection for deep brain stimulation therapy. The study [37] (Morrell, 2011) found a statistically significant reduction in seizures with treatment group vs control group, but did not report an evaluation of detection sensitivity or false detection counts. [38] (Neuropace, 2013).

## 10. Vagus connections into neural networks

The vagus nerve comprises approximately 80% afferent fibers. These fibers enter the nucleus tractus solitaries and branch bi-laterally into both sides of the brain. For the most part, both the right and left vagus seem to have very similar projections into neural networks. Both project equally into the right and left sides of the brain, although they probably have many small differences.

Krahl [39] (Krahl SE C. K., 1998) reported data that indicate the LC is involved in the circuitry necessary for the anticonvulsant effects of VNS.

Naritoku [40] (Naritoku DK, 1995) began tracing the neural networks as did Cunningham [41] (Cunningham JT, 2008).

Naritoku [40] reported that specific nuclear fos immune labeling was induced by vagus nerve stimulation (VNS) in several forebrain structures, including the posterior cortical amygdaloid nucleus, cingulate, and retrosplenial cortex, ventromedial and arcuate hypothalamic nuclei.

The immune labeling was also apparent in the vagus nerve nuclei of the brainstem, in the A5 and locus ceruleus noradrenergic nuclei and in the cochlear nucleus. The sham-operated, unstimulated control animals had no fos immune labeling of these structures.

Cunningham studied the acute and chronic effects of VNS on brain structures using c-Fos and ΔFosB immune labeling. Acutely, VNS significantly increased c-Fos staining in the nucleus of the solitary tract, paraventricular nucleus of the hypothalamus, parabrachial nucleus, central bed nucleus of the stria terminalis, and locus coeruleus, but did not increase c-Fos staining in the cingulate cortex or dorsal raphe nucleus (DRN). VNS acute stimulation did not affect ΔFosB staining in any region. Chronically, VNS significantly increased both ΔFosB and c-Fos staining bilaterally in each region affected by acute VNS as well as in the cingulate cortex and DRN.

Cunningham [41] developed the neural network schematic in Figure 3, which was presented at several conferences.

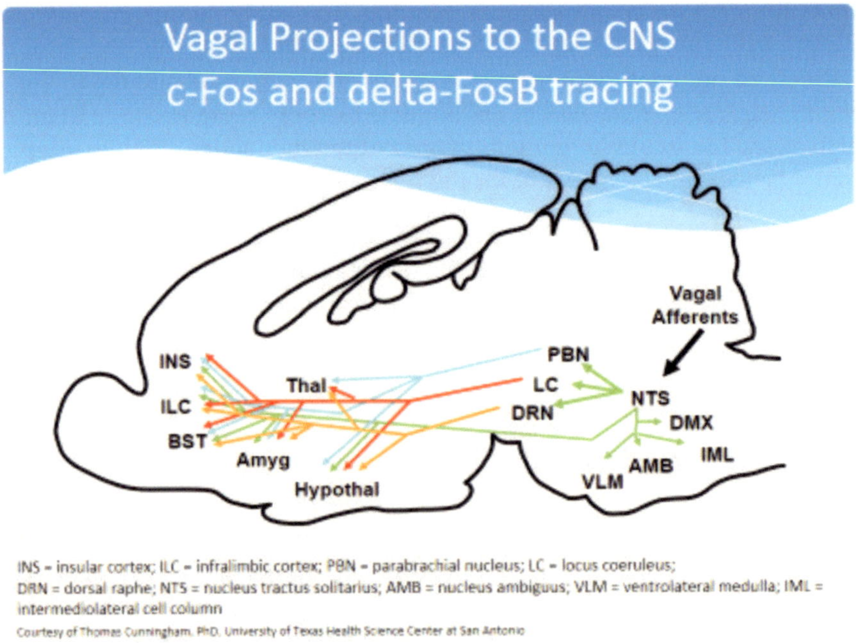


**Figure 3.** Vagal Projections to the Brain as Developed by Cunningham [41]

### 10.1. Left vs right vagus stimulation

In the clinical studies, the pulse generator was attached to the left vagus nerve for establishing safety and efficacy. Primate studies by Lockard [7] (Lockard JS, 1990) stimulated the right vagus nerve and noted no effect on heart rate nor gastric function. Several reports regarding right-sided stimulation among patients have not mentioned cardiac effects [42] (Navas M, 2010), [43] (Spuck S, 2008), [44] (Tubbs RS, 2007) (pig model), [45] (McGregor A, 2005) and [46]

(Krahl SE S. S., 2003) (rat model). Also, a positive clinical result from a pilot study of right vagus nerve stimulation for congestive heart failure has been published [47] (De Ferrari & Investigators, 2011). Although right-sided VNS may be safe, it should be used cautiously as its safety has not been established in clinical studies.

Would stimulation of both the right and left vagus nerve have an additive effect that might improve efficacy? This approach has not been studied in animal epilepsy models.

## 11. External vagus stimulation

The concept of using external stimulation has appeal because VNS requires surgical implantation, and not all patients respond. Two external vagus stimulation devices and 1 trigeminal nerve stimulator are being investigated in clinical studies. Although these devices have CE Mark approval for 1 or more indications, none have FDA approval as of February 2014.

Stefan [48] (Stefan H, 2012) described the Cerbomed NEMOS device used in a pilot study among 10 patients. Electrical stimulation was applied transcutaneously to the auricular branch of the vagus nerve (ABVN) of the left ear. Three patients left the study. Of the remaining 7 patients, 5 experienced an overall seizure reduction after 9 months of t-VNS.

DeGeorgio [49] (DeGiorgio CM, 2013) reported a 50-patient double-blind trial of the Monarch™ eTNS Trigeminal nerve stimulation device by NeuroSigma, and Pack [50] (Pack AM, 2013) commented on the study. No significant differences were found between the active group and control group in any of the 3 predefined primary end points, but a significant within-group difference improvement was seen in the responder rate over the 18-week treatment period. The responder rate was 17.8% at 6 weeks, and it increased to 40.5% at 18 weeks, providing a significant within-group improvement ($p$=0.01). Pack [50] commented that it is unclear how the responder rates at the serial observation periods were derived and concluded that although interesting, these data do not support the effectiveness of eTNS for the treatment of refractory partial seizures.

Electrocore Medical makes the gammaCore battery-powered device that is held against the neck and provides a single 180-second burst of vagus nerve stimulation. A number of presentations and publications of pilot studies have reported encouraging results for migraine, cluster headaches, asthma, and bronchial COPD. Electrocore Medical has a CE Mark for epilepsy, but no epilepsy studies have been reported as of February 2014.

## 12. Deep brain stimulation

Deep brain stimulation is attractive because it stimulates the specific part of the brain associated with the seizure. Medtronic and Neuropace have conducted clinical trials with devices that provide deep brain stimulation. Both have CE Mark approval and Neuropace received FDA approval in November 2013.

Fisher [51] (Fisher R, 2010) reported on the 110-patient double-blind trial employing the Medtronic dual-channel Model 7428 Kinetra Neurostimulator. During the final blinded month, seizure reduction was 29% greater among the stimulated group than the controls, (GEE) model (p=0.002). At 2 years, median percent reduction in seizure frequency was 56%; 54% showed seizure reduction of at least 50%, and 14 were seizure-free 6 months or more.

The Neuropace RNS® System detects seizures from the EEG and then initiates stimulation. Morrell [37] (Morrell, 2011) reported on the 191-patient double-blind study of the Neuropace System. Although seizures were significantly reduced during the blinded period for both the treatment and sham groups, the treatment group seizures were more significantly reduced (-37.9%, n=97) compared with the sham group (-17.3%, n=94; *p*=0.012). There was no difference in adverse events between groups during the blinded period. The treatment group sustained seizure reduction. The FDA approved the Neuropace RNS® System in November 2013.

## 13. Conclusion

During its first 25 years, meaningful advances have been made in VNS for epilepsy. More than 70,000 patients have been treated and more than 100,000 devices have been implanted, which includes battery replacements. Long-term clinical experience shows that efficacy continues to improve, which is different from many pharmaceutical therapies. The therapy is highly cost effective, resulting in considerable long-term savings even though most patents do not become totally seizure free. Much is still to be learned about VNS. Advances continue regarding methods of seizure detection and activation therapy. Various methods of external stimulation as well as deep brain stimulation are being developed and evaluated.

## Acknowledgements

Dr. Kiffin Penry for pioneering vagus nerve therapy in the first patients, the clinical study investigators, the clinical study patients, the basic science investigators, the Cyberonics employees, the venture capital investors who funded the clinical studies, and Jacob Zabara, PhD, who conceived and patented vagus nerve therapy. The author thanks Susan Siefert, ELS, CBC, Fellow and Past President of the American Medical Writers, for her help editing this chapter.

## Author details

Reese S. Terry Jr

Founder, Cyberonics Inc. (retired Houston, TX), Life Fellow, IEEE, USA

http://dx.doi.org/10.5772/58332

## References


[1] Corning, J. (1883). Considerations on pathology and therapeutics of epilepsy. *J Nerv Ment Dis.*, 10:243–248.

[2] Bailey, P., & Bremer, F. (1938). A sensory cortical representation of the vagus nerve (with a note on the effects of low blood pressure on the cortical electrograms. *Journal of Neurophysiology*, 405-412.

[3] Schweitzer A, W. S. (1937). Effects on the knee jerk of stimulation of the central end of the vagus and of various changes in the circulation and respiration. *J Physiol*, 459-475.

[4] Paintal, A. S. (1973). Vagal sensory receptors and their reflex effects. *Physiol Rev*, 53(1):159-227.

[5] Dell P, O. R. (1951). [Secondary mesencephalic, diencephalic and amygdalian projections of vagal visceral afferences.]fr. *C_R_Seances_Soc_Biol_Fil*, 145(13-14).

[6] Zabara, J. (1992). Inhibition of experimental seizures in canines by repetitive vagal stimulation. 1992 November/December, 33. *EPILEPSIA*, November/December, 33.

[7] Lockard JS, C. W. (1990). Feasibility and safety of vagal stimulation in monkey model. *Epilepsia.*, 31 Suppl 2:S20-6.

[8] Woodbury, W. (1991). PhD. (R. Terry, Interviewer)

[9] Uthman BM, W. B. (1993). Treatment of epilepsy by stimulation of the vagus nerve. *NEUROLOGY*, 43:1338-134.

[10] Penry JK, D. J. (1990). Prevention of intractable partial seizures by intermittent vagal stimulation in humans: preliminary results. *EPILEPSIA*, 31 Supplement 2:S40-43.

[11] Woodbury DM, W. J. (1990). Effects of vagal stimulation on experimentally induced seizures in rats. *Epilepsia.*, 31 Suppl 2:S7-19.

[12] Agnew WF, M. D. (1990). Considerations for Safety with Chronically implanted Nerve electrodes. *Epilepsia*, 31(Suppl 2):S27-S32.

[13] Cyberonics. (2013, November 17). *VNS Therapy Products*. Retrieved from VNS Therapy System Physicians's Manual (US): http://dynamic.cyberonics.com/manuals/ Per Cyberonics, this material is subject to change.

[14] Morris GL 3rd, G. D. (2013). Evidence-based guideline update: Vagus nerve stimulation for the treatment of epilepsy: Report of the Guideline Development Subcommittee of the American Academy of Neurology. *Neurology*, Oct 15;81(16):1453-9.

[15] Morris GL III, M. W. (1999). Long-term treatment with vagus nerve stimulation in patients with refractory epilepsy. The Vagus Nerve Stimulation Study Group E01-E05. *Neurology*, 53:1731–1735.

[16] Kuba R, B. M. (2009). Vagus nerve stimulation: longitudinal follow-up of patients treated for 5 years. *Seizure*, 18(4):269-7.

[17] Elliott RE, M. A. (2011). Efficacy of vagus nerve stimulation over time: review of 65 consecutive patients with treatment-resistant epilepsy treated with VNS > 10 years. *Epilepsy Behav*, Mar;20(3):478-83.

[18] Uthman BM, R. A. (2004). Effectiveness of vagus nerve stimulation in epilepsy patients: a 12-year observation. *Neurology*;63:1124–1126.

[19] Salinsky, M. (1997). Results from the Open Label Safety Study: The E04 Experience. *Vagus Stimulaton Investigators Meeting and symposium*. Colorado Springs, Colorado.

[20] Helmers SL, D. M. (2011). Clinical and economic impact of vagus nerve stimulation therapy in patients with drug-resistant epilepsy. *Epilepsy Behav*, Oct;22(2):370-5.

[21] Cukiert A, C. C.-B. (2013). A Prospective Long-Term Study on the Outcome After Vagus Nerve Stimulation at Maximally Tolerated Current Intensity in a Cohort of Children With Refractory Secondary Generalized Epilepsy. *Neuromodulation.*, Jun 5. doi: 10.1111/j.1525-1403.

[22] Wilfong AA, S. R. (2006). Vagus nerve stimulation for treatment of epilepsy in Rett syndrome. *Dev Med Child Neurol*, Aug;48(8):683-6.

[23] Elger G, H. C. (2000). Vagus nerve stimulation is associated with mood improvements in epilepsy patients. *Epilepsy Res*, Dec;42(2-3):203-10.

[24] Ben-Menachem E, H. K. (2002). Analysis of direct hospital costs before and 18 months after treatment with vagus nerve stimulation therapy in 43 patients. *Neurology.*, Sep 24;59(6 Suppl 4):S44-7.

[25] Ben-Menachem E, F. J. (2005). VNS Therapy versus the latest antiepileptic drug. *Epileptic Disord*, Jun;10(2):191.

[26] Boon P, D. M. (2002). Direct medical costs of refractory epilepsy incurred by three different treatment modalities: a prospective assessment. *Epilepsia*, Jan;43(1):96-102.

[27] Bernstein AL, H. T. (2007). Vagus nerve stimulation therapy for pharmacoresistant epilepsy: effect on health careutilization. *Epilepsy Behav.*, Feb;10(1):134-7.

[28] Smith CD, G. L. (2001). The Chronaxie and Propagation Velocity of Canine Cervical Vagus Nerve Fibers in Vivo*. *Cardiovascular Engineering,*, Vol. 1, No. 2;77-83.

[29] Helmers SL, B. J. (2012). Application of a computational model of vagus nerve stimulation. *Acta Neurol Scand.*, 126(5):336-43.

[30] Scherrmann J, H. C. (2001). Vagus nerve stimulation: clinical experience in a large patient series. *J Clin Neurophysiol.*, 2001 Sep;18(5):408-14.

[31] Alexopoulos AV, K. P. (2006). Long-term results with vagus nerve stimulation in children with pharmacoresistant epilepsy. *Seizure.*, Oct;15(7):491-503. Epub 2006 Jul 20.

[32] Shahwan A, B. C. (2009). Vagus nerve stimulation for refractory epilepsy in children: More to VNS than seizure frequency reduction. *Epilepsia.*, May;50(5):1220-8.

[33] Morris, G. (2003). A retrospective analysis of the effects of magnet-activated in conjunction with vagus nerve stimulation therapy. *Epilepsy Behav.*, Dec;4(6):740-5.

[34] Leutmezer F, S. C. (2003). Electrocardiographic changes at the onset of epileptic seizures. *Epilepsia.*, Mar;44(3):348-54.

[35] Schernthaner C, L. G. (1999). Autonomic epilepsy--the influence of epileptic discharges on heart rate and rhythm. *Wien Klin Wochenschr*, May 21;111(10):392-401.

[36] Boon P, v..K.. (2013). Vagus Nerve Stimulation Triggered by Cardiac-Based Seizure detection, A Prospective Study. *American Epilepsy Society.* Washington, DC. Abst. 1.048

[37] Morrell, M. (2011). Responsive cortical stimulation for the treatment of medically intractable partial epilepsy. *Neurology.*, Sep 27;77(13):1295-304.

[38] NEUROPACE. (2013, February 22). *RNS System for Epilepsy Neuropace, Inc.* Retrieved from FDA NEUROLOGICAL DEVICE ADVISORY PANEL: http://www.fda.gov/downloads/AdvisoryCommittees/CommitteesMeetingMaterials/MedicalDevices/MedicalDevicesAdvisoryCommittee/NeurologicalDevicesPanel/UCM340257.pdf

[39] Krahl SE, C. K. (1998). Locus coeruleus lesions suppress the seizure-attenuating effects of vagus nerve stimulation. *Epilepsia.*, Jul;39(7):709-14.

[40] Naritoku DK, T. W. (1995). Regional induction of fos immunoreactivity in the brain by anticonvulsant stimulation of the vagus nerve. *Epilepsy Res.*, Sep;22(1):53-62.

[41] Cunningham JT, M. S. (2008). Cunningham JT, Mifflin SW, Gould GG, Frazer A. *Neuropsychopharmacology*, Jul;33(8):1884-95.

[42] Navas M, N. E. (2010). Treatment of refractory epilepsy in adult patients with right-sided vagus nerve stimulation. *Epilepsy Res.*, Jun;90(1-2):1-7.

[43] Spuck S, N. G. (2008). Right-sided vagus nerve stimulation in humans: an effective therapy? *Epilepsy Res.*, Dec;82(2-3):232-4.

[44] Tubbs RS, S. E. (2007). Right-sided vagus nerve stimulation inhibits induced spinal cord seizures. *Clin Anat.*, Jan;20(1):23-6.

[45] McGregor A, W. J. (2005). Right-sided vagus nerve stimulation as a treatment for refractory epilepsy in humans. *Epilepsia*, Jan;46(1):91-6.

[46] Krahl SE, S. S. (2003). Right-sided vagus nerve stimulation reduces generalized seizure severity in rats as effectively as left-sided. *Epilepsy Res*, Sep;56(1):1-4.

[47] De Ferrari, G. C. (2011). Chronic vagus nerve stimulation: a new and promising therapeutic approach for chronic heart failure. *Eur Heart J,* Apr;32(7):847-55.

[48] Stefan H, K. G. (2012). Transcutaneous vagus nerve stimulation (t-VNS) in pharmacoresistant epilepsies: a proof of concept trial. *Epilepsia,* Jul;53(7)e115-8.

[49] DeGiorgio CM, S. J.L. (2013). Randomized controlled trial of trigeminal nerve stimulation for drug-resistant epilepsy. *Neurology,* Feb 26;80(9):786-91.

[50] Pack AM. (2013). Trigeminal Nerve Stimulation May Not Be Effective for the Treatment of Refractory Partial Seizures. *Epilepsy Curr,* Jul-Aug; 13(4): 164–165.

[51] Fisher R, S. V. (2010). Electrical stimulation of the anterior nucleus of thalamus for treatment of refractory epilepsy. *Epilepsia,* May;51(5):899-908.

Chapter 8

# Neuromodulation Therapy: Nonmedical, Nonsurgical Treatment for Intractable Epilepsy

Hyang Woon Lee

Additional information is available at the end of the chapter

http://dx.doi.org/10.5772/57441

## 1. Introduction

Neuromodulation therapy has been tried for patients with many neuropsychiatric disorders such as Parkinson's disease, tremor, obcessive-compulsive disorder, depression, intractable pain, or addiction etc [1]. In epilepsy, this kind of treatment has been applied to treat patients with intractable epilepsy for decades who are not controlled by antiepileptic drugs (AEDs) nor surgical candidates. Historically, the earliest report of the use of electrical brain stimulation to control seizures in humans showed that focal electrical cortical stimulation could stop both normal EEG rhythms and spontaneous epileptiform discharges [2].

Vagal nerve or deep brain stimulation (e.g. thalamic stimulation) is indirect neural stimulation method, which delivers high frequency electrical stimulation indirectly to the epileptic brain via vagal nerve or thalamus, regardless of seizure foci [3,4]. This kind of stimulation is called open loop system since the stimulation is delivered intermittently and regularly without external cues (e.g. seizure onset). In contrast, closed loop system has been more recently applied in which the stimulation is directly applied to the seizure focus when seizure activity actually occurs. To do this, the stimulation system is combined with early seizure detection algorithm and activated automatically to deliver the stimulation only when the seizure activity is detected [5]. Responsive neurostimulation is an example of the closed loop system, which has been tested in recent clinical trials. While these methods need surgery to implant the electrodes for stimulation, noninvasive stimulation can be tried too especially using low frequency repetitive transcranial magnetic stimulation. In this method, the stimulation has been applied directly to the seizure focus but without combining seizure detection system so far. Although the exact therapeutic mechanism is not fully understood yet, these kinds of treatments have been reported to reduce seizures in intractable epilepsy patients.

In this chapter, various kinds of neuromodulation methods are introduced with the results of treatment efficacy and side effects from previous clinical trials in intractable epilepsy patients. In addition, recent clinical and basic researches to investigate possible therapeutic mechanisms are summarized. Future study directions what should be solved to improve the therapeutic efficacy and/or reduce adverse effects are also discussed.

## 2. Types of neurostimulation

### 2.1. Open loop indirect stimulation thru the peripheral nerve: Vagal nerve stimulation

Vagal nerve stimulation (VNS) is the first and the only FDA approved neurostimulation therapy in epilepsy, which has been tried in various seizure types and epileptic syndromes [3, 4]. The most famous clinical trials using VNS are randomized controlled trials of E03 [6] and E05 [7] in 114 and 119 patients, which provided an important evidence for FDA approval of VNS therapy in 1997. The stimulation paradigm was consisted of both high and low stimulations. The high stimulation was 3.5 mA, 30 Hz, 500 μs pulsewidth, with 30 s on time and 5 min off time, whereas the low simulation was similar output currents with 1 Hz frequency rate, 130 μs pulsewidth, on time of 30 s, and off time of 180 min.

Mechanism of action has been suggested based on the animal experiments and human researches using various electrophysiological and functional brain imaging studies [8]. It is believed that VNS modulates mainly subcortical neural network that influences larger cortical areas modifying synaptic connections. However, the exact mechanism of action how VNS reduces seizures is still under investigation, which needs future studies.

### 2.2. Open loop indirect stimulation into the brain: Deep brain stimulation

Stimulation of anterior nucleus of thalamus (ANT) in epilepsy patients was reported to have therapeutic effects earlier [9], and the results from double blind multicenter clinical trials have been reported recently in 110 epilepsy patients from 17 epilepsy centers [10]. In this, so-called, SANTE trial (Clinical Trials. Gov. NCT00101933), the efficacy showed 40.4% of seizure reduction during the 3 months of blinded phase in the treatment group, for both complex partial and secondarily generalized tonic clonic seizures, compared with 14.5% in the control group. Thirty-three patients with temporal lobe epilepsy (TLE) showed better response (44% of seizure reduction during the blinded phase compared with 22% reduction in TLE patients with standard treatment). Interestingly, the seizure reduction rate was increased over time in the open label period, which showed 41% median seizure reduction at 13 months. After 2 years of seizure follow-up, treated patients showed median 56% reduction in seizure frequency and 54% of patients obtaining more than 50% seizure reduction. Although there was a concern about depression (14.8%) or memory impairment (13.0%), DBS was generally safe without serious side effects such as intracranial hemorrhage, infection or death.

Possible role of thalamic DBS in epilepsy is that the thalamus might be a relay station that inhibits or disrupts epileptic seizures spreading via thalamocortical neural network. Actually,

ANT is believed to desynchronize seizure discharges thus inhibit seizure spread from hippocampus or neocortex to other brain areas [11]. Animal studies have provided evidence that low frequency ANT stimulation increases synchronization of the cortical EEG by recruiting rhythmic thalamocortical activities whereas high frequency stimulation leads to desynchronize EEG rhythms that will be effective for reducing seizures [12].

### 2.3. Closed loop direct stimulation to the seizure focus: Responsive neurostimulation

Closed loop neurostimulation is a method that responds immediately after the seizure is detected by automated seizure detection algorithm. Early stimulation after the seizure onset is expected to stop the seizure during the early stage before the seizure propagates to remote cortical areas thus evolving to secondarily generalization. Responsive neurostimulation (RNS) is the most recently tried one, which is a combination of early seizure detection and automated neurostimulation based on detection results from real-time EEG analysis. The RNS system requires an electrode implantation to the seizure focus for both seizure detection and electrical stimulation. Once the detection algorithm detects a seizure, high frequency electrical stimulation is applied automatically to abort the seizure as early as possible.

The rationale for RNS is based on the study showing that cortical stimulation could terminate afterdischarges (ADs) during functional mapping [13]. ADs, unnecessary but inevitable events, are very similar to ictal EEG discharges during seizures although they are induced activities from high frequency electrical stimulation, not spontaneously occurred ones. Interestingly, brief bursts of electrical cortical stimulation that had induced ADs if delivered longer, could stop ADs immediately in many occasions. It worked better when the stimulation was applied briefly shorter than one second and more promptly with shorter stimulation latency within several seconds after ADs occurred [14]. After this, unblinded clinical studies were conducted in adults with refractory partial epilepsy mainly for safety and feasibility [15]. And then, a double-blind, multicenter, randomized controlled clinical trial of RNS was conducted, and the efficacy was reported recently [16].

The controlled clinical trial was performed in 191 patients from 32 epilepsy centers who were diagnosed with highly drug-resistant partial epilepsy and had one or two focal seizure foci. The stimulation parameter for RNS was the amplitude of 0.5-12 mA, the pulse width of 40-1000 μs, and the frequency of 1-333 Hz. The average seizure frequency was significantly improved with 38% of seizure reduction in the active RNS group compared to 17% in the sham control group during the initial 12 weeks of blinded phase. During the open label period, seizure frequency was reduced in the sham control group in the initial blinded phase to the level of those in the treatment group. As with ANT stimulation, there was further progressive improvement during the open label period with median seizure reduction rate up to 50% after two years [16]. Based upon these results, RNS has been approved by FDA in November 14, 2013, for patients with drug-resistant epilepsy.

Choosing the best early seizure detection algorithm is very important for successful treatment in RNS [17]. Various EEG quantification methods have been tried for seizure prediction or early seizure detection, which include correlation dimension [18-21], correlation density [22], similarity index [23-25], phase synchronization [26-28], accumulated energy [29], complexity or synchrony [30]. Improvement of early detection algorithm will be one of the most important requirements to make RNS more useful therapeutic option in treatment of intractable epilepsy.

More effective stimulation parameters have been tested in animal researches. Among the stimulation parameters, the frequency has been known as the most important factor to inhibit seizure activity better [31-33]. Several types of closed-loop brain computer interface systems have been tested in epilepsy animal model with improved seizure detection and reduced false detection rates [34].

### 2.4. Open loop direct stimulation to the seizure focus: Repetitive transcranial magnetic stimulation

Transcranial magnetic stimulation (TMS) is well established neurophysiologic study tool that has been used in neuroscience research and various clinical fields. TMS can be used in single, paired or repetitive trains, and repetitive stimulation [35]. Repetitive TMS (rTMS) is a safe and noninvasive method to alter neuronal functions thus applying to various clinical disorders such as stroke, pain, epilepsy etc. While other neural stimulation methods listed above need surgery to implant the electrodes for stimulation, TMS is a noninvasive stimulation without surgical intervention that is one of the greatest advantages when considered as a chronic treatment method clinically. In this method, the stimulation has been applied directly to the seizure focus but without combining seizure detection system so far.

Low frequency rTMS (less than 1 Hz) is known to inhibit cortical excitability [36,37] while high frequency rTMS (5-20 Hz) increases cortical excitability [38]. Low frequency rTMS, especially using 1 Hz stimulation, has been tried for refractory epilepsy. In a randomized clinical trial in 21 epilepsy patients with cortical developmental malformation, 1 Hz rTMS significantly reduced the number of seizures in the active group compared to the sham control group for at least 2 months [39]. A meta-analysis showed that low frequency rTMS has favorable antiepileptic effects, especially in patients with cortical dysplasia or neocortical epilepsy, with an effect size of 0.71 and 95% confidence interval at 0.30-1.12 [40]. Another established form of rTMS protocol is theta-burst stimulation (TBS), a burst of three 50-Hz pulses in trains repeated at 200-ms intervals. Continuous TBS (cTBS) consists of burst trains for 20-40 s that has an inhibitory effect on corticospinal excitability. On the other hand, intermittent TBS (iTBS), burst trains with a duration for 20-40 s for about 190 s, repeated in every 10 s, has a facilitating effect on corticospinal tract [41]. TBS is known to have longer effects than conventional rTMS paradigms that can be useful for clinical application, possibly including epilepsy, that will need further verification in controlled clinical trials.

## 3. Unsolved questions and future direction

In this chapter, various kinds of neuromodulation methods are introduced with a review of previous clinical trials and basic researches. We have reviewed important findings from previous clinical trials and other basic researches that contribute to our understanding of possible therapeutic mechanisms of neuromodulation for epilepsy treatment, as well as recent

http://dx.doi.org/10.5772/57441

technical notes to improve accurate and prompt seizure detection. Several important issues remain to be solved, however, such as ideal targets and stimulation parameters, and the optimization in each seizure type and/or epileptic syndrome. Investigation of the underlying therapeutic mechanisms requires more translational studies in the future that link basic researches to relevant clinical trials.

### 3.1. Targets for neurostimulation

Several targets have been tried to treat focal or generalized epilepsy patients who are refractory to medical or surgical treatment, including vagus nerve, ANT, hippocampus, and various cortical locations according to epileptic foci. Targeting the hippocampus sounds reasonable in mesial temporal lobe epilepsy, which has been tested mostly in the form of RNS [17,42].

Other brain structures have been tried as well to control seizures; for examples, the centromedian nucleus of thalamus, the subthalamic nucleus, the substantia nigra reticulata, the caudate nucleus, the cerebellum, the posterior hypothalamus, and the caudal zona incerta [43]. Subthalamic nucleus, cerebellum, and trigeminal nerve stimulations have been considered as possible targets for intractable epilepsy especially for generalized epilepsy patients. The subthalamic nucleus has been tested clinically mainly for movement disorders so far, but it is also known as a relay station in the nigral system for epilepsy control that involves in seizure propagation and secondary generalization in animal researches [44,45], which has been the rationale to suggest its usefulness in epilepsy. The cerebellar stimulation was tried earlier [46], and reevaluated recently in a double-blind, randomized controlled pilot study on five patients with medically refractory motor seizures, that showed its beneficial effects especially for generalized tonic-clonic seizures [47]. The trigeminal nerve is one of the cranial nerves that connects to the large subcortical brain areas. Early studies suggesting potential clinical benefits of trigeminal nerve stimulation for epilepsy patients have been reported [48,49].

### 3.2. Stimulus parameters

Optimization of stimulus parameters is very important to improve the efficacy of seizure control by neuromodulation. In animal experiments, low frequency electrical stimulation can decrease neural excitability and seizure activity in both in-vivo and in-vitro models of epilepsy and stimulation effects increase synaptic inhibition via long-term depression (LTD). In epilepsy patients, low frequency electrical stimulation applied to ictal onset zones reduced seizure frequency. Low frequency electrical cortical stimulation was reported to have an inhibitory effect on epileptic focus in mesial temporal lobe epilepsy [50]. However, unwanted seizures may occur as side effects of stimulation, which needs to be solved in the future studies.

High frequency electrical stimulation has been applied both indirect stimulation to the epilepsy network via both VNS and DBS, and direct stimulation to the seizure focus via RNS [4,5]. For VNS, stimulation around 30 Hz is effective to reduce seizures. For DBS, higher frequency stimulation about 120 Hz was also reported to reduce seizures. Various stimulation frequencies between 1 and 333 Hz have been used for RNS.

Intermittent versus continuous stimulation also has been discussed. Interestingly, intermittent stimulation around 1.68% of the time was reported to be effective although less than when the device was activated 50% of the time [51].

Standardization of stimulation parameters is also needed for individual seizure types and/or specific epileptic syndromes. Future studies in both basic and human researches to improve treatment efficacy based on therapeutic mechanisms will be mandatory for practical use of neuromodulation therapy in intractable epilepsy patients.

### 3.3. Unveiling therapeutic mechanism

Epileptic seizure is characterized by a brief, transient increase of abnormally excitable and synchronized activities in neural network. Interestingly, and somewhat paradoxically, the activity can be eliminated by neurostimulation, often using very high frequency especially in the VNS, DBS, and RNS.

From the observation that electrical stimulation suppressed ADs and seizures throughout the course of kindling, that indicates a strong antiepileptogenic effect. While the kindling seems very similar to long-term potentiation (LTP), electrical stimulation acts like LTD or depotentiation, which might explain how neuromodulation controls epilepsy in terms of the mechanism of action. Interestingly, LTP and kindling have many similarities although cellular mechanisms are different in many respects. On the other hand, prolonged electrical stimulation can elicit LTD that can be opposite phenomenon to LTP where the synaptic transmission is reduced. That is, electrical stimulation can depotentiate synapses that underwent LTP already. However, electrical stimulation can induce enhancement of synaptic strength like LTP, which needs further studies to verify stimulation parameters based on the therapeutic mechanisms [52].

Long lasting hyperpolarization was also suggested as a possible mechanism to reduce seizures in low frequency deep brain electrical stimulation, mediated via $GABA_B$ inhibitory postsynaptic potentials and/or slow after-hyperpolarization [53]. High frequency sinusoidal fields were reported to suppress epileptiform activity in rat hippocampal slices, which was associated with potassium efflux and following depolarization block [54].

The rationale behind chronic intermittent stimulation is to modulate the background brain activity so that epileptic seizures are less likely to occur. This could be either reducing network hypersynchronization or modulating specific pathways involving epileptic network [5]. The mechanisms are expected to be distinct from those of antiepileptic drugs in which the antiepileptic effects are mediated by alterations of cellular and/or synaptic functions. Interestingly, many clinical studies using VNS, DBS, and RNS showed that the stimulation effects were improved over time for up to two years. These observations suggest the possibility that acute seizure inhibition mechanism might be different from chronic prevention or antiepileptogenic mechanism, which is actually an ideal goal of any antiepileptic treatments.

http://dx.doi.org/10.5772/57441

## 4. Conclusion

Recent progress of technological and methodological advances in neuromodulation leads to high possibilities for using these methods more widely to treat intractable epilepsy. These technological advances have been introduced to generate more practical ways of neuromodulation methods for chronic use in clinical fields. One of the remarkable advances is the combination of neuromodulation with early seizure detection algorithm. Along with results from basic researches in animals and humans, the underlying therapeutic mechanisms have been investigated as well. In addition, studies have been performed to improve the therapeutic efficacy by optimizing stimulation parameters, and accuracy of early seizure detection.

Advances in the field of neuromodulation reviewed here, as well as general advances in neuroscience, would provide us a new insight in the treatment of intractable epilepsy. Furthermore, advances in neuromodulation in epilepsy therapy could allow us progress in related neuroscience and clinical fields, especially to investigate and modulate complex neural network based on normal and abnormal brain functions.

## Acknowledgements

This work was supported by the Ewha Global Top 5 Grant 2011 of Ewha Womans University and by the Korean Science and Engineering Foundation (KOSEF) Grant funded by the South Korean government (MOST) (R01-2007-000-11080-0), and by the Basic Science Research Program through the National Research Foundation of Korea (NRF) funded by the Ministry of Education, Science and Technology [R01-2011-0015788 and KRF-2009-006-5721].

## Author details

Hyang Woon Lee*

Department of Neurology, Ewha Womans University School of Medicine and Ewha Medical Research Institute, Seoul, South Korea

## References

[1] Krack P, Hariz MI, Baunez C, Guridi J, Obeso JA. Deep brain stimulation: from neurology to psychiatry? Trends Neurosci 2010;33(10):474-484.

[2] Penfield W, Jasper H. Electrocorticography. In: Epilepsy and the functional anatomy of the human brain. Little, Brown, Boston, pp 692-738, 1954.

[3] Ben-Menachem E. Vagus-nerve stimulation for the treatment of epilepsy. Lancet Neurology 2002;1:477-482.

[4] Ben-Menachem E. Neurostimulation – past, present, and beyond. Epilepsy Currents 2012;12:188-191.

[5] Bergey GK. Neurostimulation in the treatment of epilepsy. Experimental Neurology 2013;244:87-95.

[6] The Vagus Nerve Stimulation Study Group. A randomized controlled trial of chronic vagus nerve stimulation for the treatment of medically intractable seizures. Neurology 1995;45:224-230.

[7] Handforth A, Degiorgio CM, Schachter SC, Uthman BM, Naritoku DK, Tecoma ES, Henry TR, Collins SD, Vaughn BV, Gilmartin RC, Labar DR, Morris GL 3rd, Salinsky MC, Osorio I, Ristanovic RK, Labiner DM, Jones JC, Murphy JV, Ney GC, Wheless JW. Vagus nerve stimulation therapy for partial-onset seizures: a randomized active-control trial. Neurology. 1998 Jul;51(1):48-55.

[8] Groves DA, Brown VJ. Vagal nerve stimulation: a review of its applications and potential mechanisms that mediate its clinical effects. Neurosci Biobehav Rev 2005;29:493-500.

[9] Kerrigan JF, Litt B, Fisher RS, Cranstoun S, French JA, Blum DE, Dichter M, Shetter A, Baltuch G, Jaggi J, Krone S, Brodie M, Rise M, Graves N. Electrical stimulation of the anterior nucleus of the thalamus for the treatment of intractable epilepsy. Epilepsia 2004;45(5):346-354.

[10] Fisher R, Salanova V, Witt T, Worth R, Henry T, Gross R, Oommen K, Osorio I, Nazzaro J, Labar D, Kaplitt M, Sperling M, Sandok E, Neal J, Handforth A, Stern J, DeSalles A, Chung S, Shetter A, Bergen D, Bakay R, Henderson J, French J, Baltuch G, Rosenfeld W, Youkilis A, Marks W, Garcia P, Barbaro N, Fountain N, Bazil C, Goodman R, McKhann G, Babu Krishnamurthy K, Papavassiliou S, Epstein C, Pollard J, Tonder L, Grebin J, Coffey R, Graves N; SANTE Study Group. Electrical stimulation of the anterior nucleus of thalamus for treatment of refractory epilepsy. Epilepsia 2010;51:899–908.

[11] Graber KD, Fisher FS. Deep brain stimulation for epilepsy: animal models. In: Noebels JL, Avoli M, Rogawski MA, Olsen RW, Delgado-Escueta AV, editors. Jasper's Basic Mechanisms of the Epilepsies [Internet]. 4th edition. Bethesda (MD): National Center for Biotechnology Information (US); 2012.

[12] Mirski MA, Tsai YC, Rossell LA, Thakor NV, Sherman DL. Anterior thalamic medication of experimental seizures: selective EEG spectral coherence. Epilepsia 2003;44:355-365.

[13] Lesser RP, Kim SH, Beyderman L, Miglioretti DL, Webber WR, Bare M, Cysyk B, Krauss G, Gordon B. Brief bursts of pulse stimulation terminate afterdischarges caused by cortical stimulation. Neurology 1999;53:2073-2081.

[14] Motamedi GK, Lesser RP, Miglioretti DL, Mizuno-Matsumoto Y, Gordon B, Webber WR, Jackson DC, Sepkuty JP, Crone NE. Optimizing parameters for terminating cortical afterdischarges with pulse stimulation. Epilepsia. 2002;43(8):836-46.

[15] Fountas KN, Smith JR, Murro AM, Politsky J, Park YD, Jenkins PD. Implantation of a closed-loop stimulation in the management of medically refractory focal epilepsy: a technical note. Stereotact Funct Neurosurg. 2005;83(4):153-158.

[16] Morrell MJ, RNS System in Epilepsy Study Group. Responsive cortical stimulation for the treatment of medically intractable partial seizures. Neurology 2011;77:1295-1304.

[17] Liu C, Wen XW, Ge Y, Chen N, Hu WH, Zhang T, Zhang JG, Meng FG. Responsive neurostimulation for the treatment of medically intractable epilepsy. Brain Research Bulletin 2013;97:39-47.

[18] Lehnertz K, Elger CE. Can epileptic seizures be predicted? Evidence from nonlinear time series analysis of brain electrical activity. Physical Review Letters 1998;80: 5019–5023.

[19] Lehnertz K, Andrzejak RG, Arnhold J, Kreuz T, Mormann F, Rieke C. Nonlinear EEG analysis in epilepsy: its possible use for interictal focus localization, seizure anticipation, and prevention. Journal of Clinical Neurophysiology 2001;18:209–222.

[20] Tito M, Cabrerizo M, Ayala M, Barreto A, Miller I, Jayakar P, Adjouadi, M. Classification of electroencephalographic seizure recordings into ictal and interictal files using correlation sum. Computers in Biology and Medicine 2009;39:604–614.

[21] Rabbi, A.F., Aarabi, A., Fazel-Rezai, R., 2010. Fuzzy rule-based seizure prediction based on correlation dimension changes in intracranial EEG. Conference Proceedings: Annual International Conference of the IEEE Engineering in Medicine and Biology Society 2010;2010:3301–3304.

[22] Martinerie J, Adam C, Le Van Quyen M, Baulac M, Clemenceau S, Renault B, Varela FJ. Epileptic seizures can be anticipated by non-linear analysis. Nature Medicine 1998;4:1173–1176.

[23] Le Van Quyen M, Adam C, Martinerie J, Baulac M, Clemenceau S, Varela F. Spatiotemporal characterization of non-linear changes in intracranial activities prior to human temporal lobe seizures. European Journal of Neuroscience 2000;12: 2124–2134.

[24] Le Van Quyen M, Martinerie J, Navarro V, Boon PD, Have M, Adam C, Renault B, Varela F, Baulac M. Anticipation of epileptic seizures from standard EEG recordings. Lancet 2001;357:183–188.

[25] Navarro V, Martinerie J, Le Van Quyen M, Baulac M, Dubeau, F, Gotman J. Seizure anticipation: do mathematical measures correlate with video-EEG evaluation? Epilepsia 2005;46:385–396.

[26] Mormann F, Lehnertz K, David P, Elger CE. Mean phase coherence as a measure for phase synchronization and its application to the EEG of epilepsy patients. Physica D 2000;144:358–369.

[27] Wang L, Wang C, Fu F, Yu X, Guo H, Xu C, Jing X, Zhang H, Dong X. Temporal lobe seizure prediction based on a complex Gaussian wavelet. Clinical Neurophysiology 2011;122:656–663.

[28] Le Van Quyen M., Soss, J., Navarro, V., Robertson, R., Chavez, M., Baulac, M., Martinerie, J. Preictal state identification by synchronization changes in long-term intracranial EEG recordings. Clinical Neurophysiology 2005;116:559–568.

[29] Litt, B., Esteller, R., Echauz, J., D'Alessandro, M., Shor, R., Henry, T., Pennell, P., Epstein, C., Bakay, R., Dichter, M., Vachtsevanos, G., 2001. Epileptic seizures may begin hours in advance of clinical onset: a report of five patients. Neuron 2001;30:51–64.

[30] Jouny CC, Franaszczuk PJ, Bergey GK. Signal complexity and synchrony of epileptic seizures: is there an identifiable preictal period? Clinical Neurophysiology 2005;116:552–558.

[31] Durand DM. Control of seizure activity by electrical stimulation: effect of frequency. Conf Proc IEEE Eng Med Biol Soc 2009;2375.

[32] Nelson TS, Suhr CL, Freestone DR, Lai A, Halliday AJ, McLean KJ, Burkitt AN, Cook MJ. Closed-loop seizure control with very high frequency electrical stimulation at seizure onset in the GAERS model of absence epilepsy. Int J Neural Syst 2011;21(2): 163-173.

[33] Rajdev P, Ward M, Irazoqui P. Effect of stimulus parameters in the treatment of seizures by electrical stimulation in the kainate animal model. Int J Neural Syst 2011;21(2):151-162.

[34] Liang SF, Liao TC, Shaw FZ, Chang DW, Young CP, Chiueh H. Closed-loop seizure control on epileptic rat models. J Neural Eng 2011;8(4):045001.

[35] Dayan E, Censor N, Buch ER, Sandrini M, Cohen LG. Noninvasive brain stimulation: from physiology to network dynamics and back. Nature Neuroscience 2013;16(7): 838-844.

[36] Censor N, Cohen LG. Using repetitive transcranial magnetic stimulation to study the underlying neural mechanisms of human motor learning and memory. J Physiol 2011;589:21-28.

[37] Muellbacher W, et al. Early consolidation in human primary motor cortex. Nature 2002;415:640-644.

[38] Pascual-Leone A, Grafman J, Hallett M. Modulation of cortical motor output maps during development of implicit and explicit knowledge. Science 1994;263:1287-1289.

[39] Fregni F, Otachi PTM, do Valle A, Boggio PS, Thut G, Rigonatti SP, Pascual-Leone A, Kette D. Valente KD. A randomized clinical trial of repetitive transcranial magnetic stimulation in patients with refractory epilepsy. Ann Neurol 2006;60:447-455.

[40] Hsu WY, Cheng CH, Lin MW, Shih YH, Liao KK, Lin YY. Antiepileptic effects of low frequency repetitive transcranial magnetic stimulation: a meta-analysis. Epi Res 2011;96:231-240.

[41] Huang YZ, Edwards MJ, Rounis E, Bhatia KP, Rothwell JC. Theta burst stimulation of the human motor cortex. Neuron 2005;45:201-206.

[42] Osorio I, Overman J, Giftakis J, Wilkinson SB. High frequency thalamic stimulation for inoperable mesial temporal epilepsy. Epilepsia 2007;48(8):1561-1571.

[43] Tykocki T, Mandat T, Kornakiewicz A, Koziara H, Nauman P. Deep brain stimulation for refractory epilepsy. Arch Med Sci 2012;8(5):805-816.

[44] Gale K. Subcortical structures and pathways involved in convulsive seizure generation. J Clin Neurophysiol 1992;9(2):264-277.

[45] Deransart C, LePham BT, Marescaux C, Depaulis A. Role of the subthalamo-nigral input in the control of amygdale-kindled seizures in the rat. Brain Res 1998;807:78-83.

[46] Krauss GL, Fisher RS. Cerebellar and thalamic stimulation for epilepsy. Adv Neurol 1993;63:231-245.

[47] Velasco F, Carrillo-Ruiz JD, Brito F, Velasco M, Velasco AL, Marquez I, Davis R. Double-blind, randomized controlled pilot study of bilateral cerebellar stimulation for treatment of intractable motor seizures. Epilepsia. 2005 ;46(7):1071-1081.

[48] DeGiorgio CM, Soss J, Cook IA, Markovic D, Gornbein J, Murray D, Oviedo S, Gordon S, Corralle-Leyva G, Kealey CP, Heck CN. Randomized controlled trial of trigeminal nerve stimulation for drug-resistant epilepsy. Neurology. 2013;80(9):786-791.

[49] Pop J, Murray D, Markovic D, Degiorgio CM. Acute and long-term safety of external trigeminal nerve stimulation for drug-resistant epilepsy. Epilepsy Behav 2011;57:574-576.

[50] Yamamoto J, Ikeda A, Satow T, Takeshita K, Takayama M, Matsuhashi M, Matsumoto R, Ohara S, Mikuni N, Takahashi J, Miyamoto S, Taki W, Hashimoto N, Rothwell JC, Shibasaki H. Low-frequency electric cortical stimulation has an inhibitory effect on epileptic focus in mesial temporal lobe epilepsy. Epilepsia 2002; 43(5): 491-495.

[51] Lian J, Bikson M, Sciortino C, Stacey WC, Durand DM. Local suppression of epileptiform activity by electrical stimulation on rat hippocampus in vitro. J Physiol 2003;547(Pt 2): 427-434.

[52] Li H, Chen A, Xing G, Wei M-L, Rogawski MA. Kainate receptor-mediated heterosynaptic facilitation in the amygdala. Nat Neurosci 2001;4:612-620.

[53] Toprani S, Durand DM. Long-lasting hyperpolarization underlies seizure reduction by low frequency deep brain electrical stimulation. J Physiol 2013; Epub in advance.

[54] Bikson M, Lian J, Hahn PJ, Stacey WC, Sciortino C, Durand DM. Suppression of epileptiform activity by high frequency sinusoidal fields in rat hippocampal slices. J Physiol 2001;531(1):181-191.

# Chapter 9

# Psychogenic Non-Epileptic Spells

Batool F. Kirmani, Diana Mungall Robinson,
Jose Aceves, David Gavito, Richard Phenis and
Daniel Cruz

Additional information is available at the end of the chapter

http://dx.doi.org/10.5772/57440

## 1. Introduction

The psychogenic non-epileptic spells are the most frequent referrals to the specialized epilepsy center because of the intractable nature of the spells. Psychogenic non-epileptic spells (PNES) are also referred to as "pseudoseizures", psychogenic seizures, non-epileptic seizures or stress-related spells [1]. PNES is a form of conversion disorder but misdiagnosis and delay in diagnosis is common [2,3,4]. The reason for misdiagnosis is overinterpretation of the EEG and lack of data regarding the semiology of the spells. The gold standard is to capture these spells for definitive diagnosis under medical withdrawal in an epilepsy monitoring unit by video-EEG. The PNES are not epileptic seizures but they are extremely disabling and adversely affect quality of life [5,6].

## 2. Epidemiology of psychogenic non-epileptic seizures

The prevalence of PNES was reported by Benabadis and colleagues in a retrospective review. Benbadis et al conducted a retrospective review of available prevalence, incidence, and hospital-based data to estimate the prevalence of PNES. The prevalence was based on the following generally accepted data: prevalence of epilepsy is 0.5-1%, and proportion of intractable epilepsy is 20-55%, and 10-20% of patients referred to epilepsy centers are found to have PNES. From these data, using a prevalence of epilepsy of 0.5% to 1%, a low estimate was determined to be 1/50,000 and a high estimate was determined to be 1/3,000. The conclusion of this retrospective review was that there was a prevalence of 2 to 33 people per 100,000 [7]. The only population based study which estimated the incidence of PNES was conducted by Sigurdardottir et al. Sigurdardottir and colleagues conducted a retrospective chart review of all long-term video EEG studies made in Iceland from January 1992 to December 1996 in order to determine the incidence of PNES. All patients were aged 15 or greater and had been diagnosed with PNES at National University Hospital (Landspitalinn), Iceland. The diagnosis of PNES was determined by clinical observation and EEG, and all of these patients underwent long-term video-EEG monitoring (LVEM). A total of 14 patients met the inclusion criteria and were aged 16-54 years (mean 27.6). There were 11 female patients (78.6%). There was an average annual population of 200,191 persons aged 15 years or greater for a total of 1,000,955 persons over the entire study period. The incidence of PNES was 1.4 in a 100,000 population. The 15-24 year old age group had the highest age-specific incidence. The incidence decreased in subsequent older age groups. Half of the patients (N=7) also had epilepsy, with the majority having generalized tonic-clonic seizures only (N=3), followed by Generalized tonic clonic seizures (GTC )and myoclonus (N=2), tonic seizures (N=1), and absences (N=1). 2 of 7 patients with PNES and without epilepsy were on antiepileptic drug (AED) treatment prior to PNES diagnosis. The researchers concluded that the incidence in people aged 15-24 years was ~5% of the incidence of epilepsy and 4% of reported epilepsy from Iceland for persons aged ≥15 years [8]. Szaflarski and colleagues conducted the retrospective study which determined the incidence of PNES in a medium-sized urban community. Adult patients who underwent prolonged video and EEG monitoring (PVEM) were identified between January 1, 1995 and December 30, 1998 at the University Hospital or Veteran Administration Medical Center in Cincinnati, OH. Patients were classified into four groups: definite PNES, possible PNES, possible epilepsy, and definite epilepsy. A board-certified electroencephalographer reviewed all tracings and videos. Population characteristics were examined with univariate analysis and comparison of similar features between patients with PNES and those with epilepsy were analyzed with bivariate analysis. During the study period, 600 patients were monitored with 3 patients at the Veteran Affairs Medical Center and 174 at the University of Ohio fulfilling the residence criteria. Definite PNES was diagnosed in 77 patients and definite epilepsy was diagnosed in 85 patients. Also, there was an increase in the incidence of PNES over the study from 1.88/100,000 in 1995 to 4.6/100,000 in 1998 (mean incidence 3.03/100,000). Patients aged 25-45 years had the highest incidence of PNES (4.38/100,000). Groups of patients with PNES had a shorter average duration of illness before diagnosis and were more likely to have a history of psychiatric disorders (48.6% versus 30.4%; p=0.023) than definite epilepsy patients. Epilepsy patients were more likely to be treated with more antiepileptic drugs before admission than PNES (4.85 versus 2.53; p <0.001). No significant differences were found in the PNES compared with epilepsy groups in gender (women 73% versus 60%), age (37.2 versus 37 years), or history of febrile seizures, head trauma, or family history of epilepsy. The authors conclude that improved access to PVEM and higher clinician awareness may be related to the increasing incidence of PNES over the study period [9].

### 2.1. Psychogenic non-epileptic seizures in different population subgroups

#### *2.1.1. Psychogenic non-epileptic seizures in women*

Psychogenic seizures are more common in women. Women account for approximately 70 to 80% incidence of PNES, but the incidence varies depending on the etiology [6,10, 11,12]. There are certain factors that contribute to the increase risk of PNES. The most important factor is the history of sexual abuse seen more commonly in women. This issue has also been addressed historically by Freud based on his observations. Freud's earlier observations describe Hysteria which is now the basis of the concept of psychogenic non-epileptic seizures. Hysteria is a Greek word which means "wandering uterus" and is related to history of sexual abuse in women and repressed sexual drives [13,14]. Alper and colleagues in a case series described the history of sexual abuse in approximately 25% of the patients with PNES [12].

#### *2.1.2. Pseudoseizures in the elderly*

There are several studies which show the prevalence of pseudoseizures in the elderly. McBride et al performed a retrospective chart review in elderly patients in order to determine the utility and results of video-EEG monitoring. All patients admitted to the epilepsy monitoring unit at Columbia-Presbyterian Medical Center from January 21, 1991 to April 12, 1999 aged 60 years and older were reviewed. Reasons for admission included diagnosis of paroxysmal events, further characterization of known seizure, pre-surgical evaluation, medication adjustment or toxicity, and evaluation to rule out non-convulsive status epilepticus or subclinical seizures. Ninety-four patients were identified with 99 admissions, with five patients having two separate admissions, comprising 8% of all admissions. There were 62 females and 37 males. On average, patients were 70 years old and ranged from 60-94. The mean length of the stay was 3.8 days and ranged from 1-14 days. The most common reason for admission was to diagnose the nature of paroxysmal events (56%). A total of 118 epileptic seizures were recorded in 46 patients. Ninety-eight non-epileptic events were recorded in 27 patients. Both epileptic seizures and non-epileptic events were found in four patients. Of those with non-epileptic events, 13 patients had psychogenic seizures. There were epileptiform discharges in 26% of patients with non-epileptic events and 76% of patients with epileptic events had interictal epileptiform discharges. The authors concluded that in the majority of patients, video-EEG monitoring in elderly patients led to a definitive diagnosis. Also, they concluded that non-epileptic events are common in the elderly, including PNES, and are often misdiagnosed and mistreated as epileptic seizures [15]. Abubakr et al performed a retrospective chart review study to report the results of video/EEG recordings in patients aged 60 and older at the new Jersey Neuroscience Institute. An electronic medical record search between December 1999 and December 2001 was reviewed for all elderly patients admitted to the epilepsy monitoring unit (EMU) found 58 patients who underwent video/EEG. The elderly population accounted for 17% of EMU admissions. The reasons for video/EEG monitoring for study patients were diagnosis of events (33 patients, 57%), characterization and localization of seizure (21 patients, 36%), adjustment of medication (2 patients, 3%), and non-convulsive status epilepticus (2 patients, 3%). Study subjects were ranged between 60-91 years old and 45% were females. Six patients had psychogenic non-epileptic seizures (PNES), with five of them being women and 4 of them being greater than 70 years old. One patient presented with abdominal spasms and the others with motor symptoms. Two of six patients had a suspected diagnosis of PNES on admission. Physiologic non-epileptic seizure was the most common diagnosis and occurred in 26 patients (45%). The diagnosis of non-epileptic seizures in 27% of these patients resulted in AED discontinuation. The most common seizure type was complex partial seizures and occurred in 23 patients (40%). Six of these patients had both complex partial seizures and secondary generalization. The authors concluded that in the majority of cases, video/EEG monitoring in the elderly results in a definitive diagnosis and assists physicians with antiepileptic drug therapy management decisions [16]. Kawai et al conducted a retrospective review of video-EEG monitoring in geriatric veterans from 1999 to 2006. All patients admitted to the epilepsy monitoring unit at Michael E. DeBakey Veterans Affairs Medical Center of Houston, Texas were reviewed. Of the 440 admissions during this time, 71 of these patients were aged 60 and older, which included multiple admissions. Ninety-four percent of these were males, and the average age was 68 years. The mean duration of monitoring studies was 73.7 hours (range 2-96 hrs). Thirty-four of 71 patients (48%) had typical events, including 12 with epileptic events (35%). Nine patients (75%) had temporal lobe seizures, 2 patients (17%) had extratemporal seizures, and 1 patient (8.3%) had poorly localized seizures. The most common etiology was not identifiable (7 patients), and intracranial hemorrhage and history of tumor resection (2 patients). Non-epileptic events were seen in 22 of 71 patients (65%). Ten patients (45%) had PNES and 12 patients (55%) had physiologic non-epileptic seizures. Of the patients with non-epileptic seizures, 14 of 22 (65%) were on AEDs before video-EEG monitoring with 6 having PNES and 8 having physiologic non-epileptic seizures. The authors concluded that video-EEG monitoring in elderly patients was useful to guide physicians to the appropriate diagnosis and treatment of paroxysmal seizure-like symptoms [17].

#### *2.1.3. Psychogenic Non-Epileptic Seizures (PNES) in children*

Psychogenic non epileptic seizures (PNES) in children are transient, episodic alterations in behavior that mimic true epileptic seizures but without abnormal electrical discharges. PNES was found in 3.5 % [18] in one study and 7% [19] in another study in children that were evaluated in neurology clinics for persistent seizures. There is a paucity of literature regarding this entity in children and adolescents. In most cases, there is delay in the diagnosis of months [11,18,20]. During the delay, patients are labeled as being epileptic and are usually started on antiepileptic medication. Poor response to the medication and frequent visits to the emergency room are common. The cost of misdiagnosing PNES as epileptic seizures is very high from economic and psychosocial aspects. The distinction of PNES from epileptic seizures is difficult to be made solely on clinical grounds. The clinical manifestations of PNES vary according to age as reported by some investigators. Younger children tend to manifest more subtle motor activity which includes eye fluttering, head shaking, staring, unresponsiveness and limpness. In contrast, adolescents tend to manifest prominent motor activity which includes side to side head movements, thrashing or flailing movements of the extremities, generalized arrhythmical jerking and pelvic thrusting [11]. Triggering factors include school phobia, familial problems, social difficulties with peers or friends and sexual abuse [18,20]. Comorbid conditions include anxiety disorders and mood disorders [19,21]. The differential diagnosis of PNES is broad and includes frontal lobe seizures, vasovagal syncope with anoxic seizures, breath- holding spells, self-stimulatory behavior, gastroesophageal reflux, stereotypes, complex motor tics, parasomnias, paroxysmal kinesigenic and non-kinesigenic dyskinesias.

PNES should be strongly suspected when there are atypical clinical features, poor response to medications in spite of adequate trials and when several routine electroencephalograms have been reported within normal limits [22]. Metrick et al conducted a retrospective chart review and analyzed the records of children referred for the treatment of intractable epilepsy. A total of 222 records were found for children aged <16 years admitted to the MINCEP Epilepsy Program for Children in St. Paul, Minnesota for evaluation and treatment of refractory seizures between August 1986 and August 1988. Children with normal and abnormal intelligence were included. All children had at least 24 hours of video-EEG monitoring. Of the 222 children admitted, 27 patients (12%) had non-epileptic events on video-EEG monitoring. Study subjects were aged 7 months to 16 years (median 8.4 years) with 18 females. The study called these 4 different group pure psychogenic events (5 patients), psychogenic events plus epileptic seizures (3 patients), pure non-epileptic physiologic events (5 patients), and non-epileptic physiologic events plus seizures (14 patients). Parents or caretakers identified a history of multiple seizure types in all groups except the children with pure psychogenic seizures. Twenty-two patients (64%) had a history of status epilepticus. Twenty-five of 27 patients had a history of interictal epileptiform abnormalities on previous routine EEGs. Eight patients (30%) had their AEDs completely discontinued after the non-epileptic events were identified. Nine other patients (33%) were able to reduce the number of AEDs in the treatment regimen. The researchers concluded that in all children with refractory seizures or multiple seizure types a diagnosis of non-epileptic events should be considered [23]. Prolonged video electroencephalogram is the gold standard for diagnosis and an effort should be made to capture the typical spells that may occur spontaneously or induced by provocative measures. Definite diagnosis is made when several episodes are captured which are not associated with abnormal EEG changes. Careful consideration must be given to the fact that lack of epileptic changes in the electroencephalogram does not conclusively indicate that the episode is PNES. Frontal lobe seizures and some complex motor seizures originating from deep seated focus may manifest motor activity and normal electroencephalograms resembling PNES. Prolonged video electroencephalograms capturing several typical events may be required to make the correct diagnosis. Measuring serum prolactin levels to distinguish PNES from epileptic seizures is not routinely used in children and may have several limitations. Once the diagnosis is established, information needs to be conveyed to the family and patient in a non-judgmental tactful manner and patient needs to be referred to a mental health specialist to determine appropriate therapy which include counseling and if required psychotropic medication to treat co-morbid conditions. Prognosis in children is much better than in adults and a significant percentage were PNES free several years after the diagnosis was made [22,24].

*2.1.4. Patients with dual diagnosis of both seizures and Psychogenic Non-Epileptic Seizures*

The most complicated patients are those who have both epilepsy and Psychogenic non-epileptic seizures. The gold standard remains prolonged video–EEG monitoring in an epilepsy monitoring unit to characterize all events for definite diagnosis. Ten to 40 % patients with PNES also have true epilepsy as reported in the literature [25-29]. Interictal EEG abnormalities have also been reported in patients with PNES but they should not be interpreted as evidence of epilepsy [30].

Benbadis and colleagues performed a retrospective chart review study to determine the proportion of patients with psychogenic non-epileptic seizures (PNES) who also have evidence of epilepsy. The authors reviewed all adult patients with PNES who underwent EEG-video monitoring from January 1 to December 31, 1999. Patients were excluded if their episodes mimicked simple partial seizures or if they had a loss of consciousness. One or both board-certified electroencephalographers determined if there was evidence for epilepsy defined by epileptiform discharges, including sharp waves or spikes, spike-wave complexes, polyspikes, or any ictal pattern. Over this one-year period, 211 patients were monitored and 32 patients (15%) were diagnosed with PNES. Study patients mean age was 33.8 (range 19-72) and 20 (62%) were females. Three patients (9.4%) of the 32 patients with PNES had interictal epileptiform discharges, 20 (62%) had completely normal EEG, 6 (19%) had normal variants (three wicket spikes, three small sharp spikes), and 3 (9.4%) had mild nonspecific abnormalities (mild slowing or asymmetry). The three patients were on lower doses or no antiepileptic drugs. The authors concluded that epilepsy coexists with PNES in a small portion of patients [31]. Martin et al conducted a retrospective study to examine the frequency of epilepsy in patients with a definitive diagnosis of PNES by video-EEG monitoring from July 1, 1998 to December 31, 2002. All patients consecutively admitted to the video-EEG monitoring unit at the University of Alabama at Birmingham Hospital were reviewed. Patients were referred for characterization of paroxysmal events for undiagnosed events with uncertainty of epileptic seizures versus psychogenic evens, probable epileptic seizures with classification of seizure type, or probable epileptic seizures with localization of seizures for possible surgery. This was the first video-EEG monitoring for all patients with the average duration of 3 days (range 1-7 days). A total of 1,590 patients received a definitive diagnosis and were included in the study of 2,007 patients receiving video-EEG monitoring. PNES was diagnosed in 514 patients (32.3%) with 29 of these patients (5.3%) having both PNES and epilepsy. Other than PNES, non-epileptic diagnoses occurred in 65 patients (3.2%), including sleep disorders, migraine, panic attacks, dysautonomia, movement disorders, TIA, cough syncope, and vestibular symptoms. The authors concluded that in patients referred for video-EEG monitoring there is little overlap between epileptic seizures and PNES when strict diagnostic criteria are applied [32]. Kirmani et al conducted a retrospective chart review of all patients with epilepsy admitted to the Scott and White Hospital epilepsy monitoring unit in Temple, TX from 2008-2011. Fourteen patients who were admitted to the EMU due to increased frequency of seizures or for characterization of new kinds of spells were found to have epilepsy and concomitant PNES. The mean age of study patients was 43 years (range 21-67 years) with 10 females (71.5%) and four males (28.5%). The majority of patients had partial epilepsy (N=11, 78.6%) followed by generalized epilepsy

(N=3, 28.4%). Concurrent psychological disorders were seen in 12 of 14 patients, including depression (64%), anxiety (50%), and physical/sexual abuse (29%). On average, study patients were on 2.6 AEDs (range 1-5) and had failed 2.1 AEDs (range 0-9). Eleven patients (78.6%) improved with regular counseling. Three patients (21.4%) with mental retardation did not show improvement with regular counseling. The authors concluded that video-EEG monitoring is helpful to characterize increased frequency and new episode characterizations as well as the need for a multidisciplinary team approach between neurologists, epileptologists, psychiatrists, and psychologists to best manage these patients [33]. The reason for dual diagnosis is the development of psychiatric problems in patients with chronic long standing epilepsy or the presence of concomitant psychiatric disorders [34]. Most of the patients generally have fairly well controlled epilepsy when they develop PNES but still represent a difficult group of patients regarding management [35,36].

## 3. Clinical semiology of Psychogenic Non-Epileptic Seizures

Careful history plays a key role in the diagnosis of non-epileptic events. The history should include the seizure triggers and careful history of the semiology of the seizures from the witness. The history should also include the duration, alteration of consciousness, type of convulsive movements experienced during a seizure, presence or absence of tongue bite, urinary incontinence, autonomic symptoms, emotional symptoms like weeping or crying and eye opening and closure as these will all help in establishing the correct diagnosis.

The PNES usually occur in front of the witness or in a clinical setting [37]. The PNES occur during daytime but not during sleep. Presence of nocturnal events raise suspicion for epileptic rather than nonepileptic seizures [38]. The other finding is the frequency of seizures. Non-epileptic seizures are more common than epileptic seizures and frequency may range from daily spells to several times a week [39].

Ictal features of PNES include purposeful or semipurposeful movements, thrashing, writhing, side-to-side head jerking and pelvic thrusting which are different from synchronized tonic – clonic activity in true epileptic seizures [38-40]. Leis and colleagues conducted a retrospective chart review study to analyze ictal features in patients with psychogenic seizures undergoing video-EEG monitoring. At the Epilepsy Unit of University of Iowa, 254 patients were monitored, and 47 (18%) had psychogenic seizures and videotaped recordings to analyze their typical events. Twenty-seven patients were female (57%) and 20 were men (43%). There was a mixed seizure disorder in 11 of 47 patients (23%). The most common ictal presentation was unresponsiveness without predominant motor manifestations. The motor characteristics of out-of-phase limb movements, side-to-side head movements, and pelvic thrusting had been previously considered to distinguish psychogenic seizures, they were observed infrequently (19%, 15%, and 8% respectively). Antiepileptic drug therapy was administered to 35 patients (74%) for their spells. Of these 35 patients, 25 (69%) had pure psychogenic seizures. Six of these 25 patients (4 women, mean age 27 years) with pure psychogenic seizures were treated pharmacologically for status epilepticus entirely due to observation without even a cursory neurologic exam or chart review. Aggressive treatment of status epilepticus in 1 patient escalated to the point of respiratory arrest in 1 woman who was 2 months pregnant. A psychogenic cause to these spells was not considered in the differential diagnosis until all the patients failed to respond to pharmacologic treatment. The authors concluded that in treatment of seizures, even in the acute care of presumed status epilepticus, the diagnosis must not be based solely on inspection and should be supported by the history and physical examination findings [41]. The other study which provided detailed semiology of PNES was conducted by Seneviratne and colleagues. Seneviratne et al conducted a retrospective study of the semiology of PNES captured by video-EEG monitoring and categorize the observed patterns. From January 2002 to June 2007 the medical records were reviewed of all adult patients who underwent monitoring (mean 3, range 1-8 days) at two tertiary care epilepsy centers. Patients with PNES and no background of epilepsy were selected for the study. Sixty-one patients were identified with 330 PNES events with a mean number of 5 events recorded per patient. There were 45 females and 16 males with a mean age of 38 years (range: 16-83 years). Three types of motor manifestations were detailed on visual analysis of PNES events. 1) Rhythmic Motor PNES: 47.6% of all PNES events, rhythmic tremor, trembling, or rigor like movements of the upper limbs more commonly than the lower limbs. 2) Hypermotor PNES: 3.3 % of all PNES events had hyperkinetic or hypermotor movements with violent thrashing, punching or kicking involving the extremities or trunk. 3) Complex Motor PNES: 10% of all PNES events had complex motor movements with complex and multifocal asymmetrical movements of both proximal and distal extremities with flexion, extension, and ab/adduction. 4) Dialeptic PNES: 11.2% of all PNES events had prolonged, motionless, unresponsive patients with no motor manifestations who appeared to be in a coma-like state unresponsive to external stimuli. 5) Nonepileptic auras: 23.6% of all PNES events had various subjective sensations without any external manifestations described by the patients as "I feel weird", "zoning out", and "I am going through a trance". 6) Mixed PNES: 5.2% of all PNES events had a combination of types 1-5. Eighty-two percent of cases had the same semiologic type in a given patient. The authors concluded that the PNES patients they studied had highly stereotypic events within and across individual patients [42].

The other features including clenched mouth during a tonic seizure and injuries on the tip of the tongue rather than the sides points towards the diagnosis of PNES [43]. Ictal eye closure is also considered a sign of a psychogenic event [43,44]. Chung et al conducted a retrospective study of video-EEG monitoring data to study whether persistent ictal eye opening and closure was reliable to differentiate between ES and PNES. From July 2003 to June 2004, 234 consecutive patients (age 6-65 years) underwent long-term video-scalp EEG monitoring at the Barrow Neurologic Institute. 221 patients had a total of 938 ictal events (median number of seizures per patient=4). Fifty-two of 221 patients (23.5%) had PNES and 156 (70.6%) had ES. Seventy-five percent of patients with PNES were female. During habitual seizures, 50 of 52 PNES patients consistently closed their eyes for the entire duration of the seizure and a few who closed their eyes forcefully with facial frowning. However, 152 of 156 patients with ES had their eyes deviated to one side or were open. Rhythmic eye blinking was seen during tonic-clonic activity, followed be postictal confusion with eye closure. There was a positive predictive value of 0.943 (sensitivity of 96.2% and specificity of 98.1%) of ictal eye closure indicating a high likelihood of PNES. Conversely, true epileptic seizures had an ictal eye opening and had a high positive predictive value of 0.987 (sensitivity 98.1% and specificity 96.2%). Thus the authors concluded that careful history taking of seizure semiology may help discern between ES and PNES and home video clips of a seizure may help to make the diagnosis without long-term monitoring [44]. Autonomic symptoms are absent in PNES but weeping, ictal stuttering, partial preservation of consciousness and later recall of the ictal event also suggest PNES [38,45,46]).

## 4. Diagnosis of Psychogenic Non-Epileptic Seizures

The evaluation and diagnosis of PNES requires careful history and diagnostic testing. Ali et al conducted a literature review of PNES to make suggestions for treatment and to aid clinicians in identifying PNES episodes. The mean time between developing clinical symptoms and establishing a correct diagnosis of PNES is 7.2 years. Physicians can facilitate early diagnosis by referring patients with atypical features for video-EEG monitoring. General features that can raise a physician's suspicions include high seizure frequency with multiple emergency room visits, association with multiple other psychiatric disorders, comorbid personality disorders, abuse history, lack of response with treatment, and lack of loss of control over bladder or bowl during episodes. PNES pre-ictal features include pseudo sleep, which shows a sensitivity of 56% and specificity of 100% for pseudoseizure. PNES events typically are witnessed and begin gradually at time of stress or visual/auditory stimuli. Ictal features of PNES include asynchronous contractions, non-stereotypic movements that change during the episode, and a lack of rapid contractions with slow relaxation pattern seen in true epileptic clonic seizures. Patients may exhibit side-to-side head movements, forceful eye closure, and ictal vocalizations. Post-ictal features are easy to recognize, and physicians should watch for a short duration (~1 minute) shallow, irregular, and quiet breathing pattern, as well as the absence of confusion, headache, and fatigue. Lab findings that are not present after PNES episodes but are present after epileptic seizures are elevated serum prolactin, creatine kinase, ammonia, and white blood cell count. Video-EEG is highly recommended for patients with atypical features and in one study found that 24 percent of subjects had been misdiagnosed with epilepsy that had an accurate diagnosis of PNES. When a patient is informed of their diagnosis, the should be referred to a psychiatrist for treatment and neurologists remove AEDs, but the complete care away from neurology until the spells have decreased. The authors believe this is due to the disruption of the rapport that neurologists have with the patient, negatively affects the outcome of PNES, and premature discharge to psychiatry may increase patient resistance to accept the diagnosis. Predictors of good outcome include shorter duration of spells, presentation in children and adolescents, mild psychiatric history, identifiable acute psychological trauma, and independent living. The authors conclude that skilled clinicians can make a diagnosis based on clinical findings and their guidance can be used to help clinicians make the diagnosis of PNES [47].

Prolonged video–EEG monitoring is now considered the gold standard. Additional studies include Single positron emission computed tomography (SPECT), saline provocation during video-EEG monitoring, serum prolactin levels, and neuropsychological testing.

### 4.1. Role of video-EEG monitoring in the diagnosis of PNES

Benbadis et al conducted a retrospective chart review study of patients who underwent video-EEG monitoring at an epilepsy center in order to investigate the disposition outcomes from January to December 2012. The charts of all adults and children sent for inpatient video-EEG monitoring (>24 hours) were reviewed at University of South Florida-Tampa General Hospital. During that period of time, there were 251 patients monitored for 1-7 days (mean=2.8 days). Non-epileptic events were found in 30% of patients (N=75). Six of the 75 patients had evidence of coexisting epilepsy. Of the 69 patients with non-epileptic events without coexisting epilepsy (pure non-epileptic events), psychogenic non-epileptic seizures (PNES) were found in 61. Patients diagnosed with PNES had their AEDs gradually discontinued and were referred for mental health treatment. Fifty-eight patients with epileptic events were candidates for resective surgery, 47 with epileptic events were non-surgical candidates, and in 57 patients no events were recorded. The authors concluded that there are many possible outcomes of video-EEG monitoring, and in their case the two largest groups were PNES (30%) and surgical candidates (23%) [48].

Zhang et al conducted a retrospective study to compare the clinical outcomes after video-EEG monitoring of patients diagnosed with PNES and epileptic seizures (ES). From November 2006 and January 2008, patients were followed after admission for elective video-EEG monitoring. Sixty-two of an eligible 103 patients agreed to follow up via telephone or mail questionnaires after discharge from monitoring. Follow up occurred for 6-16 months. ES without PNES was identified in 66% of patients (N=41), followed by PNES without ES in 18% of patients (N=11), 10% (N=6) had both ES and PNES, and 6% (N=4) with indeterminate diagnoses. Improvement of overall condition was reported in ~50% of patients in each group. Both groups showed a decrease in seizure frequency and had a significant decrease in AED use at follow up. The PNES group showed a greater more sustained decrease in AED use at follow up than the ES group. The ES group reported a statistically significant improvement in Seizure Worry (P=0.003), Medication Side Effects (P<0.001), and Social Function (P<0.001) [49]. Benbadis and colleagues conducted a retrospective review to analyze the yield of short-term outpatient EEG monitoring for suspected PNES. Seventy-four adult cases of short-term outpatient EEG video monitoring were found from October 2000 to January 2003 at University of South Florida-Tampa General Hospital. Each short-term monitoring session lasted between 1-2 hours. The suspected diagnosis of PNES was confirmed in 66% of cases (N=49). No event was induced in 23 patients and 2 patients had an induced event that was not habitual type. The authors concluded that for confirmation of a suspected diagnosis of PNES, short-term outpatient video EEG monitoring obviated the need for long-term inpatient video EEG monitoring [50].

### 4.2. Role of additional diagnostic techniques in the evaluation of Psychogenic Non-Epileptic Seizures

Cragar et al conducted a literature review to analyze the possible alternatives to video-EEG for diagnosis of PNES. The literature was searched from 1967 through November 2001 using keywords in the PsychINFO database and were divided into 7 categories of alternative PNES diagnostic techniques: demographic/medical history variables, seizure semiology, provocative testing, prolactin levels, single photon emission computed tomography (SPECT), psychological testing, and neuropsychological testing. Medical history variables included history of abuse, psychiatric treatment history, frequency of seizures (not shown in four studies to have a significant difference between epilepsy an PNES groups), epileptic spells are more likely to occur during sleep and are more stereotyped, older age of onset and duration of seizure disorder, variable semiology, and length of spells (six studies all concluded that PNES spells last longer than all types of epilepsy spells). A saline induction provocation test has a 74% sensitivity, but does not always induce spells. Prolactin levels estimate the average sensitivity to be 89% and are suggestive of epilepsy, but a negative outcome is not highly predictive of PNES. The use of SPECT data to differentiate PNES from epilepsy is not recommended as a first choice due to expense, radioactive materials, and difficulty of interpretation due to muscle and movement artifact. Review of SPECT studies suggests an average sensitivity of 72% across different types of scans and is 59% specific to epilepsy when there is a presence of SPECT abnormalities. The use of the MMPI to diagnose PNES patients is 70% of epilepsy and PNES patients may be correctly diagnosed by using the Wilkus et al (1984) classification rules. The MMPI-2 may add diagnostic utility above other variables such as the medical history, but this utility was not elaborated on. Neuropsychological testing does not adequately differentiate PNES from patients with epilepsy, or both, and all 3 groups test results' suggest cognitive impairment compared to the normal population. The authors concluded that it is unlikely that the gold standard, video-EEG monitoring, will be replaced by any of the alternative techniques they reviewed, yet may be more helpful as complementary diagnostic tools [51]. Devinsky et al conducted a retrospective chart review in order to compare the clinical features of patients with epileptic seizures (ES) and nonepileptic seizures (NES) to only ES or only NES. A total of 387 consecutive admissions for video-EEG monitoring yielded 248 patients with ES (64%), 40 patients (10%) with other physiologic disorders, 99 patients (25%) with NES, and 20 patients (20%) with ES+NES. These were matched to 20 ES and 20 NES patients. These patients were 70% female with a mean age of 32 (19-57 years old). All patients underwent a saline provocation test and all ES/NES patients developed NES after ES. In patients with ES/NES, there ES seizures were similar to ES only seizures and NES were similar to NES only spells. The electrodiagnostic and neuroimaging studies in ES/NES patients were similar to ES patients, but their psychiatric interview and inventories were similar to NES patients. In ES/NES patients, the ES and NES events were different from each other, but may be stereotypic and differentiated during video-EEG recording. Once the different events are characterized, the more prevalent or disturbing types can be identified and referred for the appropriate psychiatric or AED treatment [52].

Studies of patients undergoing video-electroencephalogram (vEEG) reveal that the majority of patients with NES meet criteria for a diagnosis of conversion disorder [53,54]. The most recent version of the Diagnostic and Statistical Manual of Mental Disorders (DSM), the DSM – Fifth Edition (DSM-5), lists conversion disorder in the somatic symptom and related disorders category (American Psychiatric Association [APA], 2013) [55]. A diagnosis of conversion disorder requires a minimum of one symptom involving a change in voluntary motor or sensory function, along with evidence of "incompatibility" between known medical conditions and the symptom (e.g., vEEG capturing a paroxysmal episode). The symptom cannot be better accounted for by other medical or mental disorders, and psychosocial functioning is significantly impacted. Stress or trauma correlating with the time of symptom onset is supportive evidence for a diagnosis but, in contrast to the prior DSM, is not required [55]. Another change with the DSM-5 is that a clinician does not have to judge whether the presenting symptom is unintentionally manifested for diagnosis of conversion disorder. However, if there is clear evidence that the symptom is deliberately produced, a diagnosis of factitious disorder or malingering is more appropriate [55].

Although the majority of patients with NES are diagnosed with conversion disorder, NES may also be diagnosed as somatization disorder, dissociative disorder NOS, post-traumatic stress disorder, and undifferentiated somatoform disorder [54].Comorbid psychiatric diagnoses are common for patients with NES and, in addition to frequently diagnosed conversion disorder, consist of other somatoform disorders, posttraumatic stress disorder (PTSD), dissociative disorder, psychotic disorders, anxiety, and depression, along with the majority of patients endorsing a history of abuse [54,56,57]. It is clear that patients with NES are a heterogeneous group and pharmacological and psychological treatment should be determined by the underlying cause and psychiatric disorder [56,58,59].

## 5. Treatment

The correct diagnosis plays an important role in the management of PNES -- the earlier the diagnosis, the better the outcome [60].

Reuber and House conducted a literature review of treatment options for psychogenic non-epileptic seizures. After the diagnosis of PNES is made and communicated to the patient, they should be referred to a mental health practitioner. There is no specific treatment for PNES, but most brief psychological therapies are based on cognitive-behavioral therapy (CBT) most commonly used in patients with normal intellectual functioning. CBT tends to be less effective in patients with a history of more severe and chronic somatization and benefit from longer term contact with a clinician focusing on stress management and living with symptoms. Family therapy is also recommended but no specific recommendations were made for families of a patient with PNES. Treatment of co-existing psychiatric or neurologic disorders is recommended. The authors conclude that PNES should be diagnosed early with prompt referral for psychiatric assessment [61]. Bora et al conducted a retrospective review of the sociodemographics, clinical characteristics, and psychiatric diagnoses of patients with PNES. Data from 2000-2008 from long-term video EEG monitoring (LVEM, lasting ~5 days) from a specialized epilepsy center in Turkey was analyzed. During this period of time, 440 patients with refractory epilepsy or indeterminate diagnoses underwent LVEM and 67 patients had a diagnosis of

PNES (mean age: 30; 75% female) with 233 episodes recorded. At the time of monitoring, 56.7% were taking antidepressant medication and 100% were taking AEDs, most patients were on multiple AEDs. Six patients (9%) had concurrent epilepsy, with complex partial epilepsy being the most common (N=4). Both the neurologist and psychiatrist diagnosed all of the PNES patients with conversion disorder. Twenty-one percent of patients (N=13) were diagnosed with only a conversion disorder, while most were diagnosed with another axis I or axis II diagnosis, most commonly major depression (31%, N=19), followed by generalized anxiety disorder (15%, N=9). The authors concluded that ongoing education and cooperation between neurologists and psychiatrists are critical to properly diagnose and manage patients with PNES [62]. Drake et al conducted a retrospective review of patients with severe PNES with frequent and prolonged spells that mimicked status epilepticus in order to identify clinical and psychometric features to assist diagnosis. Twenty patients were admitted to the Epilepsy Unit of The Ohio State University Hospitals from July 1982 to July 1989. The mean age was 27.9 years old and 19 of 20 were female (95%). The clinical seizures averaged more than 2 minutes in length and were atypical with common back arching and pelvic thrusting. For the spells that continued patients received IV diazepam, phenytoin, and phenobarbital. Sixteen of the 20 patients had been previously diagnosed as epileptic due to observed seizures or abnormal EEG findings both with and without additional seizures. Five patients' PNES spells stopped spontaneously, 4 ceased with the suggestion that improvement was forthcoming, and 11 required intubation due to respiratory arrest. Four patients were cognitively impaired, 10 patients had conversion or somatization disorders, and 10 received psychiatric diagnoses of personality disorders (5 borderline and 5 mixed borderline-histrionic types). At a later date, 14 patients (70%) were found to have experienced a recent acute situational stress prior to their spell. The patients with conversion disorder had their AEDs discontinued and gradually improved, while cognitively impaired individuals were helped by situational changes, behavior modifications, or neuroleptics. Patients with personality disorders continued having attacks and eventually ceased following up [63].

Pharmacological treatment of NES consists of the use of antidepressants, particularly SSRIs [64]. In a prospective study, venlafaxine reduced the frequency of NES as well as symptoms of depression and anxiety [65]. A pilot study comparing sertraline to placebo demonstrated a lower frequency of paroxysmal events associated with sertraline but no differences were observed for quality of life and psychosocial functioning between the groups [66].

A literature review of psychological treatments of NES found that various psychological interventions are beneficial with no particular treatment being superior [67]. In a review of eye movement desensitization and reprocessing (EMDR) therapy for medically unexplained symptoms, which included a small group of NES patients in the sample, findings indicated that EMDR may be an effective treatment, especially when there is an identifiable trauma [64]. A study examining brief augmented psychodynamic interpersonal therapy for the treatment of NES indicated a significant reduction in paroxysmal event frequency, with 25% of the participants being event free for an average of 3.5 years following therapy, as well as reduced reliance on healthcare services [68]. A randomized control trial (RCT) found that the addition of cognitive-behavioral therapy (CBT) for treatment of NES resulted in a significantly greater reduction in the frequency of paroxysmal events than standard medical care alone [69]. Similarly, another study utilizing CBT demonstrated that 11 of 17 patients who completed 12

CBT sessions were episode-free by the final session and experienced improved functioning and quality of life as well as decreased psychiatric symptoms [70]. In addition, group psychotherapy for treatment of NES has demonstrated effectiveness for reducing the frequency of paroxysmal events [71,72]. Research has provided some evidence that various psychological interventions are effective for the treatment of NES, however, most studies were not well conducted and there needs to be more research conducted that utilizes RCT [68,70].

## 6. Prognosis

Most studies that have assessed the prognosis in patients after PNES diagnosis suggest that only 25 to 38 percent of patients achieve complete seizure freedom [73-77]. Children have been reported to have better prognosis than adults [60].

## Author details

Batool F. Kirmani[1,2], Diana Mungall Robinson[2], Jose Aceves[2,3], David Gavito[2,5], Richard Phenis[2,4] and Daniel Cruz[2,4]

*Address all correspondence to: bkirmani@sw.org

1 Epilepsy Center, Department of Neurology, Scott & White Neuroscience Institute, Temple, TX, USA

2 Texas A&M Health Science Center College of Medicine, Temple, TX, USA

3 Division of Pediatric Neurology, Department of Pediatrics, Scott & White Healthcare, Temple, TX, USA

4 Division of Neuropsychology, Department of Surgery, Scott & White Healthcare, Temple, TX, USA

5 Department of Psychiatry, Scott & White Neuroscience Institute, Temple, TX, USA

## References

[1] Lesser RP. Psychogenic seizures. Neurology 1996;46:1499-1507.

[2] Overintepretation of EEGs and misdiagnosis of epilepsy. Benbadis SR, Tatum WO. J Clin Neurophysiol. 2003 Feb;20(1):42-4.

[3] Benbadis SR.,Psychogenic non-epileptic seizures. In Wyllie E, editor. The treatment of epilepsy: principles and practice. 4th edition: Lippincott and Williams 2006 ; 623-630.

[4] Reuber M, Fernández G, Bauer J, Helmstaedter C, Elger CE. Diagnostic delay in psychogenic nonepileptic seizures. Neurology. 2002 Feb 12;58(3):493-5.

[5] Szaflarski JP, Szaflarski M, Hughes C, Ficker DM, Cahill WT, Privitera MD. Psychopathology and quality of life: psychogenic non-epileptic seizures versus epilepsy. Med Sci Monitor 2003; CR113-CR118.

[6] Barry E, Krumholz A, Bergey C, Alemyehu S, Grattan L. Non-epileptic posttraumatic seizures. Epilepsia 1991;32:322-328.

[7] Benbadis SR, Hauser WA. An estimate of the prevalence of psychogenic non-epileptic seizures. Seizure. 2000;9(4):280–281.

[8] Sigurdardottir KR, Olafsson E. Incidence of psychogenic seizures in adults: a population-based study in Iceland. Epilepsia. 1998;39(7):749–752.

[9] Szaflarski JP, Ficker DM, Cahill WT, et al. Four-year incidence of psychogenic nonepileptic seizures in adults in Hamilton county, OH. Neurology. 2000; 55(10):1561–1563.

[10] Meierkord H, Will B, Fish D, Shorvon S. The clinical features and prognosis of pseudoseizures diagnosed using video-EEG telemetry. Neurology 1991;41:1643-1646.

[11] Lancman ME, Brotherton TA, AsconapéJJ, Penry JK. Psychogenic seizures in adults: a longitudinal analysis. Seizure. 1993;2(4):281

[12] Alper K, Devinsky O, Perrine K, Vazquez B, Luciano D. Nonepileptic seizures and childhood sexual and physical abuse. Neurology. 1993;43(10):1950

[13] Slavney PR. Perspectives on Hysteria. Baltimore: The Johns Hopkins University Press;1990.

[14] Veith I. Hysteria : The history of a disease. Chicago: University of ChicagoPress;1965.

[15] McBride AE, Shih TT, Hirsch LJ. Video-EEG monitoring in the elderly: a review of 94 patients. Epilepsia. 2002;43(2):165–169.

[16] Abubakr A, Wambacq I. Seizures in the elderly: Video/EEG monitoring analysis. Epilepsy Behav. 2005;7(3):447–450

[17] Kawai M, Hrachovy RA, Franklin PJ, et al. Video-EEG monitoring in a geriatric veteran population. J Clin Neurophysiol. 2007;24(6):429–432.

[18] Patel H, Scott E, Dunn D, Garg BP. Non-epileptic seizures in children. Epilepsia 2007;48: 2086-2092.

[19] Wyllie E., Glazer JP, Benbadis S, Kotagal P, Wolmuth B. Psychiatric features of children and adolescents with pseudoseizures. Arch Pediatr Adolesc Med. 1999; 153: 244-248.

[20] Bathia MS, Sapra SS. Clin Pediatr 2005; 44: 617-621.

[21] Vincentiis, Valente KD, Thome-Souza S, Kuczinsky E, Fiore LA. Negrao N. Risk factors for psychogenic seizures in children and adolescents with epilepsy. Epilepsy Behav. 2006;8: 294-298.

[22] Wyllie E, Fridman D, Rothner AD, Luders H, Dudley D, Morris III H, Cruse R, Erenberg G, Kotagal P. Psychogenic seizures in children and adolescents: outcome after diagnosis by ictal video and electroencephalographic recording. Pediatrics 1990; 85: 480-484.

[23] Metrick ME, Ritter FJ, Gates JR, et al. Nonepileptic events in childhood. Epilepsia. 1991;32(3):322–328

[24] Gudmundsson O, Prendergast M, Foreman D, Crowley S. Outcome of pseudoseizures in children and adolescents: a 6 year symptom survival analysis. Dev Med Child Neurol. 2001; 43: 547-551

[25] Krumholz A, Niedermeyer E. Psychogenic seizures: a clinical study with follow-up data. Neurology. 1983 Apr;33(4):498-502.

[26] Krumholz A, Hopp J. Psychogenic (nonepileptic) seizures. Semin Neurol. 2006 Jul; 26(3):341-50.

[27] Nonepileptic seizures: diagnosis and management. Krumholz A. Neurology. 1999;53(5 Suppl 2):S76-83. Review.

[28] Evidence for epilepsy is rare in patients with psychogenic seizures. Lesser RP, Lueders H, Dinner DS. Neurology. 1983 Apr;33(4):502-4.

[29] Ramsay RE, Cohen A, Brown MC. Coexisting epilepsy and non-epileptic seizures. In Rowan AJ, Gates JR, eds. Non-epileptic Seizures. Boston : Butterworth-Heinemann; 1993:47-54

[30] Interictal EEG abnormalities in patients with psychogenic nonepileptic seizures. Reuber M, Fernández G, Bauer J, Singh DD, Elger CE. Epilepsia. 2002 Sep;43(9): 1013-20.

[31] Benbadis SR, Agrawal V, Tatum WO. IV. How many patients with psychogenic nonepileptic seizures also have epilepsy? Neurology. 2001;57(5):915–917.

[32] Martin R, Burneo JG, Prasad A, Powell T, Faught E, Knowlton R, Mendez M, Kuzniecky R. Frequency of epilepsy in patients with psychogenic seizures monitored by video-EEG. Neurology. 2003; 61(12):1791–1792.

[33] Kirmani B, Mungall D. Epilepsy and Concomittant Pseudoseizures: The Diagnostic Dilemma. The Internet Journal of Neurology. Jan 2013; 15(1).

[34] Psychological characteristics of patients with newly developed psychogenic seizures. van Merode T, Twellaar M, Kotsopoulos IA, Kessels AG, Merckelbach H, de Krom MC, Knottnerus JA. J Neurol Neurosurg Psychiatry. 2004;75(8):1175.

[35] Nonepileptic seizures after resective epilepsy surgery. Glosser G, Roberts D, Glosser DS. Epilepsia. 1999 Dec;40(12):1750-4.

[36] New-onset psychogenic seizures after intracranial neurosurgery. Reuber M, Kral T, Kurthen M, Elger CE. Acta Neurochir (Wien). 2002 Sep;144(9):901-7; discussion 907

[37] A spell in the epilepsy clinic and a history of "chronic pain" or "fibromyalgia" independently predict a diagnosis of psychogenic seizures. Benbadis SR Epilepsy Behav. 2005;6(2):264.

[38] Does the primary literature provide support for clinical signs used to distinguish psychogenic nonepileptic seizures from epileptic seizures? Avbersek A, Sisodiya S J Neurol Neurosurg Psychiatry. 2010;81(7):719.

[39] Psychogenic nonepileptic seizures: review and update. Reuber M, Elger CE Epilepsy Behav. 2003;4(3):205

[40] Ictal characteristics of pseudoseizures. Gates JR, Ramani V, Whalen S, Loewenson R Arch Neurol. 1985;42(12):1183

[41] Leis AA, Ross MA, Summers AK. Psychogenic seizures: ictal characteristics and diagnostic pitfalls. Neurology. 1992;42(1):95–99.

[42] Seneviratne U, Reutens D, D'Souza W. Stereotypy of psychogenic seizures: insights from video-EEG monitoring. Epilepsia. 2010;51(7):1159–1168.

[43] Patterns of involvement of facial muscles during epileptic and nonepileptic events: review of 654 events. DeToledo JC, Ramsay RE Neurology. 1996;47(3):621.

[44] Chung SS, Gerber P, Kirlin KA. Ictal eye closure is a reliable indicator for psychogenic nonepileptic seizures. Neurology. 2006;66(11):1730–1731.

[45] Weeping as a common element of pseudoseizures. Bergen D, Ristanovic R Arch Neurol. 1993;50(10):1059.

[46] Ictal stuttering: a sign suggestive of psychogenic nonepileptic seizures. Vossler DG, Haltiner AM, Schepp SK, Friel PA, Caylor LM, Morgan JD, Doherty MJ Neurology. 2004;63(3):516.

[47] Ali S, Jabeen S, Arain A, Wassef T, Ibrahim A. How to Use Your Clinical Judgment to Screen for and Diagnose Psychogenic Nonepileptic Seizures without Video Electroencephalogram. Innov Clin Neurosci. 2011 Jan;8(1):36-42.

[48] Benbadis SR, O'Neill E, Tatum WO, Heriaud L. Outcome of prolonged video-EEG monitoring at a typical referral epilepsy center. Epilepsia. 2004;45(9):1150–1153.

[49] Zhang YC, Bromfield EB, Hurwitz S, Nelson A, Sylvia K, Dworetzky B. Comparison of outcomes of video/EEG monitoring between patients with epileptic seizures and those with psychogenic nonepileptic seizures. Epilepsy Behav. 2009;15(3):303–307.

[50] Benbadis SR, Siegrist K, Tatum WO, Heriaud L, Anthony K. Short-term outpatient EEG video with induction in the diagnosis of psychogenic seizures. Neurology. 2004;63(9):1728–1730.

[51] Cragar DE, Berry DT, Fakhoury TA, Cibula J, Schmitt F. A review of diagnostic techniques in the differential diagnosis of epileptic and nonepileptic seizures. Neuropsychol Rev. 2002;12(1):31–64.

[52] Devinsky O, Sanchez-Villaseñor F, Vasquez B, et al. Clinical profile of patients with epileptic and nonepileptic seizures. Neurology. 1996;46(6):1530–1533

[53] Devinsky, O. (1998). Nonepileptic Psychogenic Seizures: Quagmires of Pathophysiology, Diagnosis, and Treatment. Epilepsia, 39(5), 458-462.

[54] Marchetti, R., Kurcgant, D., Neto, J., von Vismark, M., Marchetti, L., & Fiore, L. (2008). Psychiatric diagnoses of patients with psychogenic non-epileptic seizures. Seizure, 17, 247-253.

[55] American Psychiatric Association (2013). Diagnostic and Statistical Manual of Mental Disorders (5th ed.).

[56] Bowman, E. S. & Markand, O. N. (1996). Psychodynamics and psychiatric diagnoses of pseudoseizure subjects. The American Journal of Psychitary, 153(1), 57-63.

[57] Reuber, M., Howlett, S., & Kemp, S. (2005). Psychologic treatment of patients with psychogenic nonepileptic seizures. Expert Review Neurotherapeutics, 5(6), 737-752.

[58] Walzak TS, Papacostas S, Williams DT, et al. Outcome after the diagnosis of psychogenic non-epiletic seizures. Epilepsia 1995;36:1131-1137.

[59] Lesser, R. (2003). Treatment and outcome of psychogenic nonepileptic seizures. Epilepsy Currents, 3(6), 198-200.

[60] Wyllie E, Freidman D, Luders H, et al. Outcome of psychogenic seizures in children and adolescents compared to adults. Neurology 1991;41:742-744.

[61] Reuber M, House AO. Treating patients with psychogenic non-epileptic seizures. Curr Opin Neurol. 2002;15(2):207–211.

[62] Bora IH, Taskapilioglu O, Seferoglu M, Kotan OV, Bican A, Ozkaya G, Akkaya C. Sociodemographics, clinical features, and psychiatric comorbidities of patients with psychogenic nonepileptic seizures: experience at a specialized epilepsy center in Tur-

key. Seizure. 2011 Jul;20(6):458-61. doi: 10.1016/j.seizure.2011.02.007. Epub 2011 Mar 23.

[63] Drake ME Jr, Pakalnis A, Phillips BB. Neuropsychological and psychiatric correlates of intractable pseudoseizures. Seizure. 1992 Mar;1(1):11-3.

[64] van Rood,Y. & de Roos, C. (2009). EMDR in the treatment of medically unexplained symptoms: a systematic review. Journal of EMDR Practice and Research, 3(4), 248-263.

[65] Pintor, L., Bailles, E., Matrai, S., Carreno, M., Donaire, A., Boget, T., Setoain, X., Rumia, J., & Bargallo, N. (2010). Efficiency of Venlafaxine in patients with psychogenic nonepileptic seizures and anxiety and/or depressive disorders. Journal of Neuropsychiatry and Clinical Neurosciences, 22(4), 401-408.

[66] Bravo, T. P., Hoffman-Snyder, C. R., Wellik, K. E., Martin, K. A., Hoerth, M. T., Demaerschalk, B. M., & Wingerchuk, D. M. (2013). The effect of selective serotonin reuptake inhibitors on the frequency of psychogenic nonepileptic seizures: a critically appraised topic. Neurologist, 19(1), 30-33.

[67] Gaynor, D., Cock, H., & Agrawal, N. (2009). Psychological treatments for functional non-epileptic attacks: a systematic review. Acta Neuropsychiatrica, 21, 158-168.

[68] Mayor, R., Howlett, S., Grunewald, R., & Reuber, M. (2010). Long-term outcome of brief augmented psychodynamic therapy for psychogenic nonepileptic seizures: seizure control and health care utilization. Epilepsia, 51, 1162-1176.

[69] Goldstein, L. H., Chalder, T., Chigwedere, C., Khondoker, M. R., Moriarty, J., Toone, B. K., Mellers, J. D. C. (2010). Cognitive-behavioral therapy for psychogenic nonepileptic seizures: a pilot RCT. Neurology, 74, 1986-1994.

[70] LaFrance, Jr., C. & Devinsky, O. (2004). The treatment of nonepileptic seizures: historical perspectives and future directions. Epilepsia, 45(suppl. 2), 15-21.

[71] Barry, J. J., Wittenberg, D., Bullock, K. D., Michaels, J. B., Classen, C. C., & Fisher, R. S. (2008). Group therapy for patients with psychogenic nonepileptic seizures: a pilot study. Epilepsy & Behavior, 13(4), 624-629.

[72] Metin, S. Z., Ozmen, M., Metin, B., Talasman, S., Yeni, S. N., & Ozkara, C. (2013). Treatment with group psychotherapy for chronic psychogenic nonepileptic seizures. Epilepsy & Behavior, 28(1), 91-94.

[73] Outcome in psychogenic nonepileptic seizures: 1 to 10-year follow-up in 164 patients. Reuber M, Pukrop R, Bauer J, Helmstaedter C, Tessendorf N, Elger CE Ann Neurol. 2003;53(3):305.

[74] Measuring outcome in psychogenic nonepileptic seizures: how relevant is seizure remission? Reuber M, Mitchell AJ, Howlett S, Elger CESOEpilepsia. 2005;46(11):1788.

[75] Non-epileptic seizures: patients' understanding and reaction to the diagnosis and impact on outcome. Carton S, Thompson PJ, Duncan JS Seizure. 2003;12(5):287.

[76] Early outcomes and predictors in 260 patients with psychogenic nonepileptic attacks. McKenzie P, Oto M, Russell A, Pelosi A, Duncan R Neurology. 2010;74(1):64.

[77] Psychiatric and neurologic predictors of psychogenic pseudoseizure outcome. Kanner AM, Parra J, Frey M, Stebbins G, Pierre-Louis S, Iriarte J Neurology. 1999;53(5): 933.

Chapter 10

# Psychogenic Non-Epileptic Seizures in a Surgical Epilepsy Unit: Experience and a Comprehensive Review

Lorena Vega-Zelaya, Marta Alvarez, Elena Ezquiaga, Jaime Nogeiras, María Toledo, Rafael G. Sola and Jesús Pastor

Additional information is available at the end of the chapter

http://dx.doi.org/10.5772/57439

## 1. Introduction

Psychogenic non-epileptic seizures (PNES) are paroxysmal involuntary changes in behaviour, sensation, motor activity, cognitive processing or autonomic function that resemble epileptic seizures (ES). However, they have no electrophysiological correlate but instead possess a psychological origin [1, 2].

Historically known as hysteria seizures by Hippocrates and Aristotle, the French neurologist Jean-Martin Charcot recognised hysteria as a neurologically diagnosable condition. Many years after this recognition, Sigmund Freud reclassified hysteria as a psychiatric disorder.

Previously, PNES has been given various names, such as pseudoseizures, non-epileptic spells or psychogenic seizures. In general, this terminology involving the prefix 'pseudo-' has become out-dated as it implies that the seizures are not real and may suggest 'malingering'.

PNES have been commonly classified as dissociative or conversion disorders. However, PNES have also been considered to be very similar to other functional somatic symptoms and syndromes [3]. In the current and recently published DSM-V, PNES are classified as a "functional neurological symptom disorder". It appears that PNES are not a single entity and that a variety of clinical manifestations can occur. Furthermore, comorbid psychiatric disorders are common; thus, it is not surprising that a PNES diagnosis is difficult to achieve. PNES exist at the interface of neurology and psychiatry and as discussed below, constitute an important challenge in the practice of both medical specialties, because of their inherent diagnostic and therapeutic difficulties.

Misdiagnosis leads to iatrogenic effects of antiepileptic medication and aggressive or invasive procedures, such as intubations during emergency department visits [4]. After initial symptom onset, the average time to a formal PNES diagnosis is 7 to 9 years [2, 4, 5].

Video-electroencephalography (vEEG) monitoring is an indispensable tool for diagnosing PNES because it allows the simultaneous analysis of both clinical and ictal EEG findings. In some cases, provocative techniques aid in diagnosing PNES, and these are performed only when necessary and in approximately 30-50% of patients diagnosed with PNES [6]. These techniques are very useful in detecting the presence of suggestibility, which strongly supports a psychogenic mechanism.

Approximately 25-30% of epilepsy patients referred to tertiary epilepsy centres or specialised hospitals have PNES [7]. The incidence of PNES is approximately 4% of the total epilepsy incidence [8], and in fact, most patients misdiagnosed with epilepsy at epilepsy centres have PNES [9]. These patients are heavy users of emergency and nonemergency health care [10], and approximately three fourths receive inappropriate AED treatment before their diagnosis [11], which can produce adverse side effects, high costs to health care facilities, high medical utilisation rates and high personal and societal costs.

At present, PNES are considered symptoms of an underlying psychological or psychiatric disorder and are associated with impaired social function [12] and high disability [13, 14] that affect patient quality of life [15, 16]. Disability and quality of life may even be poorer than in epilepsy patients [12, 14, 17]. Additionally, PNES patients have been reported to be twice as likely to be unemployed as patients with epilepsy [12].

PNES seem to occur more frequently in women [18] but also may be more difficult to recognise in men [19]. Gender differences in clinical presentation have been reported, and males have shown higher seizure frequencies, more antiepileptic drug use, and a longer interval before PNES diagnosis [12]. Most patients experienced PNES onset at a young age, and approximately 10% of patients experienced PNES onset after the age of 60 [20].

Furthermore, patients from lower sociocultural classes are more likely to experience PNES [4]. A small number of clinical studies have observed differences in the cognitive profiles of patients with PNES versus patients with epilepsy [21].

A PNES diagnosis is a very common and important diagnosis in epilepsy units and is described as a common condition associated with epilepsy (PNES+ E). In this study, we attempted to depict and describe the characteristics of PNES patients and the difficulty of diagnosis in a national referral centre for epilepsy surgery. In light of the current literature, the physiopathological hypothesis, treatment and prognosis should be revised.

## 2. Methods

### 2.1. Patients

We retrospectively reviewed all patients with an intractable epilepsy diagnosis admitted to the National Referral Centre for Epilepsy Surgery at the Hospital Universitario de la Princesa between July 2001 and July 2013. All patients underwent prolonged video-electroencephalography (vEEG) monitoring as part of the evaluation for surgical treatment [22, 23]. The evaluation also included psychiatric, psychological and clinical evaluations, scalp EEG, 1.5T magnetic resonance imaging (MRI) and interictal single photon emission-computed tomography (SPECT) with $^{99m}$Tc-HmPAO. vEEG monitoring was conducted using 19 collodion-fixed scalp electrodes, according to the international 10-20 system. During vEEG recordings, antiepileptic drugs were progressively removed from the second to the fourth day of recordings at a rate of approximately one-third of the dose per day. The digital EEGs (NeuroWorks, XLTEK®, Oakville, Canada) were sampled at 512 Hz and filtered at 0.5-70 Hz with a notch filter of 50 Hz. Accessory electrodes were used to monitor electromyography (EMG) at different muscles in combination with an electrocardiogram (EKG).

The following criteria were used to diagnose PNES: (1) at least one single typical clinical event captured on vEEG, (2) the EEG recording did not detect any concomitant ictal activity or postictal slowing, and (3) no evidence of any alternative neurological diagnosis (e.g., movement disorders) [3]. An epilepsy diagnosis was assigned according to the International League Against Epilepsy (ILAE) classification system [24]. The provocative protocol technique with placebo was utilised in cases where discordance was observed between the clinical and electrical findings.

### 2.2. Induction techniques

We used induction techniques to support the diagnosis of PNES when considered useful [25]. The systematic steps in the provocative induction technique with placebo utilised in our epilepsy unit were [26]:

All patients received verbal information and signed an inform consent document stating the need for a test with a drug that could induce their seizures. The informed consent was accepted by the Hospital's ethical committee.

Two neurophysiologists were present and monitored the entire study.

Bioelectrical brain function and behaviour was simultaneously monitored by vEEG.

The following cardiorespiratory parameters were monitored: EKG, respiratory rate, capillary saturation oxygen ($SaO_2$) and blood pressure.

Two millilitres of a saline solution were intravenously injected. When a typical seizure appeared, a second 2-ml bolus of saline was administered to terminate the event.

Psychiatric assessment

Diagnoses were determined according to the DSM-IV-TR criteria (Diagnostic and Statistical Manual of Mental Disorders, fourth edition, text revision, Axis I Disorders).

All patients were evaluated with a non-structured clinical assessment interview according to the Protocol of Psychiatric Assessment for our Epilepsy Surgery Unit (Table 1).

| |
|---|
| The psychiatric assessment of all patients included the following: |
| Mental health exploration to identify current psychiatric disorders. |
| History of psychiatric disorders. |
| Personality traits. |
| History of drug abuse and previous and current psychopharmacologic treatments, including benzodiazepines, antidepressants and antipsychotic drugs. |
| Previous abuse or childhood trauma. |
| A neuropsychological assessment including IQ, verbal and manipulative IQ, total, verbal and visual memory and executive function tests. |

**Table 1.** Patient psychiatric assessment

### 2.3. Statistical analysis

Statistical comparisons between groups were performed using Student's t-test (two groups) or analysis of variance (more than two groups) for parametric samples. For non-parametric samples, the Mann-Whitney test (two groups) and the Kruskal-Wallis test (more than two groups) were used. The sample fit to a Gaussian distribution was assessed by the chi-square or Kolmogorov-Smirnov tests.

The SigmaStat 3.5 software (SigmaStat, Point Richmond, CA, USA) was used for the statistical analysis. The significance level was set at $p < 0.05$. The results are presented as the mean ± SEM, except where otherwise indicated.

For patients in whom differential diagnosis was especially difficult, a numerical analysis of EEG recordings was performed to differentiate between epilepsy and PNES by developing a custom program with Matlab R2008 software (MathWorks, Natic, USA).

Recordings of bioelectrical activity occurring prior to the seizure and during the epileptic event were exported as ASCII files for analysis. For every channel, a Butterworth digital filter from 0,5 to 70 Hz (notch filter, 48-52 Hz) was applied. Artefact free periods of 5 to 10 s were divided in $N$ non-overlapping windows of 1024 points each, and no software was used to remove electrooculogram signals. We calculated the fast Fourier transform (FFT) for every channel ($X_i$; i=1, 2,...19). The power spectrum ($S_i$; i=1, 2,...19) was divided into the EEG frequency bands delta (0.5-4 Hz), theta (4-8 Hz), alpha (8-13 Hz) and beta (13-30 Hz), and the total area under the curve was computed. ($S_{i,k}$; $i$=1, 2,...19; $k$=1, 2,...$N$). Cross-spectra between channels $i$ and $j$ ($S_{ij}$) were also calculated for all channel pairs. Coherence [27] for the frequency, f, and for every channel pair, $i$ and $j$, is defined according to the following expression:

$$C(f) = \frac{|\langle S_{ij}(f)\rangle_N|^2}{\langle S_{ii}(f)\rangle_N \langle S_{jj}(f)\rangle_N} \quad (1)$$

Local coherence ($c$) was obtained from neighbouring channels according to the following expression [28]:

$$c(f) = \frac{1}{p}\sum_{i=1}^{p} C_i \qquad (2)$$

Where $p$ is the number of electrodes that were computed.

Statistical comparisons between each channel before and during seizures were performed.

# 3. Results

## 3.1. Patient classifications

A total of 630 patients were included in this study. We classified the patients into 4 diagnostic groups according to the following criteria (Table 2) [26]:

- Epilepsy: clinical and bioelectrical evidence of epileptic seizures were confirmed. We did not distinguish between partial and generalised epilepsy because that classification was outside the scope of this work.
- No evidence of seizures: in these cases, the data suggested epilepsy (abnormal epileptiform activity in the vEEG), but no seizures occurred after a sufficient period of time without medication (usually more than 2 weeks).
- No evidence of epilepsy: no epileptiform activity or any clinical event was observed during vEEG recordings that supported a PNES or epilepsy diagnosis.
- PNES: at least one typical clinical event was captured on vEEG but without concomitant epileptiform bioelectric characteristics or other justifiable organic causes.

| | n | % |
|---|---|---|
| Epilepsy | 554 | 87,9 |
| No evidence of seizures | 23 | 3,7 |
| No evidence of epilepsy | 23 | 3,7 |
| PNES | 30 | 4,7 |

**Table 2.** Diagnostic classifications of patients.

We analysed the clinical features of PNES and epilepsy patients. The mean age of the PNES patients was 33,9 ± 1,7 years old, and the mean age of epileptic patients was 34,3 ± 0,5 years old ($p < 0,05$, Student's t test). The average age of seizure onset was 15,6 ± 2,0 years old for PNES patients and 11,9 ± 0,4 years old for epilepsy patients (n.s., Student's t test).

The mean seizure frequency for epileptic and PNES patients is shown in Figure 1.

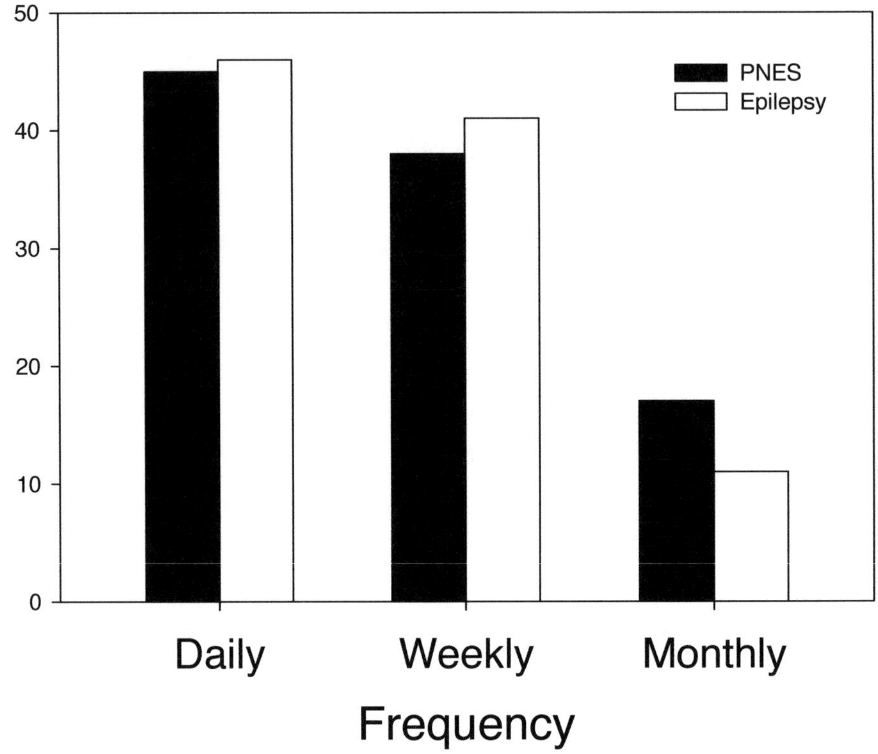


**Figure 1.** Mean seizure frequency for PNES (filled bars) and epileptic patients (empty bars).

Females represented a greater percentage of PNES patients (20/30). In addition, 70% of PNES patients presented with comorbid epilepsy (PNES+E, 12/21) or epileptiform activity (PNES +EA, 9/21), which were verified by EEG recordings (Table 3). Nine of the 30 patients were diagnosed with PNES alone.

| Patient Number | Age (years) | Age of Onset (years) | SPECT | MRI | vEEG | Engel's scale | Placebo/ Response | Psychiatric Disorder(s) |
|---|---|---|---|---|---|---|---|---|
| 1 | 18 | 18 | Normal | Normal | | _ | | PTSD |
| 2 | 31 | 7 | Normal | Normal | | _ | | DS |
| 3 | 30 | 2 | Bi-T | Tuberose Sclerosis | L-T AE | _ | Yes/Yes | DS |
| 4 | 29 | | Bi-T | Normal | Petit mal Epilepsy | _ | | Anxiety |
| 5 | 19 | 1 | Bi-T | Bi-TMS | L-T AE | _ | Yes/No | No |
| 6 | 19 | 16 | Bi-T | Multiple cavernomas | L-TLME | _ | | Dysthymia |
| | | | | is | | | | |

http://dx.doi.org/10.5772/57439

| Patient Number | Age (years) | Age of Onset (years) | SPECT | MRI | vEEG | Engel's scale | Placebo/ Response | Psychiatric Disorder(s) |
|---|---|---|---|---|---|---|---|---|
| 7 | 47 | 9 | Bi-T | Subcortical Hyperintensities | R-T AE | _ | | No |
| 8 | 37 | 8 | L-T | Normal | GE | _ | Yes/Yes | No |
| 9 | 19 | 16 | Normal | Normal | | _ | Yes/No | DS |
| 10 | 55 | 3 | R-MT | Normal | Bi-T EA | _ | Yes/Yes | No |
| 11 | 37 | 31 | Normal | Normal | L-TLE | II | | Anxiety |
| 12 | 33 | 3 | Bi-T | Surgical changes | Bi-T EA | _ | Yes/Yes | No |
| 13 | 32 | 3 | Irregular perfusion | R-F Encephalomalacia | R-FLE | | Yes/Yes | No |
| 14 | 49 | 26 | R-MT | R-MTS | R-MTLE | I | Yes/Yes | No |
| 15 | 28 | 15 | Normal | Normal | | _ | | DS |
| 16 | 35 | 30 | L-T | L-T anterior horn increase | | _ | | No |
| 17 | 49 | 35 | R-MT | R-T anterior horn increase | R-T EA | _ | | Depression |
| 18 | 41 | 28 | L-T | L-MTS | L-T EA | _ | | Depression |
| 19 | 37 | 14 | Bi-FT | Bi-FP ischemic alterations | | _ | | No |
| 20 | 40 | 1 | L-TM | Normal | G- EA | _ | | No |
| 21 | 16 | 13 | L-MT | L-MTS | | _ | Yes/Yes | No |
| 22 | 33 | 14 | R-MT | Normal | | _ | | No |
| 23 | 41 | 27 | L-FP | Normal | | _ | | No |
| 24 | 40 | 23 | Irregular perfusion | L Hippocampal dysplasia | L-T EA | _ | | Anxiety |
| 25 | 32 | 8 | L-FT | R-MTS | L-FLE | I | | Depression |
| 26 | 38 | 29 | R-MT | Normal | L-MTLE | _ | | No |
| 27 | 35 | 2 | L-MT | L-MTS | L-MTLE | I | | Anxiety |
| 28 | 48 | 23 | R-MT | Normal | L-MTLE | I | | Anxiety |
| 29 | 27 | 1 | R-MT | R-MTS | R-MTLE | _ | | No |
| 30 | 33 | 20 | Bi-F | R-F Hyper-intensities WM | L-MTLE | _ | | Depression |

**Table 3.** Characteristics of PNES patients.

Bi, bilateral; DS, dissociative disorders; EA, epileptiform activity; F, frontal; FLE, frontal lobe epilepsy; FP, fronto-parietal; FT, fronto-temporal; GE, generalised epilepsy; L, left; MT, mesial temporal; MTLE, mesial temporal lobe epilepsy; MTS, mesial temporal sclerosis; PTSD, post-traumatic stress disorder; R, right; T, temporal; TLE, temporal lobe epilepsy; WM, white matter.

The semiology of PNES was highly stereotyped in each patient and mainly consisted of *convulsive* components or bizarre bilateral motor manifestations involving the upper or lower limbs. Other manifestations lacked motor events but involved sensory feelings or unresponsiveness, although these stereotypies were less frequent.

Importantly, in the PNES+E group, structural abnormalities were evident in 11/12 neuroimaging studies. We also found that in patients diagnosed as PNES+EA, structural abnormalities associated with temporal lobe epilepsy were observed in 9 patients. Moreover, 3 patients presented with mesial temporal sclerosis in MRI studies.

When we assessed the PNES group with the same tests, we found that there were cerebral perfusion abnormalities on SPECT in 6/9 patients and that in 3/9 patients, MRI showed lesions sometimes associated with temporal lobe epilepsy (TLE).

In PNES+E patients, there were 8 cases of temporal lobe epilepsy, 2 cases of frontal lobe epilepsy, and 2 cases of generalised epilepsy. Also in this group, the following surgical interventions were utilised: temporal lobectomy (4 patients), and frontal lobectomy (1 patient) with an Engel's scale outcome of I in 4 and II in 1.

### 3.2. Provocative techniques

A provocative placebo technique was used to confirm a PNES diagnosis in 9 patients with an inconclusive diagnosis from vEEG monitoring. In 7 patients, the provocative test was positive for PNES, and in those cases, the patients experienced similar episodes to the previously recorded spontaneous episodes. However, a variable delay (of less than a minute) between saline administration and symptom onset was observed. The recorded bioelectric activity during the induced, as well as during spontaneous episodes, was completely normal, with visible muscle and movement artefacts. After a varying 30- to 60-second period after the episode started, we administered a second dose of saline after warning the patient that this drug would abort the crisis, which indeed occurred with a latency between 15 and 45 s. One example of this technique is shown in Figure 2.

Provocative induction was positive for PNES in 7 patients. However in the other two patients we did not obtain any response.

### 3.3. Diagnostic challenges

Prolonged vEEG monitoring provides a definitive PNES diagnosis in nearly all cases. However, in some situations, a definitive PNES or epilepsy diagnosis can be difficult to obtain. Here, we describe a case in which the semiology and vEEG did not confirm that the episode was psychogenic in origin. Quantitative EEG (qEEG) had been used to assess the patient by both spectral and coherence methods while recording EEGs during basal conditions and episodes.

This patient, a 29 year old female, experienced seizure onset at 9 years old. The clinical symptomatology was described as an inability to emit and understand language, with feelings of sadness and fear. Moreover, the patient reported remaining conscious during the entire

http://dx.doi.org/10.5772/57439

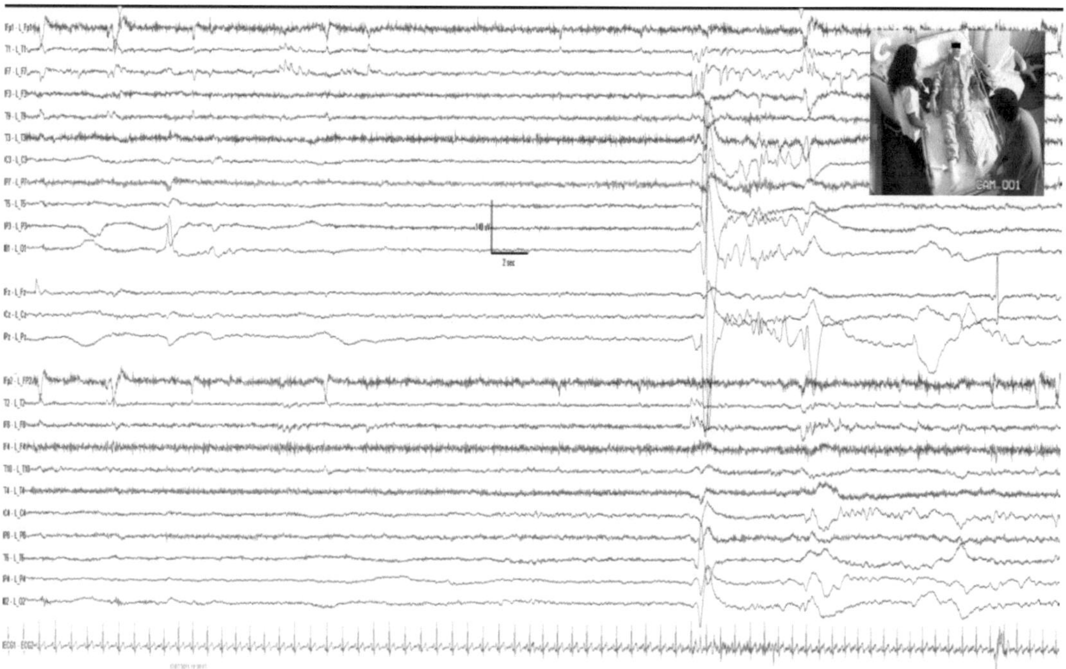

**Figure 2.** A typical recording during the provocative induction test. The image corresponds to 25 s after a 2-ml saline injection. The patient (inset) experienced myoclonus of all four limbs. Two neurophysiologist (LV-Z and JP) were monitoring the patient's response. The EEG recording showed no pathological activity, except for movement artefacts.

episode and experienced no residual amnesia. The interictal EEG showed intense and very frequent irritative activity with a burst of epileptiform discharge in the left frontal area and a generalised spike-wave complex (Figure 3).

True seizures were recorded during the spike-wave discharges, which were accompanied by an altered state of consciousness, mild to severe difficulty to speak and a clear delay in response to different orders (Figure 4).

During the episodes, no clear evidence of any ictal pattern emerged. We conducted an induction test that proved positive, reproducing the previously described patient episodes (Figure 4).

In addition to this type of seizure (of which the patient did not complain), the patient complained of severe and multi-day seizures of language emission and difficulties in understanding. The brain activity recorded during these episodes did not show a clear ictal pattern. Nevertheless, the recorded bioelectrical activity needed to be classified as a seizure or another disorder. To assess these recordings, we analysed the band coherence [*C(f)*] using to equation 1 and compared the basal and ictal activity using the *c(f)* equation. We hypothesised that a true seizure most likely increases the local coherence for any scallop region (Figure 5). However, as shown in Figure 5, the local coherence was not significantly modified from basal activity during any of the recorded paroxysmal episodes.

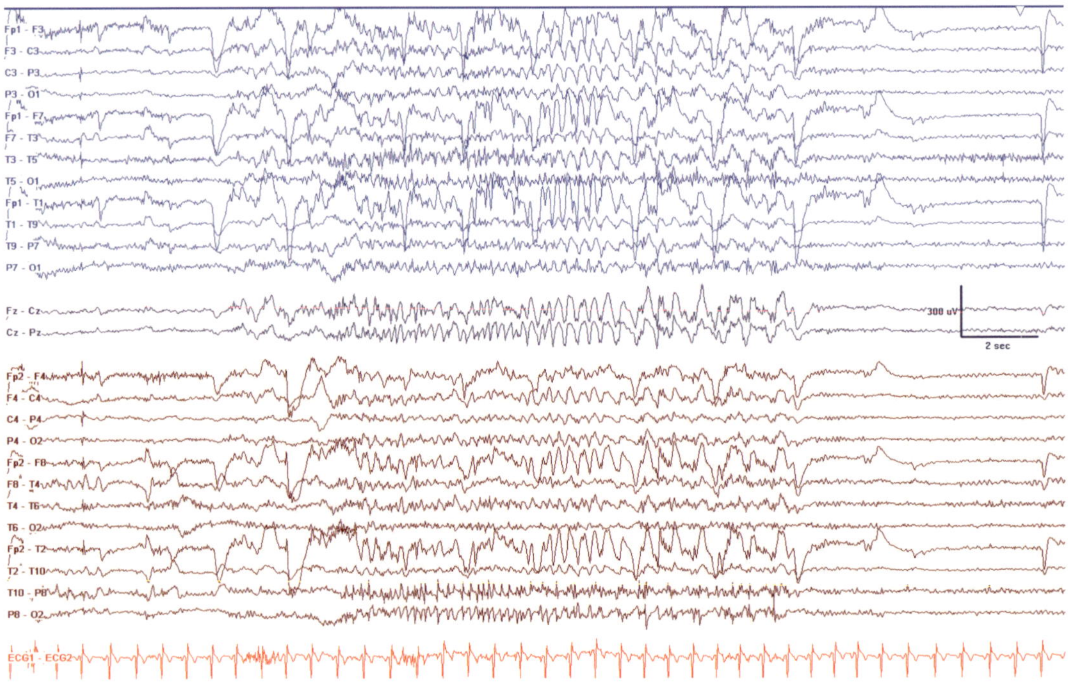


**Figure 3.** Wakeful interictal activity shows a burst of generalised spike-wave complexes.

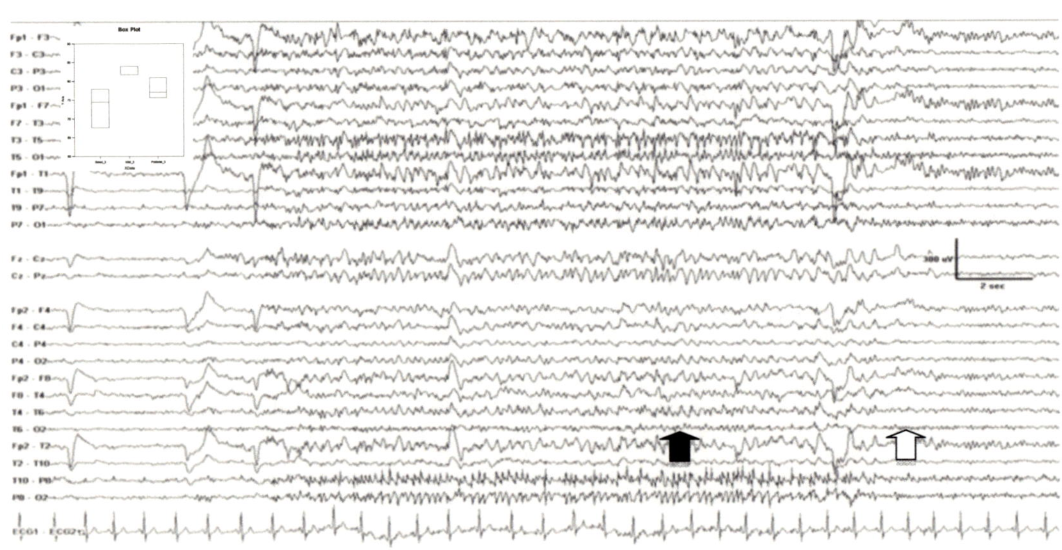

**Figure 4.** A burst of blunted spike-waves at 3 Hz shows a true alteration of consciousness. The patient was ordered to close her eyes (filled arrow), but the response was delayed by 6,4 s towards the end of the seizure. The HR (inset) increased from 72.8 ± 1.8 bpm before the burst to 83.7 ± 0.7 bpm during the episode ($p < 0.05$, ANOVA test by ranks) and returned to 78.8 ± 0.9 bpm after the episode ($p < 0.05$, ANOVA test).

http://dx.doi.org/10.5772/57439

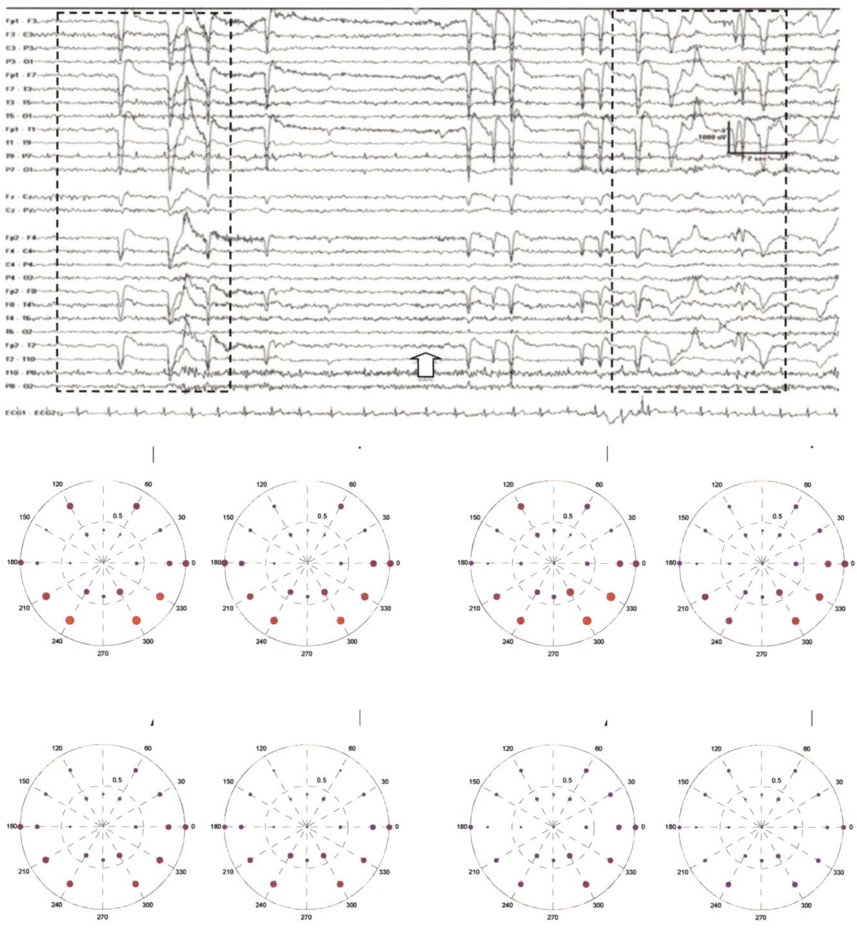


**Figure 5.** Quantitative EEG recordings assess the presence of epilepsy. A) Recording during a *typical seizure*. The empty arrow indicates when the patient pressed the patient-event button, which identifies the presence of a seizure. Dotted boxes indicate the basal (left) and *ictal* (right) periods selected for analysis. B) The EEG electrode schemes represent the 10-20 IS on the scalp. Local coherence was obtained using equation 2 for every electrode and every frequency band. Coherence is related to the diameter of each electrode, and no significant differences were observed between basal (left) and *ictal* activity (right).

Furthermore, after placebo injection, an episode was recorded that was similar to the spontaneously recorded episode, and this reinforced a PNES diagnosis.

### 3.4. Psychiatric comorbidities and patient management

The rate of psychiatric comorbidities in PNES patients was 50%. Affective disorders were the most frequent psychiatric diagnosis (10/15), in particular, depression and anxiety disorders. Dissociative disorders were observed in 4 patients. Post-traumatic stress disorder was diagnosed in the remaining patient (Table 2). Interestingly, no differences were observed in the psychiatric comorbidities between the PNES+ E and PNES patients.

Our preliminary data have confirmed reports in the literature detailing the higher disability observed in PNES patients compared with epilepsy patients. To provide proper patient management, we applied a protocol of patient management to all patients in our unit to provide the appropriate care to PNES and PNES+E patients (Table 4). We have observed that PNES patients showed good family integration, and both patients and families have adapted to the disorder, even in an overprotective way.

| |
|---|
| Neurologist refers patients to the Unit. |
| **First Step**. All patients are assessed by a:<br>Neurologist<br>or Neurosurgeon (if the patient has been referred for a neurosurgical evaluation) |
| **Second Step.** Ancillary tests and other specialists:<br>Scalp EEG<br>Single Photon Emission Computed Tomography (SPECT) with $^{99}$Tc-HmPAO<br>MRI (1.5 T) with an epilepsy protocol<br>Psychiatrist<br>Neuropsychologist<br>Clinical decisions are determined in a regular multidisciplinary clinical meeting. |
| When PNES are detected, a more complex psychiatric assessment protocol is initiated |
| In patients with PNES, it is important to differentiate between PNES alone and PNES+E patients. |
| When patients suffer from PNES alone, a joint visit from a neurologist and psychiatrist are used to inform the patient and if possible, the closest family member about:<br>The psychogenic basis of the disorder, the diagnostic process and the likelihood of individualised psycho-etiopathogenic mechanisms of the disorder. |
| The high disability associated with the disorder is also emphasised.<br>The need to treat the disorder with psychiatric and psychotherapy approaches.<br>The patient receives a complete clinical report to facilitate treatment continuation in his/her health care district.<br>The appropriate way to reduce neurological medications is supervised by a neurologist.<br>A close neurophysiological EEG survey is always recommended, which occurs during the progressive withdrawal from antiepileptic drugs.<br>Usually patients need to visit a neurologist and psychiatrist before completing care at his/her mental health centre. |
| When patients suffer from PNES+E, a joint visit from a neurologist and a psychiatrist is used to inform to the patient and if possible, the closest family member about:<br>The psychogenic basis of the disorder, the diagnostic process and the likelihood of individualised psycho-etiopathogenic mechanisms of the disorder. The high disability associated with the disorder is also emphasised.<br>The need to treat the disorder with psychiatric and psychotherapy approaches.<br>The patient is given a complete clinical report to facilitate mental health treatment in his/her health-care district.<br>An individualised and thorough study of epilepsy is crucial to before deciding on surgery. In a small number of patients, epilepsy is detected during the vEEG, because of antiepileptic drug withdrawal; however, when AEDs are reintroduced, only the PNES are clinically significant. |

**Table 4.** Protocol of Patient Management at the National Referral Centre for Refractory Epilepsy

# 4. Discussion

## 4.1. Characteristics of patients with non-epileptic seizures of psychogenic origin

We have found that 4.8% of patients referred to our epilepsy centre for pre-surgical evaluation received a diagnosis of PNES. This percentage is lower than other published studies. However, PNES was the most frequent non-epileptic condition diagnosed in our epilepsy monitoring unit. Interestingly, in 90% of PNES+E patients, we discovered alterations in brain imaging studies, and also in more than half of the PNES-alone patients, we detected alterations in brain imaging studies. These findings have been previously described, and other researchers have found MRI changes in up to 60% of PNES+ E patients and up to 10% of PNES-alone patients [29]. If epileptiform activity is also considered as marker of brain abnormality, then the percentage of PNES patients in this study with evidence of physical or functional brain abnormalities was 55%. This percentage, though lower than in previously reported studies, reinforces the theory that physical brain disease may play a role in PNES development and in other somatoform and dissociative disorders [30, 31, 32].

In this study, early diagnoses were complicated by concomitant epilepsy or epileptiform activity in 21 PNES patients; however, the semiology of the events helped differentiate PNES from epilepsy (closed eyes, non-rhythmic movements, awareness during the ictal phase and the ability of bystanders to modulate symptom intensity are more often observed in PNES). Nonetheless, no single observation provides a definitive PNES diagnosis, and no single semiological feature is shared by all PNES patients [33]. Moreover, eyewitness accounts of semiology may not be reliable in many cases. Usually, it is difficult for an experienced clinician to determine if an event is a true seizure or PNES, and generally, it is nearly impossible to rule out a diagnosis of epileptic seizures. The gold standard for ruling out epilepsy is vEEG telemetry that analyses typical seizures [34]. Observations of typical seizures combined with recorded brainwave activity lead to accurate diagnosis in up to 90% of cases [35]. In our study, 3,7% of patients did not receive a positive diagnosis because the bioelectric abnormalities or clinical manifestations typical of PNES were not observed. Thus, we could only conclude that the patient had no evidence of epilepsy, but it is possible that these patients might actually belong to the PNES group.

The diagnosis of PNES is difficult to obtain. In fact, the absence of epileptiform activity in scalp recordings during the event is necessary, and evidence of this can be extremely difficult to obtain, because a true partial seizure may have no significant changes in scalp recordings. In these cases, it is necessary to perform a more complex brain activity analysis to reach a definite and clear diagnosis. As misdiagnosis can lead to devastating consequences, all efforts should be made to determine the appropriate diagnosis, and in these cases, provocative techniques aid in providing a definitive diagnosis. We used these techniques in 9 patients to confirm a PNES diagnosis, and they were positive for PNES in 7 patients. The observed clinical manifestations were never associated with bioelectrical changes in the EEG recordings. The sensitivity to induction is unknown but estimated at 30% [25]. The key factors in PNES diagnoses are confirming that the habitual episode was indeed induced and always obtaining a simultaneous vEEG. If the recorded episode is not the habitual type typically recognised by the patient and family, then a firm conclusion is unattainable, and most likely, the patient has different types of episodes. Another important issue is the awareness of the different types of seizures that can accompany a normal ictal EEG and which types cannot.

Another confounding factor in diagnosing PNES is that both epileptic seizures and PNES may occur concomitantly. Up to a 40% of our patients experienced PNES and epilepsy. Some authors estimated the prevalence of this comorbidity as ranging from 10% to 50% [36]. In 5 patients with PNES+E, we performed resective surgery. This determination was based on the results of vEEG, psychological and psychiatric assessments that suggested the decrease or disappearance of epileptic seizures could increase patient quality of life. The outcome of these patients is good; 4 patients showed Engel's grade I and the remaining patient an Engel's grade II.

### 4.2. Psychiatric diagnosis

When considering psychiatric disorders, some authors have found no differences in the prevalence of current axis I disorders in patients with recent-onset PNES+E [37]. Nevertheless, in most studies, a high prevalence of psychiatric comorbidities has been found, and up to 90% of PNES patients met the criteria for another psychiatric disorder, axis I or axis II, according to the DSM criteria. Although a significant degree of psychopathology has been documented, there is an absence of a unique character substrate. Despite numerous studies that have linked PNES to a high prevalence of comorbid psychiatric conditions, we found a low percentage (40%) of the PNES patients exhibited these disorders. Furthermore, we did not observe differences in the psychiatric comorbid disorders of the PNES+E and PNES patient groups. However, we only detected dissociative disorder and post-traumatic stress disorder (PTSD) in the PNES-alone patients, but affective disorders were more common in the PNES+E group. Other studies have shown that approximately 60% of patients met the criteria for an affective episode [4, 5]. However, depressive symptoms are similar in PNES and epilepsy patients; thus, a negative affect could be the result of a chronic illness rather than uniquely related to PNES [38].

Other studies consider PNES as a heterogeneous entity, and to better understand it, different predisposing, precipitating and modulating factors must be considered. Patients with PNES usually present with a larger background of traumatic experiences than the general population [39], with values ranging from 84% [5, 40] to approximately 40%, in more recent studies [4, 38, 41]. These studies have shown a higher incidence of adverse childhood experiences and life events [42]. Previous trauma correlated with a high rate of psychiatric comorbidity and with strong dissociative mechanisms [43]. Additionally, a consistent history of psychiatric treatment, suicide attempts, borderline personality disorder or a history of abuse have been documented [44].

Regarding dissociative disorders, different but high prevalence rates of comorbid dissociative disorders have been reported, which range from 37% to 80% of cases [4, 5]. Not only clinically diagnosed dissociative disorders but dissociative experiences appear to be more frequent in PNES patients, as measured with the Dissociative Experience scale (DES). Patients with epilepsy and PNES scored significantly higher on the DES than patients with epilepsy without PNES and non-epileptic individuals [45]. Nevertheless, these diagnoses have not been consistently correlated with significant differences in dissociative and conversion trait scores as determined with the Minnesota Multiphasic Personality Inventory (MMPI) [4].

Somatic syndromes, such as fibromyalgia, chronic fatigue syndrome, chronic pain syndrome or tension headaches, and irritable bowel syndrome, have occurred at a higher frequency in PNES patients than in other patients with epilepsy [46, 47]. Other medical conditions, such as cancer, orthopaedic problems, and multiple surgeries, were also more frequent in the PNES group than in the epilepsy group.

In PNES patients, this high prevalence of somatic syndromes could be related to a higher risk of manifesting psychic conflicts through somatic symptoms. In fact, PNES has been considered as a somatic symptom response to a wide range of negative events, including stress in adulthood [38].

### 4.3. Psychological profile

As we analysed the psychological evaluations from our study, we realised that the amount of information from the psychological profiles and psychopathological mechanisms warrants an extended discussion of both topics. Therefore, we will discuss the psychopathological mechanisms separately in the next section.

Most PNES patients met the criteria for a diagnosis of an axis II personality disorder (around 60%) [4, 5], but patients with ES also had an increased prevalence of personality disorders. The prevalence of personality disorders in ES patients 34.3%, which was significantly higher than healthy controls.

Nevertheless, some significant differences exist in the types of personality disorders diagnosed in the PNES and ES patients. In the ES group, the most common personality disorders were cluster C personality disorders (i.e., avoidant, dependent, and obsessive-compulsive disorders) [48, 49, 50], and in PNES patients, cluster B personality disorders were more common (i.e., antisocial, borderline, histrionic and narcissistic disorders) [49, 50, 51, 52, 53,]. The presence of personality disorders is associated with a poor prognosis, diminished quality of life, and increased resource use in both PNES and ES patients [3, 4, 48, 51].

When personality traits have been assessed and specialised instruments used, a general, complex, and diverse psychological profile has been obtained, as evaluated with both neurotic and psychotic scales [4, 54]. Thus, PNES personality profiles do not reveal a specific, unique syndrome [55].

The MMPI has been the most widely used inventory to classify personality traits. The higher personality assessment scores have been in somatisation and externalisation; these patients have a strong preference for a medical explanation for their symptoms and tend to project the burden of their symptoms to their environment instead of a possible psychological explanation for their disorder [29, 55]. It is interesting to note that families of psychogenic seizure patients have also been found to be more prone to somatic problems and emotional distress than matched controls with epilepsy [56], hypochondriasis and conversion traits [4, 57].

Dissociation scores have been widely studied and, on average, are higher in PNES patients but not significantly higher [7, 45, 58, 59]. Dissociative trait scores were higher in patients with a reported trauma [7].

Additionally, an anxious, practical and perfectionist attitude with difficulties tolerating ambiguity and a strong need to control have also been reported [55]. Another study described increased shyness [7], which has been related to antisocial behaviour and a high tendency to internally control emotions [55].

PNES has been associated with an alexithymic pattern and a lack of psychological awareness of the causes and consequences of their psychological dysfunction [55]. Although alexithymia appears to be elevated in PNES, it has also been elevated in epilepsy patients [38]. PNES has been significantly correlated with symptoms of anxiety, autonomic hyperarousal, dissociation and defensive avoidance, which are all characteristic of traumatic experiences. Thus, alexithymia has been hypothesised as a deficit that results from physiological and cognitive dysregulation following post-traumatic stress disorder [60].

Another personality trait of PNES patients is difficulty with coping strategies. Elevated levels of perceived distress and a lack of coping strategies, such as a passive and/or avoidant attitude to resolve psychological problems in other ways [3, 61, 62], make it difficult to employ strategies that would normally be used to reduce the impact of a stressor [63]. Some effective minimal coping strategies, such as emotion-focused coping strategies, are more frequently used than effective coping strategies, such as task-focused strategies [60]. These features lead to difficulties in emotional experiences to daily life situations, and diminished positive emotional behaviour has also been reported [64].

Two emotion dysregulation profiles have been described [65], a high level of emotion dysregulation associated with severe psychiatric symptomatology and impaired quality of life and low emotion dysregulation that is characterised by emotional unawareness or avoidance.

Some common personality traits as well as some clinical similarities have been documented between PNES and borderline personality disorder patients. Higher prevalence rates of sexual trauma, posttraumatic stress disorder, dissociative disorders, somatoform disorders, depressive disorders, anger problems, suicide attempts, emotional instability and hostile coping styles have been reported in both conditions [49, 50, 52, 66].

Some similarities in personality traits have also been found between PNES and patients with functional somatic symptoms disorders [55].

Although the *interpersonal relationships* of PNES patients in clinical practice appear to be deeply affected, their social functioning scores remain within normal limits [4]. These patients make instrumental use of the PNES, which may contribute to the perception of being well-adapted in their relationships. Patients with higher emotional dysfunction and higher abnormal personality traits have a lower quality of life [47].

### 4.4. Etiopathogenic models of PNES

The final symptomatic expression of PNES remains unclear. Some etiopathogenic theories that comprehensively address many factors have been previously reported and have been hypothesised to explain psychogenic seizures.

Bodde y cols [55] considered factors that predispose patients to psychosomatic symptoms and increase PNES vulnerability.

Sexual abuse, other traumatic experiences, organic vulnerability (such as head trauma) and personality factors (emotional profile or neuropsychological functioning) increase PNES vulnerability. Stressful childhood events, including parental divorce and physical abuse [67], other childhood trauma and family dysfunction [68] also increase PNES vulnerability.

The observed personality traits have a tendency towards somatisation or dissociative experiences or reaction with somatic symptoms of psychological distress [7]. Alexithymia, cognitive inflexibility and hypervigilance have also been considered as PNES vulnerability factors [69].

Factors that shape patient perceptions could explain the presence of seizures and not other functional disorders, such as past epilepsy or having a close friend or relative with epileptic seizures. When examined, personally witnessing a seizure prior to their own attack occurs significantly more often in patients with PNES than in patients with epileptic seizures [67, 70].

Triggering factors, such as some psychological mechanisms or psychosocial events, might provoke seizure onset at specific moments. PNES patients report more events than epilepsy patients or patients with motor conversion disorders [67].

Thus, prolongation factors could explain the persistence of PNES over time.

The modulating factors, e.g., personality [3], patient coping style and secondary gain aspects, could also contribute. There are some criticisms of secondary gain, as this benefit can be observed in almost any illness [67].

### 4.5. PNES subtypes

In general psychopathology, somatisation, alexithymia and difficulties with most aspects of emotional regulation could be the most important factors in the etiopathogenesis of PNES [71].

Considering all the etiopathogenic factors, some studies have proposed different etiopathogenic subgroups.

Reuber and House [2] classified PNES into 3 groups based on the most prominent psychiatric comorbidity profiles: an anxiety-depressive group with depressive or anxiety symptoms, including panic and post-traumatic stress disorder (PTSD), a somatisation and abnormal illness group and a borderline personality group.

Bodde et al. [55] proposed 4 etiopathogenic groups: a psychotrauma subgroup, a high vulnerability somatisation subgroup (approximately 1/5 of patients), a sensitive personality subgroup (less than 1/5 of the patients) and a high vulnerability somatisation, low cognitive level subgroup with daily life stress.

Other authors proposed a more simplified, two-subgroup etiopathogenic classification scheme: a posttraumatic PNES group, characterised by psychogenic attacks that develop in response to acute or chronic traumatic experience exposure, and developmental PNES group, characterised by difficulties in coping with tasks and milestones on the individual's psychosocial development continuum [71].

The importance of diagnosis presentation

The first step towards symptom improvement is presenting the diagnosis to the patient and when appropriate, to his or her family. Providing patients with a PNES diagnosis, appears to reduce PNES frequency in the months following diagnosis in at least one third of patients [69, 72, 73, 74, 75, 76].

A marked reduction in health care demand after PNES diagnosis has also been established by prospective follow-up studies [10, 69, 77]. However, long-term studies examining the decrease in PNES frequency shortly after diagnosis has not been maintained longitudinally when a diagnosis is the main and sole "intervention" [3, 58].

Effectively communicating the PNES diagnosis in an understandable and acceptable way to the patient and family is critical and is considered the first therapeutic step.

The process of explaining to the patients that their disease may be psychiatric instead of neurological must be cautiously and progressively handled [16]. Patients and family must be engaged in a treatment plan very different from the previous plan.

A significant patient and family resistance to a new and non-neurologic diagnosis requires the implementation of coordinated strategies by a neurologist and a psychiatrist to manage the best outcome for these patients. When a diagnosis is given to patients and families, some of patients have a sense of relief to know that the events are not epileptic and that they will be helped in identifying sources of stress and other emotional problems responsible for triggering their events [75]. However, most patients react negatively to the diagnosis of PNES. Frequently, they feel that doctors are communicating that the symptoms are not real but invented.

Many factors may contribute to this negative, even hostile, attitude towards a psychogenic basis of their crisis [46, 52]. Some personality trait factors, such as somatisation and externalisation traits, make diagnosis acceptance difficult. Patients and families show a tendency to seek medical responsibility for these symptoms and might be very reluctant to be referred for psychological or psychiatric treatment.

An indirect benefit obtained through this behavioural pattern may also help explain their resistance to change [4].

Another factor related to the initially hostile attitude of the patients is the risk of stigma, which is inherent to psychiatric disorders. Patients fear being viewed as malingerers or attention seekers [74], and also the not infrequent stigmatisation of the doctor, who only considers neurologic diseases real diseases, may transmit this bias that a psychogenic crisis is not a real disease to patients. These stigmatising views can lead patients to believe that they do not need attentive specialised and technical care but a psychiatric clinic for insane individuals.

All these difficulties in the diagnosis and treatment of PNES patients must be considered when deliberating over the most effective strategy to provide information about a PNES diagnosis to patients previously considered neurologic patients.

A good therapeutic relationship between the neurologist and patient is crucial. One or several clinical interviews involving a neurologist, psychiatrist and patient that provide a multidisciplinary presentation of the diagnosis and treatment plan are highly recommended.

When presenting a PNES diagnosis, it is recommended to explain to patient and family the lack of ictal discharges during vEEG, the psychogenic nature of the attacks as a result of difficulties managing stressful or life situations, the acknowledgement that the attacks are real and out of the patient's control and the need for psychotherapeutic and/or psychiatric treatment to managing stressful situations and psychiatric comorbidities. Unless patients and families understand and accept the diagnosis, they typically will not comply with recommendations and will not follow-up with a psychiatrist.

A neurologist will continue to provide care, monitoring the patient outcome, and progressively decreasing antiepileptic drug doses when appropriate.

If patients present with both PNES and epilepsy and antiepileptic drugs must be maintained, patients will need to be informed and must understand both the neurologic and psychiatric treatments. However, some controversy exists about the discontinuation of antiepileptic drugs in PNES patients without epilepsy. Antiepileptic drugs could provide therapeutic benefits on mood and impulsivity and for headache or pain prophylaxis. On the contrary, toxicity, teratogenicity and medication expense need to be considered [78].

### 4.6. Treatment

The treatment plan must be individually adjusted to each patient, depending on the psychogenic factors considered to underlie the psychopathology [7, 79, 80].

For example, if difficulties in emotional reactions occur, some strategies, such as monitoring symptom triggers, re-interpreting physical sensations and increasing relaxation, could be effective in PNES patients [64].

### 4.7. Psychotherapy

Most psychotherapy techniques, such as psychodynamic interpersonal psychotherapy, group psychodynamic psychotherapy and group psychoeducation, have been evaluated in uncontrolled trials. These interventions all seem promising but require further investigation in larger scale samples and/or more rigorous methodology.

### 4.8. Cognitive-Behavioural Therapy (CBT)

CBT is based on the conceptualisation of PNES as dissociative responses to arousal when a patient is confronted with stimuli or circumstances that the patient tends to avoid, consciously or not.

Patients with PNES tend to display significantly distorted somatic beliefs and develop dysfunctional, repetitive illness behaviour patterns with associated depressive affect as well as a sudden and time-limited disturbance in controlling cognitive, emotional, and/or behavioural functions [70, 79]. CBT is intended to modify distorted beliefs and change the perception of the limitations associated with PNES.

This therapy may be useful several symptom patterns: (a) acute anxiety/panic, (b) impaired affect regulation and interpersonal skills, (c) somatisation/conversion, (d) depression, (e) PTSD, and (f) reinforced behavioural patterns [80]. These classifications were based on the most effective psychotherapeutic interventions for each patient group.

CBT is the only psychotherapeutic PNES intervention that has been studied in a randomised, controlled pilot trial and is therefore the psychotherapeutic treatment with the highest level of efficacious evidence (Class III) in this population. Goldstein et al. [81] reported that in an open-label trial for patients with dissociative seizure, a 12-session CBT intervention was significantly more effective than standard medical care (SMC) for reducing seizure frequency in PNES patients. Psychotherapy focused on the following factors: engagement in treatment, reinforcement of independence, distraction, relaxation, refocusing techniques at the earliest signs of an event, graded exposure to avoided situations, cognitive restructuring, and relapse prevention. A tendency for this benefit to persist was maintained at the 6-month follow-up. There was also an overall improvement in self-rated social functioning of the CBT group compared with SMC.

Moreover, an uncontrolled CBT study where therapy was focused on addressing cognitive distortions and promoting behavioural changes has been published [82]. Some specific objectives of the psychotherapeutic sessions included the study of the individual's context, identification of moods, situations, and thoughts, training in healthy communication and support seeking, understanding central nervous system medications and seizures, conducting a functional behavioural analysis, developing relaxation techniques, examining external stressors and internal triggers, and preparing for life after completing the time-limited intervention. Patients reduced the number of PNES and improved quality of life, family functioning, and psychosocial functioning.

In summary, the data on CBT efficacy for PNES treatment are sparse but promising.

### 4.9. Group therapy

To date, several studies have reported the use of group therapy for PNES patients.

Zaroff et al. [83] evaluated the effectiveness of group psychotherapy based on psychodynamic theories and a psychoeducational session structure. All patients included in this open-label trial participated in a one-hour weekly group psychotherapy session for a total of 10 weeks. In each session, the following relevant PNES topics were discussed: PNES, anger, trauma and abuse, depression and anxiety, somatisation tendencies, quality of life, paths toward health, stress coping techniques and topic review. At the conclusion of the 10-week program, group psychotherapy continued in a less structured and more supportive format without specific session topics. The substantial percentage of individuals whose seizures remitted following diagnosis supports the hypothesis that education about the disorder is an effective treatment.

The results of a six-month psychotherapy group, with an open, uncontrolled design [84], have been published. Over a total of 24 sessions, the discussion of feelings related to past or present events relevant to PNES was encouraged. A decrease in seizure frequency was reported in six of the nine PNES subjects (67%).

In another pilot study with a psychodynamic focus and psychoeducational structure, self-relaxation with hypnosis was also used [85]. Dissociation was considered a defence mechanism that splits traumatic memories from consciousness because they are inconsistent with the patient's self-concept. The goal of therapy was to facilitate the patient's awareness of these patterns and to thus allow change. The development of assertive coping strategies instead of passive avoidant behaviour was emphasised. The study reported an overall decrease in seizure frequency and a significant improvement in different psychometric assessments.

Recently, another uncontrolled study with group psychotherapy has been published [86]. Psychoeducation and behavioural and psychoanalytic techniques were utilised. Nine patients completed 12 weeks of psychotherapy. In this therapy, families gathering four times for 1 h during the study provided a different treatment element. Families were not included in the therapeutic process and met just before the group sessions. Families were encouraged to ask questions and talk about seizures, and behavioural management strategies were advised. A sustainable decrease in seizure frequency, which lasted from the beginning of the therapy until the 12$^{th}$ month of follow-up, was observed. All patients showed a >50% reduction in seizure frequency.

### 4.10. Paradoxical intention therapy

This technique has been used in PNES and conversion disorder patients. In a randomised trial of PNES patients, Ataoglu et al. [87] tested inpatient paradoxical intention (PI) therapy versus oral outpatient benzodiazepine. PI therapy consisted of encouraging patients to intentionally engage in their unwanted conversion symptoms. This study did not focus on crisis frequency but on conversion symptoms. Both groups recorded significantly decreased anxiety scores by the end of treatment.

### 4.11. Hypnosis

Given the theoretical importance of dissociation in PNES aetiology, hypnosis would appear to offer a useful intervention for this patient population [72]. The effectiveness of hypnosis was evaluated in a prospective study [88] where 49 patients were randomly assigned to either a hypnosis-based 10-week session treatment for conversion disorder, motor type, or a wait list control group. Only two patients in this conversion disorder study had PNES. A significant improvement in the Motor Conversion Symptoms scale (VRMC) for the treatment group was observed.

Other studies evaluated dissociation and hypnotisability in PNES patients and compared them with a nonclinical control group or epilepsy patients. In one study, PNES patients had higher levels of dissociation and hypnotisability as measured by the Dissociative Experience Scale (DES) [61]. Hypnotisability was also significantly higher for the PNES group compared with the control group, as other authors have previously reported [89].

### 4.12. Pharmacological treatment

To date, the lack of a consistent neurobiological model explaining PNES physiopathology makes it considerably difficult to identify a specific and effective pharmacological treatment. Owing to a lack of large, controlled trials evaluating the efficacy of different treatment modalities, guidelines on how to treat this complex health problem do not exist [16, 79].

Nonetheless, under the premise that some of the factors that seem to be involved in PNES etiopathogenesis, such as depressive mood and anxiety, are associated with serotonergic neurotransmission alterations, selective serotonin reuptake inhibitors (SSRIs) have been investigated as potentially useful drugs. In addition to their established efficacy for treating depression and anxiety [90, 91], SSRIs have shown promise in trials for conversion or somatoform disorders [92] and some personality disorders [93]. These disorders are frequently occurring comorbidities in PNES patients, which make SSRIs particularly attractive as a potential treatment.

LaFrance et al. conducted a randomised, double-blind, placebo-controlled clinical trial (which meets the criteria for Class II evidence) evaluating the efficacy of flexible-dose sertraline over 12 weeks at reducing PNES frequency and improving other symptom severity and psychosocial measures [94]. Patients were allowed to participate even with previous antidepressant use, except for monoamine oxidase inhibitors or sertraline at a 100 mg/day or higher dose, to maintain a constant dose of the concurrent antidepressants for the study duration. The sertraline group experienced a 45% decline in biweekly seizure rates over the 12-week course of the intervention from 22.24 to 12.18 (ratio, 0.55; 95% CI, 0.32-0.93; $p = 0.03$), while the control group showed an 8% increase from 13.38 to 14.38 (ratio, 1.08; 95% CI, 0.65-1.77; $p = 0.78$). However, the limited sample size (33 patients were included in the final analysis) implied this study was underpowered and unable to detect statistically significant differences between groups (RR, 0.51; 95% CI, 0.25-1.05; $p = 0.29$). There were no differences in secondary outcome measure changes between the groups (a series of scales assessing different psychiatric symptoms, functionality and quality of life-related measures). The evaluation of SSRIs in larger, randomised, controlled trials is therefore necessary to establish solid treatment guidelines.

Venlafaxine, a noradrenalin-serotonin reuptake inhibitor (NSRI), has also been evaluated as a potential psychopharmacological PNES treatment. An open-label, prospective, uncontrolled study evaluated the efficacy of flexible-dose venlafaxine on seizure frequency reduction and anxiety and depression disorder severity improvement [95]. All enrolled subjects had vEEG-confirmed PNES but also met the DSM-IV criteria for a unipolar depressive disorder and/or anxiety disorder. The 19 subjects who completed the 5-month follow-up experienced a statistically significant reduction in all symptom scales and monthly event frequency at the five-month assessment compared with the initial assessment; however, no differences were detected in the patient subgroup with more than ten events in the 15-day baseline pre-inclusion assessment period. The particular comorbidities of the patients included in this study prevent the extrapolation of the results to the broader PNES population, but the combined prevalence of depressive and anxiety disorders among PNES patients that may reach up to two thirds of PNES patients should not be ignored [4]. An additional caution when interpreting these results is the lack of a placebo-arm, especially if we consider the documented initial decrease in seizure frequency consistently observed during the first months after PNES diagnosis [16, 73, 75, 76]. This factor could account for some of the event frequency reductions described in this study.

Clinicians have often tried other pharmacological PNES treatments based on current knowledge of PNES-related disorders. As we have previously described, patients suffering from PNES share many underlying psychopathological characteristics with other conversion and somatoform disorders. These comorbidities may lead to the hypothesis that treatments demonstrating efficacy for these other disorders might also provide some benefit in PNES.

Evidence-based treatments for other conversion disorders have been documented in uncontrolled trials. A retrospective study on patients with psychogenic paralysis showed that repetitive Transcranial Magnetic Stimulation (rTMS) over the motor cortex contralateral to the corresponding paralysis achieved improvement in 62 of 70 patients (89%), with total recovery observed in 53 patients (76%) [96].

In an open-label, prospective study conducted on 23 patients diagnosed with chronic psychogenic movement disorder (PMD), antidepressant medications (paroxetine, citalopram and venlafaxine) showed a reduction and even remission of psychogenic movements in the subgroup of patients in which the movements were not accompanied by other somatoform disorders, such as hypochondriasis or somatisation disorders [92]. Various classes of antidepressants have also demonstrated a reduction in medically unexplained symptoms, including headache, fibromyalgia, functional gastrointestinal syndromes, idiopathic pain, tinnitus, and chronic fatigue [97].

In summary, no strong evidence exists to date that establishes the efficacy of any particular medication for the treatment of PNES. Despite this fact, preliminary studies suggest that sertraline and venlafaxine could be beneficial therapeutics. The data investigating treatments for PNES-related disorders also support this hypothesis

## 5. Conclusions

PNES is one of the most frustrating clinical pathologies. Diagnosis is sometimes difficult and can take several years to reach a definitive PNES diagnosis. Moreover, treatment can be even more difficult. However, PNES is a severely debilitating illness for the patient, family and society; thus, we must continue to identify and improve patient quality-of-life. Nevertheless, a significant percentage of patients referred to epilepsy surgery units present with PNES, where the necessary assessments, such as vEEG, can provide a correct diagnosis; therefore, the actual magnitude of PNES diagnoses may be larger than currently detectable.

Hence, when assessing patients with epilepsy, the PNES pathology should be kept in mind and patients should be sent to a specialised unit, even under the minimal suspicion of PNES, to provide a correct diagnosis in a timely manner. These steps should minimise medication side-effects and consumption of health care resources and offer the patient the most efficacious treatment.

## Acknowledgements

This work was supported by a grant from the Plan Nacional de Investigación Científica, Desarrollo e Innovación Tecnológica (I+D+I), Instituto de Salud Carlos III, Subdirección General de Evaluación y Fomento de la Investigación PI12/02839.

## Author details

Lorena Vega-Zelaya[1], Marta Alvarez[2], Elena Ezquiaga[2], Jaime Nogeiras[2], María Toledo[3], Rafael G. Sola[4] and Jesús Pastor[1*]

*Address all correspondence to: jesus.pastor@salud.madrid.org

1 Clinical Neurophysiology, National Referral Centre for Refractory Epilepsy, Hospital La Princesa, Madrid, Spain

2 Psychiatry, National Referral Centre for Refractory Epilepsy, Hospital La Princesa, Madrid, Spain

3 Neurology, National Referral Centre for Refractory Epilepsy, Hospital La Princesa, Madrid, Spain

4 Neurosurgery, National Referral Centre for Refractory Epilepsy, Hospital La Princesa, Madrid, Spain

## References

[1] Bowman ES, Coons PM. The differential diagnosis of epilepsy, pseudoseizures, dissociative identity disorder, and dissociative disorder not otherwise specified. Bull Menninger Clin. 2000;64(2) 164-80.

[2] Reuber M, House AO. Treating patients with psychogenic non-epileptic seizures. Curr Opin Neurol. 2002;15(2) 207-11.

http://dx.doi.org/10.5772/57439

[3] Bodde NM, Brooks JL, Baker GA, Boon PA, Hendriksen JG, Mulder OG, Aldenkamp AP. Psychogenic non-epileptic seizures – definition, treatment and prognostic issues: a critical review. Seizure 2009;18(8) 543–53.

[4] Bailles, E, Pintor L, Torres X, Fernández-Egea E, de Pablo J, Arroyo S. Psychiatric disease in patients with psychogenic non-epileptic seizures referred from an epilepsy unit in a general hospital. Actas Esp Psiquiatr 2004;32(2): 76-81.

[5] Bowman ES, Markand ON. Psychodynamics and psychiatric diagnoses of pseudoseizure subjects. Am J Psychiatry. 1996;153(1) 57-63.

[6] Ribai P, Tugendhaft P, Legros B. Usefulness of prolonged Video-EEG monitoring and provocative procedure with saline injection for the diagnosis of non epileptic seizures of psychogenic origin. J. Neurol 2006;253(3) 328-32.

[7] Bodde NM, van der Kruijs SJ, Ijff DM, Lazeron RH, Vonck KE, Boon PA, Aldenkamp AP. Subgroup classification in patients with psychogenic non-epileptic seizures. Epilepsy Behav. 2013;26(3) 279-89.

[8] Reuber M, Elger CE. Psychogenic nonepileptic seizures: review and update. Epilepsy Behav 2003;4(3) 205–16.

[9] Lobello K, Morgenlander JC, Radtke RA, Bushnell CD. Video/EEG monitoring in the evaluation of paroxysmal behavioural events: duration, effectiveness, and limitations. Epilepsy Behav 2006;8(1) 261-6.

[10] Razvi S, Mulhern S, Duncan R. Newly diagnosed psychogenic nonepileptic seizures: health care demand prior to and following diagnosis at a first seizure clinic. Epilepsy Behav. 2012;23(1) 7-9.

[11] Reuber M. The etiology of psychogenic non-epileptic seizures: toward a biopsychosocial model. Neurol Clin. 2009;27(4) 909-24.

[12] O'Sullivan SS, Spillane JE, McMahon EM, Sweeney BJ, Galvin RJ, McNamara B, Cassidy EM. Clinical characteristics and outcome of patients diagnosed with psychogenic nonepileptic seizures: a 5-year review. Epilepsy Behav. 2007;11(1) 77-84.

[13] Reuber M, Baker GA, Gill R, Smith DF, Chadwick DW. Failure to recognize psychogenic nonepileptic seizures may cause death. Neurology. 2004;62(5) 834-5.

[14] Szaflarski JP, Szaflarski M, Hughes C, Ficker DM, Cahill WT, Privitera MD. Psychopathology and quality of life: psychogenic non-epileptic seizures versus epilepsy. Med Sci Monit. 2003;9(4) CR113-8.

[15] Testa SM, Schefft BK, Szaflarski JP, Yeh HS, Privitera MD. Mood, personality, and health-related quality of life in epileptic and psychogenic seizure disorders. Epilepsia. 2007;48(5) 973-82.

[16] Baslet G. Psychogenic nonepileptic seizures: a treatment review. What have we learned since the beginning of the millennium? Neuropsychiatr Dis Treat. 2012;8 585-98.

[17] Breier JI, Fuchs KL, Brookshire BL, Wheless J, Thomas AB, Constantinou J, Willmore LJ. Quality of life perception in patients with intractable epilepsy or pseudoseizures. Arch Neurol. 1998;55(5) 660-5.

[18] Dixit R, Popescu A, Bagić A, Ghearing G, Hendrickson R. Medical comorbidities in patients with psychogenic nonepileptic spells (PNES) referred for video-EEG monitoring. Epilepsy Behav. 2013;28(2) 137-40.

[19] Noe KH, Grade M, Stonnington CM, Driver-Dunckley E, Locke DE. Confirming psychogenic nonepileptic seizures with video-EEG: sex matters. Epilepsy Behav. 2012;23(3) 220-3.

[20] Behrouz R, Heriaud L, Benbadis SR. Late-onset psychogenic nonepileptic seizures. Epilepsy Behav. 2006;8(3) 649-50.

[21] Hermann BP. Neuropsychological assessment in the diagnosis of non-epileptic seizures. In: Rowan AJ, Gates JR, (ed) Non-epileptic Seizures. Stoneham, MA: Butterworth-Heinemann; 1993. p. 221–32.

[22] Pastor J, Hernando-Requejo V, Domínguez-Gadea L, De Llano I, Meilán-Paz ML, Martínez-Chacón JL, et al. Impacto de la experiencia sobre los resultados quirúrgicos en la epilepsia del lóbulo temporal. Rev Neurol 2005;41(12) 709-16.

[23] Sola RG, Hernando-Requejo V, Pastor L, García-Navarrete E, De Felipe J, Alijarde MT, Sánchez A, Domínguez-Gadea L, Martín-Plasencia P, Maestú F, Ramón-Cajal S, Pulido-Rivas P. Epilepsia farmacorresistente del lóbulo temporal. Exploración con electrodos del foramen oval y resultados quirúrgicos. Rev Neurol. 2005;41(1) 4-16.

[24] Commission on Classification and Terminology of the International League Against Epilepsy. Proposal for Revised Classification of Epilepsies and Epileptic Syndromes. Epilepsia. 1989;30(4) 389–99.

[25] Benbadis SR. Provocative techniques should be used for the diagnosis of psychogenic nonepileptic seizures. Epilepsy Behav 2009;15(2) 106-9.

[26] Gallego I, Ezquiaga E, Betancor D, Sola RG, Pastor J. Crisis no epilépticas de origen psicógeno en una unidad de cirugía de la epilepsia. Rev Neurol 2011;52(8) 449-56.

[27] Shaw JC. An introduction to the coherence function and its use in EEG signal analysis. J Med Engineering & Technology. 1981;5(6) 279–288.

[28] Ortega GJ, Menendez de la Prida L, Sola RG, Pastor J. Synchronization clusters of interictal activity in the lateral temporal cortex of epileptic patients: intracranial analysis. Epilepsia 2008;49(2) 269–280.

[29] Reuber M, Fernández G, Helmstaedter C, Qurishi A, Elger CE. Evidence of brain abnormality in patients with psychogenic nonepileptic seizures. Epilepsy & Behavior 2002;3(3) 249–254.

[30] Devinsky O, Mesad S, Alper K. Nondominant hemisphere lesions and conversion nonepileptic seizures. J Neuropsychiatry Clin Neurosci 2001;13(3) 367–73.

[31] Glosser G, Roberts D, Glosser DS. Nonepileptic seizures after resective epilepsy surgery. Epilepsia 1999;40(12) 1750–4.

[32] Labate A, Cerasa A, Mula M, Mumoli L, Gioia MC, Aguglia U, Quattrone A, Gambardella A. Neuroanatomic correlates of psychogenic nonepileptic seizures: a cortical thickness and VBM study. Epilepsia. 2012;53(2) 377-85.

[33] Leis AA, Ross MA, Summers AK. Psychogenic seizures: ictal characteristics and diagnostic pitfalls. Neurology 1992;42(1) 95–9.

[34] Alsaadi TM, Thieman C, Shatzel A, Farias S. Video-EEG telemetry can be a crucial tool for neurologists experienced in epilepsy when diagnosing seizure disorders. Seizure 2004;13(1) 32-4.

[35] LaFrance WC Jr. Psychogenic nonepileptic seizures. Curr Opin Neurol. 2008;21(2) 195-201.

[36] Turner K, Piazzini A, Chiesa V, Barbieri V, Vignoli A, Gardella E, Tisi G, Scarone S, Canevini MP, Gambin O. Patients with epilepsy and patients with psychogenic nonepileptic seizures: Video-EEG, clinical and neuropsychological evaluation. Seizure 2011;20(9) 706–710.

[37] Binzer M, Stone J, Sharpe M. Recent onset pseudoseizures--clues to aetiology. Seizure. 2004;13(3) 146-55.

[38] Tojek TM, Lumley M, Barkley G, Mahr G, Thomas A. Stress and other psicosocial characteristics of patients with psychogenic nonepileptic seizures. Psychosomatics. 2000;41(3) 221-6.

[39] Gumnit RJ, Gates JR. Psychogenic seizures. Epilepsia. 1986;27(2) S124-9.

[40] Arnold LM, Privitera MD. Psychopathology and trauma in epileptic and psychogenic seizure patients. Psychosomatics 1996;37(5) 438-43.

[41] Myers L, Perrine K, Lancman M, Fleming M, Lancman M. Psychological trauma in patients with psychogenic nonepileptic seizures: trauma characteristics and those who develop PTSD. Epilepsy Behav. 2013;28(1) 121-6.

[42] Stone J, Binzer M, Sharpe M. Illness beliefs and locus of control: a comparison of patients with pseudoseizures and epilepsy. J Psychosom Res. 2004;57(6) 541-7.

[43] Hingray C, Maillard L, Hubsch C, Vignal JP, Bourgognon F, Laprevote V, Lerond J, Vespignani H, Schwan R. Psychogenic nonepileptic seizures: characterization of two

distinct patient profiles on the basis of trauma history. Epilepsy Behav. 2011;22(3) 532-6.

[44] Stewart RS, Lovitt R, Stewart RM. Are hysterical seizures more than hysteria? A research diagnostic criteria, DMS-III, and psychometric analysis. Am J Psychiatry. 1982;139(7) 926-9.

[45] Ito M, Adachi N, Okazaki M, Kato M, Onuma T. Evaluation of dissociative experiences and the clinical utility of the Dissociative Experience Scale in patients with coexisting epilepsy and psychogenic nonepileptic seizures. Epilepsy Behav. 2009;16(3) 491-4.

[46] Mökleby K, Blomhoff S, Malt UF, Dahlström A, Tauböll E, Gjerstad L. Psychiatric comorbidity and hostility in patients with psychogenic nonepileptic seizures compared with somatoform disorders and healthy controls. Epilepsia. 2002;43(2) 193-8.

[47] Myers L, Lancman M, Laban-Grant O, Matzner B, Lancman M. Psychogenic non-epileptic seizures: predisposing factors to diminished quality of life. Epilepsy Behav. 2012;25(3) 358-62.

[48] Devinsky O. Psychiatric comorbidity in patients with epilepsy: implications for diagnosis and treatment. Epilepsy Behav. 2003;4(4) S2-10.

[49] Harden CL, Jovine L, Burgut FT, Carey BT, Nikolov BG, Ferrando SJ. A comparison of personality disorder characteristics of patients with nonepileptic psychogenic pseudoseizures with those of patients with epilepsy. Epilepsy Behav. 2009;14(3) 481-3.

[50] Direk N, Kulaksizoglu IB, Alpay K, Gurses C. Using personality disorders to distinguish between patients with psychogenic nonepileptic seizures and those with epileptic seizures. Epilepsy Behav. 2012;23(2) 138-41.

[51] Lacey C, Cook M, Salzberg M. The neurologist, psychogenic nonepileptic seizures, and borderline personality disorder. Epilepsy Behav. 2007;11(4) 492-8

[52] Bodde NM, Brooks JL, Baker GA, Boon PA, Hendriksen JG, Aldenkamp AP. Psychogenic non-epileptic seizures-diagnostic issues: a critical review. Clin Neurol Neurosurg. 2009;111(1) 1-9.

[53] Galimberti CA, Ratti MT, Murelli R, Marchioni E, Manni R, Tartara A. Patients with psychogenic nonepileptic seizures, alone or epilepsy-associated, share a psychological profile distinct from that of epilepsy patients. J Neurol. 2003;250(3) 338-46.

[54] Kalogjera-Sackellares D, Sackellares JC. Analysis of MMPI patterns in patients with psychogenic pseudoseizures. Seizure. 1997;6(6) 419-27.

[55] Bodde NM, Bartelet DC, Ploegmakers M, Lazeron RH, Aldenkamp AP, Boon PA. MMPI-II personality profiles of patients with psychogenic nonepileptic seizures. Epilepsy Behav. 2011;20(4) 674-80.

[56] Thompson N, Connelly L, Peltzer J, Nowack WJ, Hamera E, Hunter EE. Psychogenic nonepileptic seizures: a pilot study of a brief educational intervention. Perspect Psychiatr Care. 2013;49(2) 78-83.

[57] Griffith NM, Szaflarski JP, Schefft BK, Isaradisaikul D, Meckler JM, McNally KA, Privitera MD. Relationship between semiology of psychogenic nonepileptic seizures and Minnesota Multiphasic Personality Inventory profile. Epilepsy Behav. 2007;11(1) 105-11.

[58] Reuber M, House AO, Pukrop R, Bauer J, Elger CE. Somatization, dissociation and general psychopathology in patients with psychogenic non-epileptic seizures. Epilepsy Res. 2003;57(2-3) 159-67.

[59] Lawton G, Baker GA, Brown RJ. Comparison of two types of dissociation in epileptic and nonepileptic seizures. Epilepsy Behav. 2008;13(2) 333-6.

[60] Myers L, Fleming M, Lancman M, Perrine K, Lancman M. Stress coping strategies in patients with psychogenic non-epileptic seizures and how they relate to trauma symptoms, alexithymia, anger and mood. Seizure. 2013;22(8) 634-9.

[61] Goldstein LH, Drew C, Mellers J, Mitchell-O'Malley S, Oakley DA. Dissociation, hypnotizability, coping styles and health locus of control: characteristics of pseudoseizure patients. Seizure. 2000;9(5) 314-22.

[62] Frances PL, Baker GA, Appleton PL. Stress and avoidance in Pseudoseizures: testing the assumptions. Epilepsy Res. 1999;34(2-3) 241-9.

[63] Testa SM, Krauss GL, Lesser RP, Brandt J. Stressful life event appraisal and coping in patients with psychogenic seizures and those with epilepsy. Seizure. 2012;21(4) 282-7.

[64] Roberts NA, Burleson MH, Weber DJ, Larson A, Sergeant K, Devine MJ, Vincelette TM, Wang NC. Emotion in psychogenic nonepileptic seizures: responses to affective pictures. Epilepsy Behav. 2012;24(1) 107-15.

[65] Uliaszek AA, Prensky E, Baslet G. Emotion regulation profiles in psychogenic nonepileptic seizures. Epilepsy Behav. 2012;23(3) 364-9.

[66] Skodol AE, Gunderson JG, Pfohl B, Widiger TA, Livesley WJ, Siever LJ. The borderline diagnosis I: psychopathology, comorbidity, and personality structure. Biol Psychiatry. 2002;51(12) 936-50.

[67] Dickinson P, Looper KJ, Groleau D. Patients diagnosed with nonepileptic seizures: their perspectives and experiences. Epilepsy Behav. 2011;20(3) 454-61.

[68] Reuber M, Howlett S, Khan A, Grünewald RA. Non-epileptic seizures and other functional neurological symptoms: predisposing, precipitating, and perpetuating factors. Psychosomatics. 2007;48(3) 230-8.

[69] Baslet G. Psychogenic non-epileptic seizures: a model of their pathogenic mechanism. Seizure. 2011;20(1) 1-13.

[70] Bautista RE, Gonzales-Salazar W, Ochoa JG. Expanding the theory of symptom modeling in patents with psychogenic nonepileptic seizures. Epilepsy Behav. 2008;13(2) 407-9.

[71] Brown RJ, Bouska JF, Frow A, Kirkby A, Baker GA, Kemp S, Burness C, Reuber M. Emotional dysregulation, alexithymia, and attachment in psychogenic nonepileptic seizures. Epilepsy Behav. 2013;29(1) 178-83.

[72] Walczak TS, Papacostas S, Williams DT, Scheuer ML, Lebowitz N, Notarfrancesco A. Outcome after diagnosis of psychogenic nonepileptic seizures. Epilepsia. 1995;36(11) 1131-7.

[73] Reuber M, Mitchell AJ, Howlett S, Elger CE. Measuring outcome in psychogenic nonepileptic seizures: how relevant is seizure remission? Epilepsia. 2005;46(11) 1788-95.

[74] Kanner AM, Parra J, Frey M, Stebbins G, Pierre-Louis S, Iriarte J. Psychiatric and neurologic predictors of psychogenic pseudoseizure outcome. Neurology. 1999;53(5) 933-8.

[75] Farias ST, Thieman C, Alsaadi TM. Psychogenic nonepileptic seizures: acute change in event frequency after presentation of the diagnosis. Epilepsy Behav. 2003;4(4) 424-9.

[76] Iriarte J, Parra J, Urrestarazu E, Kuyk J. Controversies in the diagnosis and management of psychogenic pseudoseizures. Epilepsy Behav. 2003;4(3) 354-9.

[77] Jirsch JD, Ahmed SN, Maximova K, Gross DW. Recognition of psychogenic nonepileptic seizures diminishes acute care utilization. Epilepsy Behav. 2011;22(2) 304-7.

[78] LaFrance WC Jr, Alper K, Babcock D, Barry JJ, Benbadis S, Caplan R, Gates J, Jacobs M, Kanner A, Martin R, Rundhaugen L, Stewart R, Vert C; NES Treatment Workshop participants. Nonepileptic seizures treatment workshop summary. Epilepsy Behav. 2006;8(3) 451-61.

[79] LaFrance WC Jr, Barry JJ. Update on treatments of psychological nonepileptic seizures. Epilepsy Behav. 2005;7(3) 364-74.

[80] Rusch MD, Morris GL, Allen L, Lathrop L. Psychological Treatment of Nonepileptic Events. Epilepsy Behav. 2001;2(3) 277-283.

[81] Goldstein LH, Chalder T, Chigwedere C, Khondoker MR, Moriarty J, Toone BK, Mellers JD. Cognitive-behavioral therapy for psychogenic nonepileptic seizures: a pilot RCT. Neurology. 2010;74(24) 1986-94.

[82] LaFrance WC Jr, Miller IW, Ryan CE, Blum AS, Solomon DA, Kelley JE, Keitner GI. Cognitive behavioral therapy for psychogenic nonepileptic seizures. Epilepsy Behav. 2009;14(4) 591-6.

[83] Zaroff CM, Myers L, Barr WB, Luciano D, Devinsky O. Group psychoeducation as treatment for psychological nonepileptic seizures. Epilepsy Behav. 2004;5(4) 587-92.

[84] Prigatano GP, Stonnington CM, Fisher RS. Psychological factors in the genesis and management of nonepileptic seizures: clinical observations. Epilepsy Behav. 2002;3(4) 343-349.

[85] Barry JJ, Wittenberg D, Bullock KD, Michaels JB, Classen CC, Fisher RS. Group therapy for patients with psychogenic nonepileptic seizures: a pilot study. Epilepsy Behav. 2008;13(4) 624-9.

[86] Metin SZ, Ozmen M, Metin B, Talasman S, Yeni SN, Ozkara C. Treatment with group psychotherapy for chronic psychogenic nonepileptic seizures. Epilepsy Behav. 2013;28(1) 91-4.

[87] Ataoglu A, Ozcetin, A, Icmeli C, Ozbulut O. Paradoxical Therapy in Conversion Reaction. J Korean Med Sci 2003;18(4) 581-584.

[88] Moene FC, Spinhoven P, Hoogduin KA, van Dyck R. A randomised controlled clinical trial on the additional effect of hypnosis in a comprehensive treatment programme for in-patients with conversion disorder of the motor type. Psychother Psychosom. 2002;71(2) 66-76.

[89] Barry JJ, Atzman O, Morrell MJ. Discriminating between epileptic and nonepileptic events: the utility of hypnotic seizure induction. Epilepsia. 2000;41(1) 81-4.

[90] Cipriani A, La Ferla T, Furukawa TA, Signoretti A, Nakagawa A, Churchill R, McGuire H, Barbui C. Sertraline versus other antidepressive agents for depression. Cochrane Database Syst Rev. 2010;14 (4).

[91] Ipser JC, Kariuki CM, Stein DJ. Pharmacotherapy for social anxiety disorder: a systematic review. Expert Rev Neurother. 2008;8(2) 235-57.

[92] Voon V, Lang AE. Antidepressant treatment outcomes of psychogenic movement disorder. J Clin Psychiatry. 2005;66(12) 1529-34.

[93] Lieb K, Völlm B, Rücker G, Timmer A, Stoffers JM. Pharmacotherapy for borderline personality disorder: Cochrane systematic review of randomised trials. Br J Psychiatry. 2010;196(1) 4-12.

[94] LaFrance WC Jr, Keitner GI, Papandonatos GD, Blum AS, Machan JT, Ryan CE, Miller IW. Pilot pharmacologic randomized controlled trial for psychogenic nonepileptic seizures. Neurology. 2010;75(13) 1166-73.

[95] Pintor L, Baillés E, Matrai S, Carreño M, Donaire A, Boget T, Setoain X, Rumia J, Bargalló N. Efficiency of venlafaxine in patients with psychogenic nonepileptic seizures and anxiety and/or depressive disorders. J Neuropsychiatry Clin Neurosci. 2010;22(4) 401-8.

[96] Chastan N, Parain D. Psychogenic paralysis and recovery after motor cortex transcranial magnetic stimulation. Mov Disord. 2010;25(10) 1501-4.

[97] O'Malley PG, Jackson JL, Santoro J, Tomkins G, Balden E, Kroenke K. Antidepressant therapy for unexplained symptoms and symptom syndromes. J Fam Pract. 1999;48(12) 980-90.

Chapter 11

# Quality of Life Issues in Epilepsy

Jane McCagh

Additional information is available at the end of the chapter

http://dx.doi.org/10.5772/58689

## 1. Introduction

There is overwhelming evidence that people with epilepsy (PWE) have a number of psychosocial difficulties which impact greatly on their quality of life [1]. To this end the chapter will summarise some definitive clinical features of the disorder and then go on to provide an overview of the types of psychosocial deficits that PWE experience and the influence of epilepsy related variables on these factors. The chapter will conclude by considering interventions that may alleviate the burden of psychosocial problems.

### 1.1. Epilepsy & psychosocial functioning

The themes addressed in this chapter emerged from observing the symptoms and behaviours exhibited by patients with epilepsy who were attending a tertiary referral unit in the UK. The majority of patients have intractable epilepsy (difficult to manage seizures) and a number of these patients were being assessed to see if they were viable candidates for surgery. A recurring problem reported by epilepsy patients attending the unit was that they experienced a number of difficulties in relation to social functioning. Such difficulties have been evident in the wider epilepsy population where PWE often report difficulties such as low self-worth, stigma, social isolation, difficulties with interpersonal relationships and in gaining and maintaining employment [1].

In reviewing the literature it is evident that quality of life in PWE is determined by the interplay of a number of multifaceted biopsychosocial factors. Psychosocial factors include comorbid anxiety and depression, cognitive deficits, the impact of stigma, low self-esteem, reduced opportunities for social interaction, difficulties in both intimate and non-intimate relationships and employability [1-15]. The impact of these factors on the quality of life of people with epilepsy in relation to epilepsy related variables will be discussed throughout the chapter.

## 2. Epilepsy

Epilepsy is the most common serious neurological disorder affecting people across the world with approximately 50 million people living with the condition [16]. The highest incidence and prevalence rates of epilepsy are in the early and later years of life. The largest majority of PWE (approx. 80%) live in developing countries with limited access to treatment [17]. This can be due to economic constraints but is most commonly a consequence of misconceptions about the origins of epilepsy. Some cultures believe that the condition is not organically based and therefore not treatable by medical intervention. Such misconceptions have been reflected historically where epilepsy was regarded as a 'sacred disease' representing possession by evil spirits or retribution by the gods.

Epilepsy is a complex disorder with many different seizure types and syndromes. An epileptic seizure is defined as 'a transient occurrence of signs and/or symptoms due to abnormal excessive or synchronous neuronal activity in the brain' [19, p 471]. Definitions of epilepsy often focus on the unprovoked and recurrent nature of seizures as not all individuals who experience a seizure will be given a diagnosis of epilepsy. Some individuals have a lower threshold to seizure activity [18]. Single isolated seizures and those which stem from systemic changes in the body, illness or an insult to the head do not constitute a diagnosis [20]. Epilepsy can often be comorbid with or caused by other neurological disorders such as cerebrovascular accidents, brain tumours, brain infection, congenital defects, exposure to toxic agents, degenerative disorders, head injury and birth complications. The aetiology of epilepsy varies in accordance with age.

The remainder of the chapter will go on to discuss how epilepsy related variables interrelate with psychosocial factors and effect quality of life.

## 3. Stigma & misconception

Historical and media misrepresentations of epilepsy have been overwhelmingly negative and consequently the condition has been clouded by misunderstanding, superstition and discrimination present over many centuries and across different cultures. This in turn increases the experience of stigma in PWE. Contagion beliefs existed up until the 18$^{th}$ century and are still apparent in some African cultures [21-23]. Laws prohibiting marriage in both the UK (until 1970) and America (1956) have only been revoked in the past sixty years [23-24] and more recently over a third of PWE were refused one or more types of insurance in the UK [25]. Cross culturally the impact of stigma on PWE has caused them to be excluded from important social roles, under achieve in school, to find it more difficult to gain employment, have intimate relationships or maintain family relations [1, 21, 26-28]. Consequently negative attitudes and lack of family support render the individual as being more vulnerable to psychosocial difficulties [2]. Conversely recent research suggests that attitudes towards PWE have become more positive [29-31].

Misconception, myths and stereotypes are still prevalent in media portrayals of epilepsy despite advances in education [32]. These misconceptions contribute to the psychosocial limitations and stigma experienced by PWE. The idea that epilepsy is defined by the symptoms of a tonic clonic seizure is often perpetuated by the media. Media portrayals are often misguided by ancient myths that see the person with epilepsy as being possessed by demons, frothing at the mouth, violent and in need of urgent medical care [32-33]. Consequently the media can hinder educating people about epilepsy by broadcasting inaccurate portrayals or misinformation [23]. Research suggests that misconceptions are more evident in individuals who do not know someone with epilepsy [32].

Negative attitudes in society have created fear and stigma. The stigma of epilepsy can often be more debilitating for PWE than presenting symptoms. Subsequently individuals with epilepsy conceal their condition or are isolated within their family in an attempt to lessen the social stigma associated with the disorder and to increase marital prospects [2,12]. Psychosocial limitations in relation to employment and relationships are more likely to be a consequence of stigma in society than symptoms of epilepsy [23]. Withdrawing from society in order to reduce the experience of stigma can greatly impact on the quality of life of PWE by making them feel more socially isolated, have less social opportunities and smaller social support networks. This in turn can reduce their ability to cope with epilepsy. The uncertainty of having a seizure in public without warning can increase the likelihood of withdrawal. This is supported by research findings where fear of having a seizure and the uncertainty associated with epilepsy (such as if seizures will ever be controlled) were primary concerns of individuals living with the condition [7]. This uncertainty can increase vulnerability in social situations and greatly impact on quality of life [1, 34].

The presence of stigma is complex and researchers have established two different types of stigma affecting PWE. Enacted stigma is when PWE experience discrimination because of their condition and felt stigma is apparent when PWE fear being subjected to enacted stigma [35]. Often there is a difference between the two, with felt stigma being more prevalent [36-37]. In a large cross cultural study of with over 5,000 participants, 51% of PWE felt stigma as a consequence of having the condition [3]. Perceived stigma is subjective and will differ across individuals. Felt stigma has been shown to depend on whether the individual feels that they have been discriminated against in the work place or constrained in their day to day life as a consequence of having epilepsy [38]. Personality, coping ability and the perceived impact of epilepsy in gaining and maintaining employment can greatly influence felt stigma [39] and increased time spent in education has been shown to reduce felt stigma [38]. The severity of the condition in early life has been found to impact on psychosocial sequalae and treatment interventions at this stage can be effective in reducing felt stigma [12, 40].

Felt and enacted stigma are influential in getting work and keeping it and stigma has far reaching consequences that impact upon interpersonal relationships, employability, health and quality of life [12; 41]. Children may feel stigma because they are subjected to an overprotective parental style which in turn may compromise their self-esteem when they are older [42-44].

Particular clinical features of epilepsy can influence feelings of stigma and psychosocial functioning. Severe and frequent seizures, a chronic form of epilepsy and comorbidity of other conditions can increase psychosocial risk [2]. Becoming seizure free has been reported as the most important epilepsy related factor in relation to quality of life [45]. Seizure type, frequency, severity, age of onset and duration of epilepsy can greatly impact on stigma. PWE who experience more frequent seizures or whose seizures are not well controlled report more stigma, and tonic clonic seizures are more likely to induce stigma due to their dramatic nature and the attention they draw to the individual [3, 9].

## 4. Psychological factors: Anxiety and depression

Both anxiety and depression are common comorbid features in PWE yet psychopathology is undertreated and underdiagnosed [17]. Forty to sixty percent of PWE display symptoms of depression and more than 40% present with anxiety [8, 46]. Depression is 4 to 5 times more likely in PWE than in the general population [16] and anxiety is the most common form of psychopathology reported [4, 9]. Suicide is also far more common in PWE than in the general population [47-48]. There is a reciprocal relationship between anxiety, depression and epilepsy such that those with anxiety and depression are more likely to experience epilepsy and those with epilepsy are more likely to experience anxiety and depression [46, 49-51]. The exact mechanism that underpins this relationship as yet is not fully understood.

Depression and seizure worry were found to be the most influential predictors of quality of life in people with intractable epilepsy [56]. Surgery has been shown to significantly reduce anxiety and increase quality of life in patients who have become seizure free [57]. Seizure related variables such as age of onset, seizure type, frequency, severity, duration, side effects of AED's, underlying aetiology and difficult to control seizures all impact on the prevalence of depression and anxiety [9, 11, 58-61].

Seizure activity in itself can be anxiety inducing especially if medial temporal lobe structures such as the amygdala are involved. Anxiety can occur before, during and after a seizure [46]. PWE may fear having a seizure which will in turn raise anxiety thresholds as will the associated stigma of having a seizure in public [7]. To this end PWE can often isolate themselves so as not to have a seizure at an inopportune moment [62]. Consequently social isolation is very common in PWE [2]. Increased feelings of anxiety and depression can be further compounded by reduced social opportunities, lack of social support, poor self-esteem, a reduced sense of mastery, stigma and discrimination, an overprotective parental style in childhood or vocational disability [35, 52-54, 63-70]. Self-esteem and sense of mastery have been inversely related to higher scores on measures of depression and anxiety in PWE [71].

## 5. Dispositional factors: Self-esteem and sense of mastery

Unpredictability is central to living with epilepsy. Individuals may not know when a seizure will occur and often have difficulty accepting living with a chronic condition that may or may not improve. This can cause PWE to feel that they lack control over their life, can lower mood and heighten feelings of anxiety [6, 52-54]. This has been reflected in research where PWE demonstrate a lower sense of mastery in relation to healthy controls [55].

Seizure severity and frequency have been shown to be the most influential factors in determining self-esteem and sense of mastery in PWE [61, 72]. Knowledge has been found to mediate the impact on self-esteem. Adolescents with more awareness of their condition report higher levels of self-esteem and educational interventions aimed at increasing knowledge have been successful at enhancing self-esteem in PWE [72-73].

Low self-esteem is commonly reported in PWE and is often mediated by employability, with those being unemployed reporting poorer self-esteem [4, 7, 9, 39, 74-76]. Research suggests that PWE with an increased sense of mastery are less likely to be affected by stigma and more likely to adhere to drug therapy [77]. Over protective parental styles can work to reduce a sense of mastery and consequently make PWE less independent in adulthood [43].

## 6. Social factors: Social isolation and relationships

Epilepsy has been shown to reduce opportunities for social interaction and PWE can isolate themselves for fear of having a seizure in public and the injury this may cause [7, 10]. Parents may have been over protective in early life which has impeded important independent living skills and confidence in social settings [2, 78]. PWE may also find it harder to gain employment and so are not afforded the same social networks as people without the condition. A reduction in social life has many confounding consequences in terms of social support. Social support can be a protective factor in aiding PWE to cope with living with a chronic condition. The opportunity to develop relationships and maintain them can be compromised and this is evident in the reduced marital status and likelihood of parenting a child in PWE [4, 7, 62, 79-81]. Seizure frequency and age of onset have been negatively related to marital status with those who have more frequent seizures and being younger at the age of onset being less likely to marry [9].

A research study investigated the employment, marital, social and educational status of 343 PWE [81]. The marital status of males and females over the age of twenty was much lower than that of the general population (males 33% compared to 65% and females 46% compared to 73%). Poor seizure control was indicative of people with poor social status. A similar disparity was found in another study where 42% of PWE in comparison to 71% without epilepsy were married [4].

Parental beliefs may be self-fulfilling, one study found that parents who believe their child will experience stigma and experience limitations reported more behavioural problems in their children than parents who did not hold such beliefs 82]. This study also found that children who report their parents as over controlling had more behavioural problems than children with epilepsy who did not. The authors concluded that seizure type and frequency did not predict behavioural problems but parents perceived stigma, perceived limitations and extent of control did.

Young people with epilepsy report social isolation as the most influential factor in determining their quality of life and find it especially difficult to take part in social activities outside of their home environment and to make friends [62]. PWE find it very hard to develop friendships [2], perceived and enacted stigma, lower self-esteem, lack of employment and reduced social opportunities may well account for this.

Epilepsy can have consequences for the whole family. There is an increased likelihood of mental health problems, stress, reduced social opportunities, marital problems and lower self-esteem in families of PWE [83]. Parental anxiety can also reduce quality of life in children with epilepsy and carers of PWE report being discontent due to reduced social and personal opportunities as a consequence of their role [84-85].

## 7. Education & employment

Children with epilepsy have been shown to underachieve at school in comparison to their peers and are more prone to educational difficulties [7, 34, 62]. These difficulties may arise as a consequence of a number of factors. Drug therapy and post ictal confusion may slow cognitive functioning and impact on children's capacity to learn. Children who experience absence seizures in class can often be mistaken for daydreaming, consequently they may not take in all the material taught to them or their attentiveness and behaviour in class may be misinterpreted by the teacher and result in adverse consequences. Children with more severe seizure types may miss time off school and may also be more prone to stigma if they have a seizure in class. Seizures can impair storage of learned information and consequently frequent seizures are more likely to interfere with educational progress [22].

PWE and those who have a history of epilepsy are prohibited by law from a variety of occupations [86]. Practices that discriminate against people with epilepsy are another major contributing factor to unemployment. Worldwide PWE are unemployed and underemployed in relation to the general population [23, 87]. Seizure related variables contribute to this, PWE who experience frequent seizures have less chance of employment and so do those who experience tonic clonic seizures [9, 88-89]. Adequate seizure control, early age of onset, stigma, side effects of AEDs, poor self-efficacy, poor social skills, education level, social isolation, cognitive deficits, negative attitudes of family members, employers and teachers have all been linked to unemployment and underemployment [90-95].

Internal work beliefs have been highlighted as an important factor in the successful inclusion of people with epilepsy into the workplace [91]. Self-worth, worry about safety at work, perceptions of the likelihood of injury in relation to self and others at work and attitudes of family members were primary factors in work status. A major barrier is stigma which in turn may lower self-worth and discourage PWE from seeking employment, conversely being part of the workforce is also likely to increase self-worth [62, 75 & 91]. As well as individual feelings of stigma, professional stigma may affect employability in the work place [97]. Perceptions of stigma in the workplace and experience of stigma have been found to be of a similar magnitude [40, 98]. Discrimination is apparent in relation to the availability of employment and employers who recruit PWE and in those who lose their jobs after diagnosis [44, 96]. Whilst the PWE may worry about being discriminated against in the workplace, employers also hold negative attitudes about employing PWE [97-98].

The IBE Employment Commission [cited in 77] conducted a cross cultural study investigating factors that contributed to unemployment. PWE attributed their employment difficulties to; employers having stigmatised views, lack of self-worth, missing school and training, the uncertainty of whether epilepsy will cause a problem in the workplace and not getting the job they want. They felt that laws against discrimination, vocational assessment and epilepsy associations working together with employers would help solve the problem [77].

## 8. Strategies to enhance quality of life

Clearly improvements in diagnostic procedures and medical intervention will have important consequences for management, prognosis and psychosocial outcome of epilepsy and in turn improve quality of life. For example, seizure severity is the most influential clinical feature that impacts on felt stigma so seizure management is fundamental in reducing psychosocial consequences in relation to stigma. Two main avenues to increase quality of life in PWE are promoting attitude change in society and in the individual.

Many of the psychosocial difficulties experienced by people with epilepsy stem from the society within which the individual lives which can determine how restricted they are as a consequence of their condition. Misconception about epilepsy is still prevalent in today's society and is influenced by inaccurate perceptions of the disorder. Such representations of epilepsy are disseminated by the media who perpetuate myths and stereotypes that consequently maintain stigma. Cultural differences in how epilepsy is perceived can influence whether PWE receive adequate treatment, the only way to resolve this is to educate society to increase understanding and reduce stigma.

Felt stigma is reported to be the main obstacle for the individual with epilepsy and impacts greatly on social networks, relationships and employability. PWE who have developed efficient coping techniques, have high self-worth and foster more positive attitudes towards their condition are at less risk of psychosocial dysfunction [41, 99-103]. Therefore interventions that focus on increasing self-worth, developing effective coping strategies and positive cognition will help the individual accept their diagnosis, reduce psychosocial deficits and ultimately enhance quality of life. Cognitive behavioural therapy is likely to be particularly useful to this end and is also effective in managing the comorbid impact of anxiety and depression which are common in PWE.

Interventions that educate and integrate PWE more into society and increase social opportunity would reduce feelings of isolation and enhance coping strategies. Education both in schools and the wider community will help reduce stigma which in turn will enhance social prospects. The impact of education is apparent in adolescents with more knowledge about their condition who report higher levels of self-esteem [73]. Social anxiety has been related to knowledge of the condition such that PWE who have more knowledge are less likely to feel socially anxious [73]. Individual and family counselling may also work to enhance self-esteem in PWE [12].

Support groups for both the individual and their family can increase social support and facilitate coping strategies [104-105]. Educating families and PWE will enable better self-management of the condition and reduce seizure activity which may be effective in reducing psychosocial sequalae and felt stigma. Non adherence to medication has been found to be the main reason for a seizure on PWE, also lack of awareness of the condition can increase felt stigma in PWE [106]. Stress also plays a significant role in seizure activity so stress management techniques may prove to be useful in aiding PWE to reduce seizure triggers.

Respite, social support groups and increased education for families of children with epilepsy would be useful in highlighting how parental behaviour can impact on the child [85]. This would help reduce the impact that parental anxiety and overprotectiveness can have on the self-esteem, independence and overall quality of life of the child now and in later life.

Vocational interventions could increase employability and confidence in dealing with epilepsy related issues in the work place especially in relation to disclosure of the disorder and how best to manage this. PWE report being very concerned and unsure about disclosing their condition when applying for jobs [107]. Role play may be useful in helping to increase the confidence of PWE in being able to disclose their condition and aid them in explaining to employers and colleagues what they should do in the event of a seizure [108]. This may have another added benefit by increasing feelings of control and sense of mastery in PWE and by encouraging positive coping strategies. Another strategy to improve employability is to make vocational training more available to PWE [97].

## 9. Chapter summary

This chapter has reviewed the main psychosocial difficulties PWE experience as consequence of epilepsy and its treatment. The main psychosocial issues stem from felt and enacted stigma which can reduce social opportunities in a number of ways. Stigma impacts on self-esteem, sense of control, the ability to gain and sustain employment, to form and maintain relationships and can increase feelings of social isolation. PWE may have a reduced sense of autonomy as a consequence of being exposed to over protective parental styles and underachieve in the education system due to ill health and absence. PWE are more commonly unemployed or underemployed than the general population which may well be a consequence of stigma or disruptions to academic engagement. This in turn can increase feelings of isolation which may also be compounded by the fear of having a seizure in public. Reduced autonomy and social opportunity can impact on building friendships and relationships which is evident in the reduced marital status of PWE in relation to the general public.

Psychiatric comorbidities such as anxiety and depression are often present in PWE, these can precede and be an outcome of having the condition and impinge on psychosocial outcomes.

http://dx.doi.org/10.5772/58689

Psychosocial factors are influenced by a number of clinical features of epilepsy such as seizure frequency, type and severity, duration of the condition, age of onset, aetiology and comorbid conditions. Seizure activity can impair cognitive dysfunction as can AED therapy. Memory deficits are the most commonly reported deficit and seizure activity can disrupt memory consolidation, learning and information storage which can interfere with academic progress consequently PWE may underachieve in education which in itself can reduce psychosocial opportunities, employability and financial status.

It is worth noting that many of the psychosocial difficulties discussed are pertinent to individuals with intractable epilepsy where much of the research has been conducted as the severity of the disorder has more profound effects on psychosocial functioning so may not be relevant to all PWE [2].

## Author details

Jane McCagh*

Liverpool Hope University, England

## References

[1] McCagh J., Fisk J.E., Baker G.A. Epilepsy, psychosocial and cognitive functioning. Epilepsy Research, 2009; 86 1-14.

[2] Austin K. and deBoer H. Disruptions in social functioning and services facilitating adjustment for the child and adult. In: Engel Jr. J., Pedley, T.A. (eds.) Epilepsy: A Comprehensive Textbook). Philadelphia: Lippincott-Raven; 1997 p2191–2201.

[3] Baker G A., Brooks J., Buck D., Jacoby A. The stigma of epilepsy: a European perspective. Epilepsia 2000; 41 98–104.

[4] Collings J. A. Psychosocial well-being and epilepsy: an empirical study. Epilepsia 1990; 31, 418-426.

[5] Corcoran R., Thompson P. Epilepsy and poor memory. Who complains and what do they mean? British Journal of Clinical Psychology 1993; 32 199-208.

[6] de Souza.E.A.P. and Salgado P.C.B. A psychosocial view of anxiety and depression in epilepsy. Epilepsy &Behavior 2006; 8 232-8.

[7] Fisher R.S., Vickrey B.G., Gibson P., Hermann B., Penovich P., Scherer A. et al. The impact of epilepsy from the patient's perspective II: Views about therapy and health care. Epilepsy Research 2000; 41 53–61.

[8] Grabowska-Grzyb A., Jędrzejczak J., Nagańska E., Fiszer U. Risk factors for depression in patienst with epilepsy. Epilepsy &Behavior 2006; 8 411-417.

[9] Jacoby A., Baker G.A., Steen N., Potts P., Chadwick D.W. The clinical course of epilepsy and its psychosocial correlates: findings from a U.K. community study. Epilepsia 1996; 37 148–61.

[10] Mittan R.J. Fear of Seizures. In: Whitman S. and Hermann B. (eds.). Psychopathology in Epilepsy: Social Dimensions. Oxford: Oxford University Press; 1986.

[11] Mensah S.A., Beavis J.M., Thapar A.K., Kerr M.P. A community study of the presence of anxiety order in people with epilepsy. Epilepsy &Behavior 2007; 7 438–46.

[12] Morrell M.J. Stigma and epilepsy. Epilepsy & Behaviour, 2002; 3 S21–5.

[13] Scicutella A.B., Ettinger A. Treatment of anxiety in epilepsy. Epilepsy &Behavior 2002; 3 (5S) 10-12.

[14] Suurmeijer T.P., Reuvekamp M.F, Aldenkamp B.P. Social functioning, psychological functioning and quality of life in epilepsy. Epilepsia 2001; 42 1160–8.

[15] Thompson P.J. and Corcoran R. Everyday memory failures in people with epilepsy. Epilepsia 1992; 33 (suppl.6.) S18-S20.

[16] Brodie M.J. and Schachter S.C. Epilepsy, (2nd ed.). Oxford: Health Press; 2001.

[17] de Boer H.M., Mula M., Sander J.W. The global burden and stigma of epilepsy. Epilepsy &Behavior 2008; 12 540-546.

[18] McCagh J. Epilepsy: classification, aetiology and management. In R. E. D., Bautista (ed.), Epilepsy: A century of discovery. New York: Nova Science Publishers; 2012 p31-40.

[19] Fisher R.S., van Emde Boas W., Blume W., Elger C., Genton P., Lee P., Engel Jr J. Epileptic seizures and epilepsy: definitions proposed by the International League Against Epilepsy (ILAE) and the International Bureau for Epilepsy (IBE). Epilepsia 2005; 46 470-471.

[20] Baker G.A. and Jacoby A. (Eds.). Quality of life in epilepsy beyond seizure counts in assessment and treatment. Amsterdam: Harwood Academic Publishers 2000.

[21] Carod-Artal F.J., Vazquez-Cabrera C.B. An anthropological study about epilepsy in native tribes from Central and South America. Epilepsia 2007; 48 886-893.

[22] deBoer H.M. Epilepsy and society. Epilepsia 1995; 36 S8-11.

[23] International League Against Epilepsy. The history and stigma of epilepsy. Epilepsia, 2003; 44 (suppl. 6): 12-14.

[24] Epilepsy Foundation of America. The legal rights of persons with epilepsy, (6th ed). Landover: MD; 1992.

[25] Jacoby K., and Jacoby A. Epilepsy and insurance in the UK: an exploratory survey of the experiences of people with epilepsy. Epilepsy & Behavior2005; 5 884–94.

[26] McCagh J. Epilepsy, myths, stereotypes & stigma. Brain Research Journal 2010; 3 75-88.

[27] ILAE/IBE/WHO. "Out of the shadows": global campaign against epilepsy. Annual Report. Geneva: WHO; 1999.

[28] Wiebe S., Bellhouse D.R., Fallahay C., Eliasziw M. Burden of epilepsy: the Ontario Health Survey. Canadian Journal of Neurological Science 1999; 26 263–70.

[29] Jacoby A., Gorry J., Gamble C., Baker G.A. Public knowledge, private grief: a study of public attitudes to epilepsy in the United Kingdom and implications for stigma. Epilepsia 2004; 45 1405–15.

[30] Jacoby A., Gorry J., Baker G.A. Employers' attitudes to employment of people with epilepsy: still the same old story? Epilepsia 2005; 46 1978-1987.

[31] Jacoby A., Snape D., Baker G.A. Epilepsy and social identity: the stigma of a chronic neurological disorder. Lancet Neurology 2005; 4 171–8.

[32] Baxendale S. and O'Toole A. Epilepsy myths: alive and foaming in the 21st century. Epilepsy and Behavior 2007; 11 192-196.

[33] Krauss G.L., Gondek S., Krumholz A., Paul S., Shen F."The scarlet E": the presentation of epilepsy in the English-language print media. Neurology 2000; 54 1894–8.

[34] International League Against Epilepsy. Living with epilepsy: Employment. Epilepsia 2003; 44, (suppl. 6): 49-50.

[35] Scambler G. Sociological aspects of epilepsy. In: Hopkins A. (ed.) *Epilepsy*. New York: Demos; 1987 p 497-510.

[36] International League Against Epilepsy. Living with epilepsy: Epilepsy education within statutory school system. Epilepsia 2003; 44 (suppl. 6): 45-48.

[37] Jacoby A. Felt versus enacted stigma: a concept revisited. Evidence from a study of people with epilepsy in remission. Social Science Medicine 1994; 38 269–74.

[38] Ryan R., Kempner K., Emlen A.C. The stigma of epilepsy as a self concept. Epilepsia 1980; 21 433–44.

[39] Lee S. What we confront with employment of people with epilepsy in Korea. Epilepsia 2005; 46 57–8.

[40] Chaplin J.E., Yepez Lasso R., Shorvon S.D. and Floyd M. National general practice study of epilepsy: the social and psychological effects of a recent diagnosis of epilepsy. British Medical Journal 1992; 304 1416-1418.

[41] Smeets V.M.J., van Lierop B.A.G., Vanhoutvin J,P.G., Aldenkamp A.P., Nijhuis F.J.N. Epilepsy and employment: Literature review. Epilepsy &Behavior 2007; 10, 354–362.

[42] Goodyer I. The influence of epilepsy on family functioning. In: Hoare P. (ed.). Epilepsy and the Family: A Medical Symposium on New Approaches to Family Care. Manchester: Sanofi UK Ltd; 1998.

[43] McCollum A.T. The chronically ill child: A guide for parents and professionals. New Haven, CT: Yale University Press; 1981.

[44] Scambler G. and Hopkins A. Social class, epileptic activity and disadvantage at work. Journal of Epidemiology and Community Health 1980; 34 129–33.

[45] Jacoby A. Epilepsy and the quality of everyday life: findings from a study of people with well controlled epilepsy. Social Science & Medicine 1992; 43 657-666.

[46] Munger Clary M. Anxiety and epilepsy: what neurologists and epileptologists should know. Current Neurology & Neuroscience Reports, 2014; 14 445.

[47] Pompili M., Girardi P., Tatarelli R. Death from suicide versus mortality from epilepsy in the epilepsies: A meta-analysis. Epilepsy & Behaviour 2006; 9 641-8.

[48] Pompili M., Girardi P., Ruberto A., Tatarelli R. Suicide in the epilepsies: a meta-analytic investigation of 29 cohorts. Epilepsy &Behavior 2005; 7 305–10.

[49] Hesdorffer D.C., Hauser W.A., Annergers J.F., Cascino G. Major depression is a risk factor for seizures in older adults. Annals of Neurology 2006; 47 246-249.

[50] Forsgren L. and Nystrom L. An incident case referent study of epileptic seizures in adults. Epilepsy Research 1999; 6 66-81.

[51] Hesdorffer D.C., Hauser W.A., Olafsson E., Ludvigsson P., Kjartansson O. Depression and suicide attempt as risk factors for incident unprovoked seizures. Annals of Neurology 2006; 59 35–41.

[52] Brier J.I., Fuchs K.L., Brookshire B.L. Quality of life perception in patients with intractable epilepsy or pseudoseizures. Archives of Neurology 1998; 55 660-5.

[53] Salgado P.C.B. and Souza E.A.P. Qualidade de vida em epilepsia e percepcao de controle de crises. Arq Neuropsiquiatria 2001; 19 537-40.

[54] Vasquez B. and Devinsky O. Epilepsy and anxiety. Epilepsy &Behavior 2003; 9 S20–5.

[55] De Vellis R.F., De Vellis B.M., Wallston B.S., Wallston K.A. Epilepsy and learned helplessness. Basic and Applied Social Psychology 1989; 1 241-253.

[56] Loring D.W., Meador K.J., Lee G.P. Determinants of quality of life in epilepsy. Epilepsy &Behavior 2004; 5 976-980.

[57] Taft C., Sager Magnusson E., Ekstedt G., Malmgren K. Health-related quality of life, mood, and patient satisfaction after epilepsy surgery in Sweden—A prospective controlled observational study. Epilepsia 2014 (in press) 1-8.

[58] Kanner A.M. and Balabanov A. Depression and epilepsy: how closely related are they? Neurology 2002; 58 S27-39.

[59] Mendez, M.F., Cummings, J.L., Benson, D.F. Depression in epilepsy: Significance and phenomenology. Archives of Neurology 1986; 43 766-70.

[60] Smith D.F., Baker G.A., Dewey M. Jacoby, A and Chadwick D.W. Seizure frequency, patient perceived seizure severity and the psychosocial consequences of intractable epilepsy. Epilepsy Research 1991; 9 231-241.

[61] Thompson P.J. Psychosocial aspects of epilepsy. In: Duncan J.S., Sisodiya S.J.J., Smalls J.E. (eds.). Epilepsy 2000 from science to patient. I.L.A.E.; 2000.

[62] Chaplin J.E., Yepez R., Shorvon S.D. Floyd M. A quantitative approach to measuring the social effects of epilepsy. Neuroepidemiology 1990; 9 151-8.

[63] Dell J. Social dimensions of epilepsy: stigma and response: In Whitman S. and Hermann B. Psychopathology in epilepsy: social dimensions. New York: Oxford University Press; 1986.

[64] Ferrari M., Matthews W.S., and Barabas G. The family and child with epilepsy. Family Process 1983; 22 53-59.

[65] Hecimovic H., Goldstein J.D., Sheline Y.I., Gilliam F.G. Mechanisms of depression epilepsy from a clinical perspective. Epilepsy Behaviour 2003; 4 (suppl. 3.) S25-30.

[66] Hermann B.P. and Wyler A.R. Depression, locus of control, and the effects of epilepsy surgery. Epilepsia 1989; 30 332-8.

[67] Jacoby A. Felt versus enacted stigma: a concept revisited. Evidence from a study of people with epilepsy in remission. Social Science Medicine 1994; 38 269–74.

[68] Mungas D. Interictal behaviour abnormality in temporal lobe epilepsy. Archives of General Psychiatry, 1982; 39 108-11.

[69] Baker G A. The psychosocial burden of epilepsy. Epilepsia 2002; 43 (suppl.6.) 26-30.

[70] Baker G.A., Smith D.F., Dewey M., Jacoby A., Chadwick D.W. The initial development of a health-related quality of life model as an outcome measure in epilepsy. Epilepsy Research, 1933; 16 65-81.

[71] Baker G.A., Spector S., McGrath Y., Soteriou, H. Impact of epilepsy in adolescence: A UK controlled study. Epilepsy &Behavior 2005; 6 556-62.

[72] Snead K., Ackerson J., Bailey K., Schmitt M.M., Madan-Swain A., Martin R.C. Taking charge of epilepsy : the development of a structured psychoeducational group inter-

vention for adolescents with epilepsy and their parents. Epilepsy &Behavior 2004; 5 547-56.

[73] Jahoda M. Employment and unemployment: a social psychological analysis. Cambridge: Cambridge University Press; 1982.

[74] Levin R., Banks S., Berg B. Psychosocial dimensions of epilepsy: a review of the literature. Epilepsia 1988; 29 805–16.

[75] Chaplin J. Vocational assessment and intervention for people with epilepsy. Epilepsia 2005; 46 S1 55–6.

[76] Schachter S.C.Improving quality of life beyond seizure control. Epileptic Disorders 2005; 7 (suppl. 1.), S34-S38.

[77] Lothman D.J., Pianta R.C., Clarson S.M. Mother-child interaction in children with epilepsy: relations with child competence. Journal of Epilepsy, 1990;3 157-163.

[78] Arnston P., Drodge D., Norton R. and Murray E. The perceived psychosocial consequences of having epilepsy. In: Whitman S. and Hermann B.P. (eds.) Psychopathology in epilepsy: social dimensions. New York: Oxford University Press; 1986.

[79] Zahn C.A., Morrell M.J., Collins S.D., Labiner D.M., Yerby M.S. Management issues for women with epilepsy: a review of the literature. American Academy of Neurology Practice Guidelines. Neurology 1998; 51 949–56.

[80] Callaghan N., Crowley M., Goggin T. Epilepsy and employment, marital, education and social status. Irish Medical Journal 1992; 85 17–19.

[81] Carlton-Ford S., Miller R., Mealeigh N., Sanchez N. The effects of perceived stigma and psychological over-control on the behavioural problems of children with epilepsy. Seizure 1997; 6 383-391.

[82] Ellis N., Upton D., Thompson P. Epilepsy and the family: A review of current literature. Seizure 2000; 9 22-30.

[83] Thompson P.J. and Upton D. The impact of chronic epilepsy on the family. Seizure 1992; 1 (1) 43-48.

[84] Williams J., Steel C., Sharp G.B. Parental anxiety and quality of life in children with epilepsy. Epilepsy &Behavior 2003; 4 (5) 483-6.

[85] Jacoby A. and Baker G.A. Psychosocial handicaps associated with the epilepsies. In: Vinken P.J. and Bruyn G.W. (eds.). Handbook of Clinical Neurology. Amsterdam, Elsevier Press; 2000.

[86] Chaplin J.E., Wester A., Tomson T. Factors associated with the employment problems of people with established epilepsy. Seizure 1998; 7 299–303.

[87] International League Against Epilepsy. Pharmacological treatment. Epilepsia 2003; 44 (suppl 6): 33-34.

[88] Yagi K. Epilepsy: comprehensive care, quality of life, and factors preventing people with epilepsy from being employed. Clinical Therapy, 1998;20 (suppl. A) A19–29.

[89] Rätsepp M., Õun A., Haldre S., Kaasik A. Felt stigma and impact of epilepsy on employment status among Estonian people: exploratory study. Seizure 2000; 9 391–401.

[90] Bishop M. Determinants of employment status among a community-based sample of people with epilepsy: implications for rehabilitation interventions. Rehabilitation Counseling Bulletin 2004; 47 112-120.

[91] Clarke B.B., Upton A.R.M., Castellanos C. Work beliefs and works status in epilepsy. Epilepsy & Behavior 2006; 119-125.

[92] Devinsky O. Vickery B., Hays R., Perinne K., Hermann B., Meador K et al. Quality of life in epilepsy:QOLIFE-89 instrument development. Neurology 1994; 44 (suppl.2) A141.

[93] Seidenberg M., and Clemmons D.C. Maximising school functioning and the school-to-work transition. In: Engel, J. Jr., Pedley. T.A. (Eds.). Epilepsy: a comprehensive textbook. Philadelphia: Lippincott–Raven; 1988 p2203-2209.

[94] Thorbecke R., and Fraser R.T. The range of needs and services in vocational rehabilitation. . In: Engel J. Jr., Pedley T.A. (eds.). Epilepsy: a comprehensive textbook Philadelphia: Lippincott–Raven; 1997 p2211-2226.

[95] International League Against Epilepsy. Epidemiology. Epilepsia 2003; 44 (suppl. 6): 17-18.

[96] Thompson P.J. and Oxley J. Socioeconomic accompaniments of severe epilepsy. Epilepsia 1988; 29 (suppl.1.) : S9-18.

[97] Hauser W.A. and Hesdorffer D.C. Epilepsy: frequency, causes and consequences. New York: Landover, MD, Epilepsy Foundation of America; 1990.

[98] Bautista R.E.D. and Wludyka P. Factors associated with employment in epilepsy patients. Epilepsy & Behavior 2005; 10 89–95.

[99] Cooper M. Epilepsy and employment: employer's attitudes. Seizure 1995; 4 193-9.

[100] Oosterhuis A. Coping with epilepsy: the effects of coping styles on self-perceived seizure severity and psychological complaints. Seizure 1999; 8 93-6.

[101] Livneh H., Wilson L.M., Duchesneau A., Antonak R.F. Psychosocial adaptation to epilepsy: the role of coping strategies. Epilepsy & Behavior 2001; 2 533–44.

[102] May T.W., Pfafflin M. The efficacy of an educational treatment program for patients with epilepsy (MOSES: results of a controlled, randomized study). Epilepsia 2002; 43 539–49.

[103] Dilorio C., Shafer P.O., Letz R., Henry T.R., Schomer D.L., Yeager K. et al. Project EASE : a study to test a psychosocial model of epilepsy medication management. Epilepsy & Behavior 2004; 5 926-36.

[104] Kobau R., Diloro C. Epilepsy self-management: a comparison of self-efficacy and outcome expectancy for medication adherence and lifestyle behaviours among people with epilepsy. Epilepsy & Behavior 2004; 4 217-25.

[105] Kurtz L.F., Powell T.J. Three approaches to understanding groups. Social Work Groups 1987; 10 3 69-80.

[106] Trostle J.A., Hauser W.A., Sharbrough F.W. Psychological and social adjustment to epilepsy in Rochester, Minnesota. Neurology 1989; 36 633.

[107] Baker G.A. People with epilepsy: what do they know and understand, and how does this contribute to their perceived level of stigma? Epilepsy & Behavior 2002; 3 S26-32.

[108] Bishop M. Barriers to employment among people with epilepsy: Report of a focus group. Journal of Vocational Rehabilitation 2002; 17 281-286.

Chapter 12

# Marriage of Epileptic Patients

Azra Alajbegovic, Jasminka Djelilović-Vranić,
Salem Alajbegović and Lejla Alajbegović

Additional information is available at the end of the chapter

http://dx.doi.org/10.5772/58635

## 1. Introduction

Epilepsy is a common syndrome. It occurs equally in all countries and in all nations represents a significant social and medical problem [1].

Human beings and their environment are constantly affecting each other, while under the term environment we include social, physical and economic milieu. The main need of people in general, as well as patients with epilepsy is to lead a normal life. The patient must be accepted as a whole person with the inevitable limitations imposed by the disease. Epileptic patients during their life are exposed to a series of related specific social relations from early childhood including relationships with parents, relationships during education, relation to vocational rehabilitation, the ability to drive, and for a variety of activities throughout life. Yet the oldest and biggest handicap is the dilemma of patients with epilepsy related to marriage and parenthood [1,2].

The existence of the disease always results in response of the patients to disease, most often with non-psychotic character such as: mood disorders, certain personality changes that are closely related to relationship patient-society through the following parameters: public attitudes toward epilepsy, social situation of patients outside hospitals, rehabilitation of patients with epilepsy, education of patients and the immediate environment, the problems of vocational guidance, aspects of military doctrine, the ability to drive, ability to be involved in some sports, etc. Therefore, patients with epilepsy, to which is imposed the need "to live with epilepsy", often react with depression. Dominian et al (1963) found that depression is the most common psychiatric symptom in patients with epilepsy [3].

Depressive reactions are usually of reactive nature andmore often present in patients with temporal lobe epilepsy than in patients with other types of seizures. In addition to depression, in response to the disease, also are seen neurotic syndromes and anxiety disorders.

These physiological changes in patients are the result of the complexity of psychosocial factors in relationship patient-environment, while is more significant relationship between the environment to disease and patient than relationship of patients to the disease [4,5].

In addition to responses to the disease that is of reactive character in patients with epilepsy, there is often "accompanying personality disorder" which is usually seen in patients with temporal lobe epilepsy. Changes which are encountered in this group of patients are: reduced ability to adapt, some "stickiness" in behavior, slowed thinking, redundancy in speech, formalism and pedantry, hypertrophic modesty, lability of mood and frequent hypochondria. In emergence of this specific personality profile are associated many factors: the occasional temporal lobe lesions, antiepileptic therapy, social and psychological impacts.

The specificity of the social status of people with epilepsy and attitude of the environment to the epilepsy can be best seen in the context of the attitude of the environment to some other, more serious and severe diseases with significantly greater disability than epilepsy and their better social acceptance than the acceptance of epilepsy. Such patients can, and sometimes do, discover the secret of their illness and from the knowledge of the disease have a certain social, material and other benefits [2,3].

On the other side the patient with epilepsy keeps the disease as "a shameful secret". Fear of seizures in public places restrains their freedom of thought and behavior which leads to avoidance of many life activities with suspicion in regard to marriage and parenthood. Family as a primary community is where the attitude of patients towards themselves, their illness and society is determined [6]. If the family accept the disease and makes an emotional basis for all life activities with frequent monitoring and the possibility of seizure control by medications a patient with epilepsy have a chance to lead a normal life with all its amenities [7].Kocijan-Hercegonja et al. examined the association between adverse psychosocial factors and variations in the behavior of patients with epilepsy and found that the social status of the family is one of the most important factors in the forming of the personality, as well as the emergence of mental aberrations. Results showed that children with mental aberrations significantly more frequent originate from families from the countryside, from parents farmers and parents of low economic status [8].

The analysis showed that demographic factors: employment and education, as well as treatment with antiepileptic drugs, with accompanying depressive disorder are variables that are significantly associated with QOLIE-31 total score ($p<0.01$) which makes 64.8% of the variance in QOLIE-31 overall evaluation including seizure severity, comorbid depression and seizure frequency [5,6].

An important factor is the relationship of psychosocial environment and seizures.Reduction in the frequency of seizures can lead to better social adjustment, on the other hand better social opportunities can lead to a reduction in frequency of seizures [9,10,11]. Fear of the consequences of seizures can by itself lead to lowering of the threshold for seizures [12,13].

In relation to marriage and parenthood better psychosocial adjustment in many ways increases the possibility of a patient suffering from epilepsy to get married and make the decision about the offspring [14].

http://dx.doi.org/10.5772/58635

Alstrom (1950) showed that eugenic laws which prohibited a patient suffering from epilepsy to marry and have children did not discouraged mentally healthy patients to marry and have children. Alstrom's study showed that the following parameters significantly reduce the possibility of marriage and fertility:

1. Early onset of the seizures, seizures which began before the age of ten;
2. Mental disorders related to the illness or incurred as a reaction to the disease in terms of organic deterioration of mental function.

In the U.S., Lennox and Mchram in 1963 found that the specific nuptiality in patients with epilepsy is lower than in the general population, especially in patients whose seizures began before the age of ten. In Germany, the eugenic law was declared in 1935. Essen and Moler (1955) determined that the establishment of eugenic laws in Germany prevented patients to establish marital community, but did not significantly influence the fertility of women with epilepsy [7].

Andersen (1972) proved that for most patients with epilepsy hereditary factor is not higher than hereditary factor for other diseases that have a variety of factors in their emergence, such as diabetes considering that it in the same way predetermines marriage and parenthood in patients with epilepsy as for those suffering from diabetes. In Quebec, where there were no eugenic laws that would affect the patients with epilepsy, there are few reports on the social conditions of people with epilepsy. Tomasi and Davidson (1949) found a low rate of married men (33%) among 160 patients. Pond and Gudmston (1966) found that the reduced rate of marriages of male patients compared to women with epilepsy in comparison with the general population. At the Clinic of Neurology at Montreal Dansky Linda, Eva Andermann and Frederic Andermann (1978) found a reduced specific nuptiality in patients with epilepsy. Social handicap of men with epilepsy is higher because woman can stay in the house, living with their illness separated from the environment. A man has to deal with the problems of employment, to obtain the material basis for himself and his family. Therefore, the ability of men suffering from epilepsy to devise the marriage and to get married are significantly lower than in women with epilepsy as it is more unlikely that men with epilepsy will have offspring [14].

It's common cohabitation of women with epilepsy, as well as the readiness of having children in these communities. The diversity of male and female pattern is partly conditioned by psychosocial factors, in part probably this has a background in often found and described disturbed sexuality of patients with epilepsy. After discontinuation of bromide therapy it was observed that many patients with temporal lobe epilepsy continue to have reduced sex drive in terms of hypopotence and impotence, as well that in patients with temporal lobe epilepsy after temporal lobectomy sexual activity and sexual interest is restored. In some patients is also described hyper sexuality, with other changes in the quality of sex drive like transsexualism, fetishism and homosexuality [15,16,17].

## 2. Goal

The goals of this research were:

1. Determine the marital status in the random sample of patients with epilepsy with special reference to the marriage in relation to the age when the illness started and determine how the future spouses were informed about the disease in cases when the disease occurred before marriage.
2. Determine the difference in relation to marriage status between men and women with epilepsy.
3. Determine correlations between significant social parameters of marriage and epilepsy as well as patient's profession, marital status, level of education, also marital status according to EEG findings, marital status according to the therapeutic responsiveness, marriage duration according to the type of epileptic events, marriage duration according to therapeutic responsiveness, duration of marriage according to the psychological changes.

## 3. Material and methodology

Material for our work represented a group of 98 patients suffering from idiopathic epilepsy with different epileptic manifestations that were treated on outpatient basis at the Neurology Clinic, Clinical Center of the University of Sarajevo. The diagnosis of epilepsy was certain, previously confirmed during clinical treatment. Patients were followed as outpatients for years through Dispensery for epilepsies. For each patient was made a special questionnaire with 40 questions to obtain targeted information about epilepsy, marriage and parenting.

### 3.1. Marriage of epileptic patients

**i.** Basic patient data

1. Last name, father's name, name ______________
2. Place of birth ______________
3. Place of residence (ZIP code and address) ______________

______________

1. Sex: Male (1), Female (2) ______________ ☐
2. Year of birth ______________ ☐☐
3. Education: without formal education (1), grammar school (2), high school (3), higher (4), college (5), unknown (6) ______ ☐
4. Profession: workman (1), craftsman (2), clerk (3), farmer (4), pupil (5), student (6), retired (7), housewife (8), unknown (9) ______________ ☐
5. Employed: in profession (1), other workplace (2), unknown (3) ______ ☐
6. Employment duration ______________ ☐☐

http://dx.doi.org/10.5772/58635

## ii. Basic data on epilepsy

7. Type of seizures: primary generalized seizures (1), partial seizures with or without generalization (2), seizures with complex symptomatology (3), combined seizures (4), unknown (5) ............ ☐

8. Specific heredity: mother (1), father (2), father's relatives (3), mother's relatives (4), unknown (6) ............ ☐

9. Etiology: idiopathic (1), reliable data on birth trauma (2), CNS infectious diseases (3), data on head injury with or without loss of consciences (4), existence of so called "macro factors" (5), unknown (6) ............ ☐

10. First epileptic manifestation: at age of ............ ☐☐

11. Last epileptic manifestation: at age of ............ ☐☐

12. Constitution: leptos (1), dysplastic (2), athletic (3), picnic (4), combined (5) ............ ☐

13. Neurology findings: with focal disturbances (1), without focal disturbances (2),unknown (3) ............ ☐

14. Mental changes: personality changes (1), signs of organic psychosyndrome (2), neurotic disturbances (3), oligophrenic (4), psychoses (5), without mental disturbances (6), unknown (7) ............ ☐

15. EEG finding: not recorded (1), focal specific changes (2), generalized specific changes (3), normal (4), unknown finding (5), unknown whether it was recorded (6) ............ ☐

16. Frequency of epi manifestation: daily (1), weekly (2), monthly (3), every couple months (4), annually (5), in several years (6), unknown (7) ............ ☐

17. Treatment: on regular basis (1), inordinately (2), not treated (3), unknown (4) ............ ☐

18. Therapy: barbiturates (1), hydantoins (2), barbiturates and hydantoins combined (3), diones (4), ethosuximid (5), Primidon (6), Tegretol (7), Hydanphen and Tegretol combined (8), unknown (9) ............ ☐

19. Therapeutic reactivity: good – without seizures (1), average – rare crises (2), poor – crises frequent as without therapy (3) ............ ☐

20. Family attitude: interested (1), not interested (2), denial (3), unknown (4) ............ ☐

### iii. Marriage

21. Marital status: married (1), single (2),
divorced (3), widowed (4),
extramarital community (5), unknown
(6)

22. Previous marriages: one (1), two (2), three (3), extramarital
community (4), unknown (5)

23. Age of marriage:

24. Marriage: before illness is diagnosed (1), after the illness was diagnosed
(2)

25. Spouse knew about the disease: yes (1), no (2), unknown
(3)

26. Spouse: health (1), have epilepsy (2), organic somatic or
neurology patient (3), suffers from some other
mental disorder (4), unknown
(5)

27. Did the spouse mothers or fathers relatives or parents
have epilepsy: yes (1),
no (2), unknown (3)

28. Duration of marriage: in years:

29. Emotional relationships in marriage: harmonious (1), average (2), poor (3),
unknown (4)

30. Household members: only married couple (1), with children (2),
with parents (3), single (4), unknown
(5)

After data collection we performed the statistical analysis.

## 4. Results

The results are presented in tabular form.

| AGE OF FIRST EPI EVENT | Average age when patient got married | Standard deviation (S.D.) |
|---|---|---|
| Under 20 years | 21.2 yrs. | 2.95 |
| 20 – 29 years | 22.5 yrs. | 3.70 |
| 30 – 39 years | 23.0 yrs. | 3.69 |
| 40 and older | 25.0 yrs. | 7.36 |

**Table 1.** Average age when patient got married according to age of the first epi event

http://dx.doi.org/10.5772/58635

| FREQUENCY OF EPI EVENTS | TOTAL ALL PATIENTS | | MARITAL STATUS | | | | | |
|---|---|---|---|---|---|---|---|---|
| | | | MARRIED | | SINGLE | | OTHER | |
| TOTAL | 100 | 100% | 65 | 100% | 24 | 100% | 11 | 100% |
| 1. Daily | 10 | 100% | 6 | 92% | 1 | 42% | 3 | 273% |
| 2. Weekly | 14 | 140% | 5 | 77% | 7 | 292% | 2 | 182% |
| 3. Monthly | 21 | 210% | 19 | 292% | 1 | 41% | 1 | 91% |
| 4. Every couple months | 19 | 190% | 12 | 185% | 5 | 208% | 2 | 182% |
| 5. Annual | 8 | 80% | 5 | 77% | 3 | 125% | - | - |
| 6. Several years | 27 | 270% | 17 | 262% | 7 | 292% | 3 | 272% |
| 7. Unknown | 1 | 10% | 1 | 15% | - | - | - | - |

**Table 2.** Frequency epi events and marital status

| MARITAL STATUS | Gender | No. of patients | Average age of first EPI event, years | Standard deviation (S.D.) |
|---|---|---|---|---|
| TOTAL | Total | 98 | 25.1 | 12.98 |
| | Male | 52 | 26.9 | 13.77 |
| | Female | 46 | 23.1 | 11.71 |
| MARRIED | Total | 63 | 27.7 | 11.63 |
| | Male | 37 | 28.2 | 11.80 |
| | Female | 26 | 27.0 | 11.56 |
| SINGLE | Total | 24 | 14.0 | 6.38 |
| | Male | 11 | 13.7 | 6.31 |
| | Female | 13 | 14.2 | 6.69 |
| OTHER | Total | 11 | 34.1 | 15.81 |
| | Male | 4 | 50.3 | 7.95 |
| | Female | 7 | 24.9 | 11.05 |

**Table 3.** Average age of patients when they had first epi event

| AGE OF FIRST EPI EVENT | TOTAL | | *MARITAL STATUS* | | | | | |
|---|---|---|---|---|---|---|---|---|
| | | | MARRIED | | SINGLE | | OTHER | |
| TOTAL | 100 | 100% | 65 | 100% | 24 | 100% | 11 | 100% |
| 1. 0 – 9 yrs. | 11 | 11.0% | 2 | 3.1% | 8 | 33.3% | 1 | 9.1% |
| 2. 10 – 19 yrs. | 31 | 31.0% | 17 | 26.2% | 12 | 50.0% | 2 | 18.2% |
| 3. 20 – 29 yrs. | 25 | 25.0% | 19 | 29.2% | 4 | 16.7% | 2 | 18.2% |
| 4. 30 – 39 yrs. | 11 | 11.0% | 10 | 15.4% | - | - | 1 | 9.1% |
| 5. over 40 yrs. | 20 | 20.0% | 15 | 23.1% | - | - | 5 | 45.4% |
| 6. Unknown | 2 | 2.0% | 2 | 3.0% | - | - | - | - |

**Table 4.** Age of first epi event and patient marital status

| MARITAL STATUS | TOTAL | | Male | | Female | |
|---|---|---|---|---|---|---|
| | N | % | N | % | N | % |
| TOTAL | 100 | 100 | 53 | 100 | 47 | 100 |
| 1. MARRIED | 65 | 65.0 | 38 | 71.7 | 27 | 57.5 |
| 2. SINGLE | 24 | 24.0 | 11 | 20.8 | 13 | 27.7 |
| 3. DIVORCED | 9 | 9.0 | 4 | 7.5 | 5 | 10.6 |
| 4. WIDOWED | 1 | 1.0 | - | - | 1 | 2.1 |
| 5. UNKNOWN | 1 | 1.0 | - | - | 1 | 2.1 |

**Table 5.** Sample according to gender and marital status

| MARITAL STATUS | Gender | N | Average age in years | Standard deviation (S.D.) | Student t test of differenceamong average age of men and women |
|---|---|---|---|---|---|
| TOTAL | Total | 100 | 39.1 | 11.18 | t = 0.171 |
| | Male | 53 | 39.0 | 11.30 | not significant |
| | Female | 47 | 39.4 | 11.05 | |
| MARRIED | Total | 65 | 41.2 | 9.31 | t = 0.463 |
| | Male | 38 | 40.8 | 9.76 | not significant |
| | Female | 27 | 41.9 | 8.63 | |
| SINGLE | Total | 24 | 31.8 | 11.13 | t = 2.021 |
| | Male | 11 | 27.4 | 5.47 | Significant at |
| | Female | 13 | 35.5 | 13.16 | p < 0.01 |
| OTHER | Total | 11 | 43.9 | 14.36 | |
| | Male | 4 | 54.0 | 8.39 | |
| | Female | 7 | 38.1 | 13.88 | |

**Table 6.** Average age of the patients according to gender an marital status

| Age of first epi event | TOTAL | | *Age when got married* | | | | | |
|---|---|---|---|---|---|---|---|---|
| | | | up to 22 years | | 23 -27 years | | 28 years and older | |
| YEARS | 75 | 100 | 43 | 57,3 | 25 | 33,3 | 7 | 9,4 |
| 0 – 9 years | 3 | 100 | 2 | 66,7 | 1 | 33,3 | - | - |
| 10 – 19 years | 19 | 100 | 12 | 63,2 | 7 | 36,8 | - | - |
| 20 – 29 years | 20 | 100 | 12 | 60,0 | 6 | 30,0 | 2 | 10,0 |
| 30 – 39 years | 11 | 100 | 6 | 54,5 | 4 | 36,4 | 1 | 9,1 |
| 40 and older | 20 | 100 | 10 | 50,0 | 6 | 30,0 | 4 | 20,0 |
| Unknown | 2 | 100 | 1 | 50,0 | 1 | 50,0 | - | - |

**Table 7.** Age of first epi event and years when patient got married

| Frequency of epi seizures | TOTAL | *Marriage duration in years* | | | | | |
|---|---|---|---|---|---|---|---|
| | | 0 – 1 | 2 – 6 | 7 - 11 | 12 - 16 | 17 and more | Unknown |
| TOTAL | 75 | 11 | 9 | 10 | 5 | 37 | 3 |
| 1. Daily | 9 | 4 | - | 1 | 1 | 3 | - |
| 2. Weekly | 6 | 2 | - | - | - | 4 | - |
| 3. Monthly | 20 | 1 | 3 | 3 | 2 | 9 | 2 |
| 4. Every couple months | 14 | 2 | 1 | 3 | - | 8 | - |
| 5. Annual | 5 | - | - | 1 | - | 4 | - |
| 6. Several years | 20 | 2 | 5 | 2 | 2 | 8 | 1 |
| 7. Unknown | 1 | - | - | - | - | 1 | - |

**Table 8.** Frequency of epileptic events

| Type of epi events | TOTAL | *Marriage duration in years* | | | | | |
|---|---|---|---|---|---|---|---|
| | | 0 - 1 | 2 - 6 | 7 – 11 | 12 – 16 | 17 and more | Unknown |
| TOTAL | 75 | 11 | 9 | 10 | 5 | 37 | 3 |
| 1. Primary generalized seizures | 42 | 3 | 5 | 7 | 5 | 20 | 2 |
| 2. Partial seizure with or without generalization | 7 | 1 | 2 | - | - | 4 | - |
| 3. Seizures with complex symptomatology | 10 | 3 | - | 1 | - | 6 | - |
| 4.Combined seizures | 14 | 4 | 2 | 2 | - | 5 | 1 |
| 5. Unknown | 2 | - | - | - | - | 2 | - |

**Table 9.** Marriage duration and type of epi events

The difference between the average age of the first epileptic event between married [27.7] and total bachelors/unmarried (14 years) was statistically highly significant. Value of t-test is: t=7.00, p<0.01. The result shows that the unmarried were significantly younger than married when they had first epileptic event.

The difference between men and women, depending on the average age of life where they had first epileptic seizuredwas not significant withthe value of t-test t=1.506.

T-test of the difference significance between the average age of men and women in the group of married is t=0.463, and is not significant. In the group of bachelors/unmarried in the total sample, we had 24 (24%) patients with a mean age of 31.8±13.11 years. There was 11 men in this group at average age of 27.4±5.47 years.

Significance of the difference between the average age of men and women in a group of bachelors/unmarried measured by Student's t test show result of $t=2.021$ which was significant at the level $p<0.01$.

The difference between the average age of the married (41.2 years) and a group of bachelors/ unmarried (31.8 years) was statistically highly significant, or group of bachelors/unmarried was significantly younger than the group of married. Value of t-test is: $t=3.716$, at the confidence level of $p=0.01$.

We tested the difference between marital status and age at which respondents first had epileptic manifestation in life and came to the result that these differences were highly significant ($\chi 2$ test values: 18.801, $p<0.01$).

Based on the above data we tested the difference between men and women according to "marital status", the difference is not statistically significant ($\chi 2$ test value is 0.920), which means that between men and women there is no significant difference according to marital status.

We tested the difference between married and unmarried according to the type of epileptic manifestations which was statistically significant. The value of $\chi 2$ test is: $\chi 2=9.313$, $p<0.05$, DF=3)

The difference between men and women according to the type of epileptic manifestations was not significant. The value of$\chi 2$ test was 3.839, DF=3.

The test of significance between specific focal EEG changes in relation to marital status between married and unmarried is 1.008 which was not significant.

Contingency test between EEG generalized specifically altered and marital status was 0.341 which is also not significant.

According to the results of all three groups of respondents, regardless of marital status, the largest number of patients have generalized specifically altered EEG findings. Testing the significance of differences between married and unmarried according to the proportion of patients with generalized specifically altered EEG findings indicated that the difference was not statistically significant. Value of Sudent's t-test is: $t=1.008$. Thus there is no significant difference in the proportion of patients with focal specifically altered EEG findings. Value of t-test is: $t=0.341$.

Having in this manner calculated average duration of marriage in relation to the type of epileptic manifestations, we tested the significance of differences in the average duration of marriage of patients with primary generalized seizures (type 1) in relation to other types of epileptic events by Student's t-test and came to the following results:

The difference in the average duration of the marriage between patients with primary generalized seizures (17.4 years) and patients with partial seizures with or without generalization (13.9 years) was not statistically significant (t-test value is: $t=0.850$).

Also, there is no significant difference in the average duration of marriage between patients with primary generalized seizures and patients with seizures of complex symptomatology (Value of t-test is: $t=0.768$)

The difference between patients with primary generalized seizures and patients with combined seizures is significant, but at a lower level of reliability, $t=1.908$, $p<0.10$.

## 5. Discussion

Progress in medicine over the last century has been accompanied by advances in social and human attitudes in relation to man in general, while it is especially pronounced in some groups of patients, for example, in patients who have epilepsy [18].

The main motive for taking care of these patients is functioning of whole personality which implies besides proper diagnosis, treatment and therapy,also complete social rehabilitation of these patients. It should be noted that "living with epilepsy" means to live with a series of restrictions aimed at protecting these patients. It first refers to the sphere of education, relationship to the patient, education, career choice, limitations in daily life that the disease inevitably brings and restrictions on marriage and parenthood [19,20].

Advice whether the patient with epilepsy should get married or not depends solely on individual decisions. Regardless of epilepsy form the spouse must be informed about the disease and be familiar with it, and make a decision about marrying a person who has epilepsy. It is best that future spouse meet with the doctor who treated his/hers future spouse, and that with enough patience doctor informboth the patient as well as the future spouse on the character of the disease and the limitations that it brings, the possible situations in which they may found themselves, with a special emphasis on providing accurate and useful information in terms of genetic factors [21,22,23].

The existence of laws which was prohibiting marriage to patient with epilepsy has shown that the establishment of such legislation did not discourage mentally healthy patients to marry and have children [14].

Alstrom (1950) proved that in spite of eugenic laws patients with epilepsy get married in Sweden. Lennox and Mchram (1963) found that in the U.S., patients with epilepsy more rarely get married in relation to the general population [14].

Goodmsson (1965) presented the results made in Iceland on marriage of patients with epilepsy and low rate of marriages in patients where the seizures began early in life, in which the seizures were frequent and when personality changes occurred [18].

In our sample, we had 100 respondents, 65 married, 24 unmarried/single, 9 divorced, 1 widowed and one whose marital status was unknown.

In the group of married we had 38 men and 27 women.

Kantardzic et al (1990) on a sample of 5076 respondents, obtained results that differ from ours. In that study 64.4% of the sample were unmarried/single, 13.9% married, 2.3% divorced, 28% widowers, 10.1% in common law marriage and 8.8% unknown [24,25].

The diversity of the results can be explained by the fact that Kantardzic and associates conducted a wide social study at the territory of the Republic of Bosnia and Herzegovina, and that sample covered all age populations of patients in all regions including residents of the Institute for the mentally retarded. In our sample, we had patients who have good social adjustment in relation to the disease and whose treatment trough the Dispensary for epilepsy was aimed at increasing the quality of life and social adaptation [24,25].

Our results agree with the results of Schupf and Ottman (1996) which in their study on 863 patients treated by voluntary organizations in cooperation with Neurological Clinic in Montreal, presented findings that male respondents were more likely to be married than women. It should be noted that all subjects in study by Schupf and Ottman were married [26,27].

In respect of occupation among a group of married we had 75 patients: 26 laborers, 4 craftsmen, 20 clerks, 16 retirees and 8 housewives.

According to Kantardzic and colleagues (1990) on a sample of 7698 respondents in the highest percentage wererepresented students, then laborers, followed by housewives, craftsmen and clerks [24,25].

Also this difference in results can be explained by the size and sample specificity in our study and study conducted by Kantardzic and associates (1990).

In relation to marriage in our results we should indicate that more than half of the respondents were married before the age of 22 years. The exceptions were clerks who entered into marriage after age of 22 years.

Dansky Linda, Anderman Eva and Frederic Andermann (1980) examined the marital status of epileptic patients and fertility of women with epilepsy compared with the fertility of women whose husbands suffering from epilepsy and obtain the results that were previously described in the literature: there is a direct correlation between marital status and age when the first epileptic event in life occurred, and as the first epileptic event occurs later in life the chance that such a person get married and remains in marriage is higher. [14]

Their results are as follows: if the seizures began in the first decade of life the married status among male patients was reduced to 32% from expected with a very significant difference. In the female sample the same authors also found reduced specific nuptiality directly related to the age of forstepileptic event occurrence by 58% from expected, which is also important [28,29].

In our sample, we had only 3 patients with the occurrence of the first epileptic event in the first decade of life, of which 2 got married before 22 years of age, and one at age from 23-27 years.

Number of patients with the disease onset in the first decade of life in our sample is so small that these results could not be compared with the results of Dansky, Andermann andAndeemann [14,3].

For patients with onset of the seizures in the second decade of life the men had significantly reduced specific nuptiality, their ratio was 47% of expected. In our sample of 19 patients which had a first epileptic manifestation in the second decade of life, married was all 19 respondents after age of 22 years, so after the first epileptic event. First epileptic event after age of 20 in our sample had 51 patients, 38 is married before 22 years of age, 16 at the age from 23-27 years and 7 after age of 28 years.

In our study, we came to the conclusion that there is a significant difference in the average age of the respondents at age when they get married and age of the first epileptic event. The linear correlation coefficient between the age of the first occurrence of epileptic event and age of marriage amounts to 0.435 and means that for the possibility of marriage an important parameter is the absence or presence of disease. So there is a significant correlation between age of marriage and the occurrence of the first epileptic event.

Dansky and Andermann (1980) determined that in patients with the first seizure after 20 years of age specific nuptiality is not much different than the general population [14].

Our results also show that there is a significant difference between the average age of a total number of married (41.2 years) compared to the average age of the total number of unmarried, or the group of unmarried was significantly younger than the group of married.

Gudmuston (1966) stated that male patients more rarely get married. Dansky andAndermann (1980) showed that the difference between male and female patients is equal. Our results agree with the results of the Dansky and Andermann considering that the difference between men and women according to marital status was not statistically significant, with chi square value for this parameter of 0.920. These results can be explained by significant advances in medicine and epileptology in general, and the changes of attitudes towards patients with epilepsy who do not show mental changes in terms of alteration of personality within nosology entity or the disease with consequent changes in personality [14,18].

Thus, from the current results and literature we have data that women get married more often, or those patients in whom the first epileptic event occurred after 20 years of age, or to say, before the first epileptic event [30,31].

In our sample, we had 53 men and 47 women. The average age of men was 39.0 years and of women 39.4 years.

We had 38 married men, with a mean age of 40.8 years, and 27 married women, with a mean age of 41.9 years.

Dansky and Andermann (1980) tested 100 patients with epilepsy. The average age of the male patients in their sample was 39.3±1.7 years and average age of female was 42.3±1.6 years.

Our results in relation to the age of the male and female patients are comparable with the sample from study conducted by Dansky andAndermann (1980) at the Neurological Clinic in Montreal [14].

In our sample, as well as in a sample from comparative literature there is no significant difference between the average agebetween genders.

In relation to the etiology the recent division into genuine (idiopathic, cryptogenic) and symptomatic is already largely abandoned. Epilepsy in many ways is idiopathic while the genuine epilepsy can manifest if the hereditary tendency is influenced by some other symptomatology. Basically pathophysiological cause of epilepsy unclude all that can overcome the polarization balance in groups of nerve cells located at the main afferent-efferent pathways of the brain. On inherited tendencies, level of maturity, location, size, and material cells in focus depends on whether one cause will initiate the crisis-epileptic form of seizure, or a single disorder of polarization balance will cause the second and so on, until the first primarily functional, and then structurally "biochemical focus" does not lead to such balance disorder of irritant and inhibitory processes in the brain that are clinically manifested as epilepsy [32,33,34].

In the group of exogenous factors in our study, we considered:

Birth trauma (reliable data), infectious diseases of the CNS, head injury with or without loss of consciousness, and the existence of "macro factors". Positive heredity in epilepsy in our sample was present in 6.88% of patients, negative in 54 and unknown in 16 patients.

Our results in relation to specific heredity agree with results from the literature: In case of petit mal 15.3%, 2.6% in case of psychomotor epilepsy (Lennox), 3% (Gastaut), Gibbs 4.5%.

Jears et al. (1975) at the Children's Hospital in Ljubljana found among 525 patients positive specific heredity in 12.98%. Jears and associates had sample of 809 children between 3 and 15 years of age. In our sample the youngest patient was 17 years old, the oldest 53 and the positive result of a specific heredity is considered to be in accordance with results from the literature [22,30].

Hajnsek (1980) presented the results of the age group from 20-50 years in testing the etiology of epilepsy on a sample of 572 patients and came to the following results: trauma 40.7%, 35.4% of unknown etiology, inflammatory CNS processes 11.4%. Positive heredity for epilepsy in his study was 10%.

In our sample we found: trauma 6% (tested reliable data of birth trauma), infectious CNS diseases 6%, and head injury with or without loss of consciousness 17% and 58% with unknown data. Positive heredity in the total sample was 13% and among married 8.8%.

We can conclude that our results agree with the results of Hajnsek, while certain differences can be explained by age variations of the samples.

Classification of epilepsy is an open problem. International classification of diseases in the latest version from the 1981 simplifies the main categories of epilepsy into a) generalized and b) partial seizures. Hercegovac (1976) deepened earlier given classification by Gastaut (1969) which divided epileptic events into the following groups:

I.

**a.** Partial seizures with simple symptomatology.

**b.** Partial seizures with complex symptomatology.

c. Partial seizures with secondary generalization.

II. Generalized seizures:

**a.** Generalized seizures without local onset,

**b.** Generalized seizures with local onset.

III. Unilateral or predominantly unilateral seizure

IV. Unclassified seizures.

In our study we use the following classification of seizures: primary generalized seizures 51%, partial seizures with or without generalization 10%, seizures with complex symptomatology 12%, combined seizures 25% and unknown in 2%.

According to the results of Jears (1975) from the Clinical Center in Ljubljana on a sample of 525 children-aged between 3 and 15 years, using the division on focal and generalized seizures there was 31.3% cases of generalized epilepsy and 68.7% of focal epilepsy.

Our results agree with the results of Kantardzic and associates who on a sample of 250 patients from the territory of Sarajevo, Tuzla, Mostar and Livno (1990) found that most respondents had generalized seizures, followed by partial with complex symptomatology, then partial seizures with simple symptomatology and combined seizures.

In relation to marriage in our sample of 65 respondents more were married (65.0%) compared to unmarried (35%). 38 patients had generalized seizures, 7 partial seizures with simple symptomatology. Our result was also that the group of unmarried had in much larger number of cases 41.7% the combined seizures than married 15.4%.

Study by Schupf and Ottman (1996) on a sample of 863 patients treated by voluntary organizations in cooperation with the Clinic of Neurology in Montreal showed partial onset in 82% of patients, which is different compared to our study.

These authors state in their discussion diversity compared with previous studies. Weber also states that specific nuptiality in the group of patients with epilepsy with partial seizures is higher than in group suffering from generalized seizures. The average duration of marriage in 40 patients with primary generalized seizures was 17.4 years, patients with partial seizures with or without generalization 13.9 and patients with seizures of complex symptomatology 14.3 years.

Our conclusion is that the type of epileptic event does not affect the duration of the marriage, which we did not compared with the results from other authors as these results in the available literature were not found.

In relation to EEG results significance of difference between married and unmarried according to proportions of generalized specifically altered EEG, and the results of difference significance between married and unmarried according to specific proportions of focal EEG changes in terms of the absence of statistical significance, we also did not compare with the earlier studies because for us such results in the literature were not known.

Understanding the neurological aspects of epilepsy significantly improved during last decades. Psychiatric aspects have for many years been neglected. Lately in the project to improve the quality of life of patients with epilepsy more attention is focused on the psychological aspects of life of patients in order to achieve better social functioning and good social reintegration.

Quality of life of patients with epilepsy was one of the leading topics of the 29$^{th}$ Congress on Epilepsy (Tokyo 1995] and the Second European Congress of epileptologists (Hague 1996).

The most important contribution in this regard was the European Quality of Life study conducted on 5000 adult patients in 15 European countries. The study collected clinical and demographic data on psychosocial functioning. Survey was filled by the patients. The questions focused on: types of seizures, the frequency of seizures, injuries related to the seizures, the side effects of anti-epileptic therapy, problems in social functioning in relation to the family, workplace and the environment in general. Results are as follows: 1/5 of the respondents felt that the seizures are not under good control, 1/3 considered to have frequent seizures. The most common side effects of antiepileptic drugs that patients listed were: fatigue, memory problems and focusing difficulties. In our country in the last two years special attention is given.in light of the global trend of epileptology psychological changes,to patients with epilepsy. Gavranovic (1996) and associates all psychological changes within epilepsy divided into two large groups:

1. Epilepsy and intellectual functions (mental retardation, cognitive disorders, dementia or psychosindrome) and;
2. Psychiatric disorders (epilepsy personality changes, psychosis, vacillating moods-dysphoria, depression, psychoreactive disorders).

According to Kantardzic (1997) study on the prevalence of conduct disorder shows that these disorders occurs in 12-95% of cases and depends on the study population. They are lowest in the normal population, in schools and general practice (12-23%), in the out-patient clinics for epilepsy 50% and 95% in inpatient psychiatric hospitals [24,25].

In our sample psychological changes that have been investigated are: character changes 4%, signs of organic mental disorder 23%, neurotic disorders 37%, 13% oligophrenic, without psychiatric disorders 21%, unknown 2% and 0% of psychosis.

Neurotic reaction in our sample occurred in 37% of patients, twice as often is present neurotic reactions among women with epilepsy than in men. Our result is comparable with some results from the literature like ones by Currie and colleagues, Betts, Ramesch et al., Fenton and colleagues, and Bingley [12].

Signs of organic mental disorder we found in 23 patients, 17 were in the group of married, 16 men and 7 women. Our results are compatible with the results of Broxn and Abeyasinghe

(1984) fromMausdley and Kings College Hospital, which have proven that the intellectual deterioration as the ultimate outcome of epilepsy occurs much less frequently than previously thought at the beginning of the 20th century. Trough comparative results we can say that the mental deterioration occurs in a certain number of patients, but not in majority of cases, as previously thought. In both studies it was concluded that the mental deterioration occurs more frequently in men than women. This can be explained by the long history of the disease in terms of early onset of the seizures, poor therapeutic reactivity and the possible effects of antiepileptic drugs, especially barbitone and diphetoin. Confirmation of toxic effects on mental functioning of AEDs have was proven Thomson and Trimbler, De Niegri et al., Andrewes et al., Calandre et al., Thompson and Trimble [14].

In our sample, we had 21 patients without mental changes, 14 men and 7 women. In relation to marrital status 14 respondents in this group was married with different marriage duration.

These results can be correlated with good psychosocial functioning of patients on which there are many studies: Salajpal and Ristovic (1986) have shown an adverse effect of some social factors in the course of the disease and psychological status: poverty, deficient education within the family, poor emotional family relationships, duration of conflict family situations with an aggressive and rejecting attitude towards the patient [27].

In the same context Kocijan-Hercegonja (1986) and colleagues have concluded that the social status of the family is the most important factor in forming the patient's personality, and for the presence or absence of psychological changes in patient [22].

Loewenson et al (1980) in terms of socioeconomic status, interviewed 298 patients: all was older than 18 years and none of them was mentally retarded. The average duration of disease was 15 years at mean age of 30 years. The results were: There was no significant correlation between marital status, employment, good income with the frequency of seizures, type of seizures and psychological problems of reactive nature [24,25].

All respondents without psychiatric disorders in our sample were of good social status. Seventeen respondents who were married had a harmonious relationship with the spouse.

Slodnjak et al (1986) evaluated the overall psychosocial functioning of patients with epilepsy and found poor prognosis in 26% of boys and only 7% girls. Our result agrees with the result of Slodnjak et al (1986), the women in our sample had more disorders of mental sub normality without disorder of psychosocial functioning [30].

Our results on mental changes depending on the marriage duration are the following: mental changes have occurred after 12 or more years of marriage in 64.7% of cases: neurotic disorders 71% and 21.4% was without problems.

Our results in terms of the correlation between the marriage duration and psychological changes can be explained by multiple factors: disease factors, factor of antiepileptic therapy, psychosocial dynamics factor in marriage not related to the disease, problems that marriage brings in relation to the provision of the financial basis for the family as a whole and the like.

In relation to gender differences in psychological problems that we found, they also may be explained by the above-mentioned factors that we discussed, as well as socio cultural specific factors of the environment and climate of our respondents.

When we talk about psychological changes in our patients we must emphasize the fact that we tested those patients who had been treated through the Dispensary for epilepsy, and that the triage of these patients in terms of nonpsychotic disorders by sample selection was already made.

Our results on psychological changes agree with the other results from the literature. According Pekovic (1990) neurotic problems are encountered in 32% of patients, character personality disorders in 22%, signs of organic mental disorder at an early form in 15% and 32% of respondents had no psychological disorders. Our results are parallel with this.

In relation to treatment there are strict principles with the aim of successful treatment. According to Gavranovic (1988) principles of epilepsy treatment are:

1. Diagnosis of epilepsy must be reliable.

2. When the diagnosis is certain the treatment should be started as soon as possible because is well-known fact that the neurons damage by repeated seizures is much more severe than by other factors.

3. Treatment should always begin with monotherapywhilepolytherapy should be avoided. Reynolds et al (1976, 1981) reported that 80% of seizures can be treated by monotherapy. When introducing AEDs start with small doses. Each patient is an individual in terms of treatment and the determination of AEDs dose is highly complex depending on: type of seizure, seizure frequency, age, general condition, individual drug tolerance, the environment and the attitude of the environment in relation to the disease and socioeconomic factors.

4. Selection of antiepileptic drugs depends on the type of epileptic manifestations. Partial seizures with elementary symptomatology and generalized seizures are treated with phenobarbitone, carbamazepine, and diphenyl-hydantoin. Partial complex seizures are the most common drug-resistant seizures, they are treated with carbamazepine, diphenyl-hydantoin and primidone. In case of generalized non convulsive seizures of simple type the drug of choice is ethosuccimide, then sodium valproate, or these two drugs in combination.

5. In refractory cases when we are forced to use polytherapy we must respect the principle: never combine identical antiepileptics, avoid combinations of toxic drugs, the choice of antiepileptic drugs and its concentration should be within the therapeutic dose spectrum.

6. It is necessary to control the drug tolerance which in clinical practice means to control the skin, lymph nodes, liver, spleen, mental state and possible ataxia.

7. A special task is to educate patients in terms of knowledge about the disease and precipitating factors of potential seizures such as: reduced sleep, alcohol use, stress, hormonal changes.

http://dx.doi.org/10.5772/58635

In our sample we found the following therapeutic reactivity: 27% of our patients had one seizure in several years, 19 every couple of months, 8 once per year, 21 once a month, 14 weekly and 10 daily seizures, so complete or very satisfactory therapeutic reactivity was found in 54% of our respondents.

Remission rates of patients with different types of seizures in the literature varies: in patients with tonic seizures,they vary according to Gavranovic (1988) from 60-80%, according to Sofijanov (1982) remission rate as simple absence seizures is 70-80% [24,25].

Satisfactory therapeutic reactivity in our sample we found in 54% of patients which is consistent with those authors.

In relation to marrital status in our sample we had the following results: the group of married and group of unmarried in approximately same percentage had epileptic events during few years before the study, while a group of unmarried usually have weekly frequency of seizures (29.2%).

Comparison of therapeutic responsiveness in relation to marriage we did not conducted because in the available literature we did not find such results [31,32].

Explanation for better therapeutic responsiveness of married in our sample we can find in following facts: our sample was respondents from the Dispensary for treatment of epilepsy, so the patients who in previous period were treated and whose social status (employed in the profession 55, at another job 8) requested functionality within the job, environment and family. The purpose of their treatment through the Dispensary is "best possible therapeutic reactivity" in relation to the type of disease, EEG changes, disease duration, age and choice of antiepileptic drugs.

Our results in terms of patient's employment patients agree with the results of Kantardzic and associates (1990) as they found the overall employment rate of 43.2% (28% in the profession and 15.2%at another job).

Hydanphen received 61 patients, 4 patients were taking barbiturates, Tegretol 9, 22 combinations of Tegretol and Phenobarbitone.

## 6. Conclusion

1. Epilepsy affects the possibility of marriage and is in the direct correlation with the age at which the first epileptic event occurred and the type of the epileptic event.
2. Early occurrence of the disease reduces the possibility of entering into marriage, if the disease occurs later in life chances of conceiving marriage are significantly higher.
3. Patients who are married have less frequent epileptic seizures.

## Author details

Azra Alajbegovic[1], Jasminka Djelilović-Vranić[1], Salem Alajbegović[2] and Lejla Alajbegović[3]

1 Neurology Clinic, Clinical Center of Sarajevo University, Sarajevo, Bosnia and Herzegovina

2 Cantonal Hospital Zenica,Zenica, Bosnia and Herzegovina

3 Berlin-Chemie Sarajevo, Representative office Bosnia and Herzegovina, Germany

## References

[1] Al-Khateeb JM, Al-Khateeb AJ Research on psychosocial aspects of epilepsy in Arab countries: A review of literature Epilepsy Behav. 2013 Nov 5. doi:pii: S1525-5050(13)00499-X.

[2] Helmstaedter C, Droege F, Witt JA Assessing activities of daily living in patients with epilepsy).FortschrNeurolPsychiatr. 2013 Aug;81(8):452-8. doi: 10.1055/ s-0033-1335993. Epub 2013 Aug 12.

[3] Dominian J, Serafetinides EA, Dewhurst MA. Follow-up study of late-onset epilepsy.Br Med J 1963;1:431-5

[4] TodorovaK.S.KaprelyanA.G.Influence of deprtessive mood on quality of life rating og women with epilepsy of childbearing age.Folia Med.2013.Jan-Mar;55(1):70-5.

[5] CastanoMonsalvie.Antidepressants in epilepsy.RevNeurol 2013 Aug 1;57(3):117-22.

[6] Helmstaedter C, Witt JA. Epilepsia. Multifactorial etiology of interictal behavior in frontal and temporal lobe epilepsy 2012 Oct;53(10):1765-73. doi: 10.1111/j. 1528-1167.2012.03602.x. Epub 2012 Aug

[7] Andermann E, Dansky L, Kinch RA. Complications of pregnancy.Labour and delivery in epileptic women.Janz D, Dams M, Richens A, Bossi L, Helge H, Schmidt D, eds, Epilepsy, pregnancy and the child. New York, Raven Press1982:61-74.

[8] Aicardi J. The Prevention of Epilepsy in Children.D.Janz (Ed).Epileptology, George Time Publishers, Stuttgart, pp. 2-9, 1976.

[9] American Academy of Pediatrics Committee on Drugs. Anticonvulsant and pregnancy, Pediatrics, 63:331-335.1979.

[10] Baker GA, Jaciby A, Smith DF, Dewey ME,Chadvick DW. The development of novel scale to assess life fulfillment as part of the further refinement ofa quality of life model for epilepsy.Epilepsia 1994; 35:591-6.

[11] Baker GA, Smith DF, Dewey M, Jacoby A, Chadwick DW. The initial development of healt-related quality of life model as an outcome measure in epilepsy.Epilepsy Res 1993; 16:65-81.

[12] Betts TA. Depression, anxiety and epilepsy.Reynolds EH, Trimble MR, eds. epilepsyand psychiatry. New York, ChurcillLovingstone, 1981.

[13] Britten N, Morgan K, Fenwick PBC, Britten H. Epilepsy and handicap from birth to age thirty-six. Dev Med Child Neurol 1986; 28:719-28.

[14] Dansky LV, Andermann E, Andermann F. Mariageet fertility in epilepticpatients. Epilepsia 1980;261-71.

[15] Dodrill CB, Batzek LW. Patterns of Psychosocial Problems in Chronic Epilepsy-acta-NeurolScand 62 (suppl-79) 9:31-32,1980

[16] Ercegovac D, Savić Č, Mitrović D. Psihosocijalniaspektiepilepsije. X Jugoslovenski-simpozij o epilepsijama, Knjiga sa`etaka,Zadar,1984.

[17] Goodridge D, Shorvon SD. Epileptic seizures in a population of 6000:Treatment and prognosis. Br Med J 1983;287:645-7.

[18] Gudmunsson G. Epilepsy in Iceland.A clinical and epidemiological investigation.ActaNeurologicaScandinavica (supplementum 25), Vol. 43, 1966.

[19] Hauser WA, Hersdorffer DA. Epilepsy: frequency, causes and conse-quences. Epilepsy Foundation of America. New York: Demos 1990.

[20] Jacoby A, Baker GA, Smith DF, Dewey M, Chadwick DW. Measuring the impact ofepilepsy:the development of a novel scale. Epilepsy Res 1993; 16:83-8.

[21] Jacoby A, Graham Jones S, Baker GA, RatoffR,HaynesJ,Chadvick DW. A general practice audit of care for people with epilepsy (In preparation)

[22] Janko M. Liječenjeepilepsijeantiepilepticima.Psihifarmakoklogija, 1990.

[23] Jensen P, Jensen SB, Sorenson PS, et al. Sexual dysfunction in male and female patients with epilepsy: A study of 86 patients Arch Sex Behavior 1990;19:1-14.

[24] KantardićDž, et al. Epilepsija-analizasociomedicinskihpokazatelja u grupiod 4071 pacijenata u BiH, 1989.

[25] KantardžićDž, Gavranović M, Jovanović M. Socijalna traumau provokacijiepileptičnihmanifestacija, Medicinski arhiv,33,105,1979.

[26] Reproduction Among Individuals with Idiopathic (Cryptogenic Epilepsy) Risk Factors for Reduced Fertility in Marriage Epilepsia 37 (9): 833-840,1996 Nicole Scupl and Ruth Ottman

[27] Salajpal T. RisovićDž. Utjecajpsihosocijalnihfaktoranatokepilepsije-Soc.Psihijatrija 15 (2):125-135.1987.

[28] Self –reported Sexual Function and Sexual Arousability in Women withepilepsy, Marta J Morrell and Gregory T GuldnerEpilepsia 37 (12) 1204-1210.1996.

[29] Sexual disfunction in epilepsy.Epilepsia 1991;32:S38-45. Reutens DC, Howell RA, Gebert KE, et al. Validation of sexual dysfunction in women: results of a survey study of 329 women in aoutpatient gynecological clinic.J sex Marital Ther 1993;19:171-88.

[30] Slodnjak V, Goriček M, Ravnik I et al Pokušajopredjeljenjadijagnostičnoznačajnihmjerilapsihosocijalnogfunkcionisanjadjece i omladinesaepilepsijama. XI jugosl.simpozij o epilepsijama,SavezligezaborbuprotivepilepsijeJugoslavije (Ed),Dom štampe,1986.

[31] The clinical Course of epilepsy and Its Psychosocial Correlates: Findings from a. c.k.Community Study Ann Jacoby, Gus A.Baker,NickSteen,PaulinePotts,and David W,Chadwick Epilepsia,37(2):148-161,1996

[32] Toone BK, Edeh J, Nanjee MN, et al. Hyposexuality and epilepsy: a coomunity survey of hormonal and behavioural changes in male epileptics, psychological Medicine, 19, 937-943, 1989.

[33] Zielinski JJ. Social prognosis in epilepsy.Epilepsia 1972;13:133-40.

[34] Zielinski JJ. Social Prognosis of E pilepsy-Epilepsia 13:133-140,1972.